Harrison Learning Centre
City Campus
University of Wolverhampton
St Peter's Square
Wolverhampton
WV1 1RH
Telephone 0845 408 1631
Online renewals:
www.wlv.ac.uk/lib/myaccount

Telephone Renewals: 01902 321333
Please RETURN this item on or before the last date shown above.
Fines will be charged if items are returned late.
See tariff of fines displayed at the Counter. (L2)

TOURISM AND THE ENVIRONMENT

REGIONAL, ECONOMIC, CULTURAL AND POLICY ISSUES

Revised Second Edition

ENVIRONMENT & ASSESSMENT

VOLUME 6

The titles published in this series are listed at the end of this volume.

Tourism and the Environment

Regional, Economic, Cultural and Policy Issues

Revised Second Edition

Edited by

Helen Briassoulis

*University of the Aegean,
Mytilini, Lesvos, Greece*

and

Jan van der Straaten

*Tilburg University,
Tilburg, The Netherlands*

KLUWER ACADEMIC PUBLISHERS

DORDRECHT / BOSTON / LONDON

A C.I.P. Catalogue record for this book is available from the Library of Congress.

ISBN 0-7923-6136-9

Published by Kluwer Academic Publishers,
P.O. Box 17, 3300 AA Dordrecht, The Netherlands.

Sold and distributed in North, Central and South America
by Kluwer Academic Publishers,
101 Philip Drive, Norwell, MA 02061, U.S.A.

In all other countries, sold and distributed
by Kluwer Academic Publishers,
P.O. Box 322, 3300 AH Dordrecht, The Netherlands.

Revised second edition

Printed on acid-free paper

CONTENTS

vi

TOURISM AND THE ENVIRONMENT: AN OVERVIEW[1]

HELEN BRIASSOULIS
Department of Geography
University of the Aegean
Karantoni 17, Mytilini, Lesvos 81100
Greece

JAN VAN DER STRAATEN
Department of Leisure Studies
Tilburg University
PO Box 90153, 5000 LE Tilburg
The Netherlands

1. Introduction

Seven years ago, in 1992, the first edition of the present volume appeared with the title *Tourism and the Environment: Regional, Economic and Policy Issues*. For the most part, it contained modified versions of papers presented at the 30th European Congress of the Regional Science Association, which had taken place in Istanbul, Turkey, in August 1990. This was one of the first occasions where conference sessions were devoted to the tourism-environment relationship. Also non-conference papers were included as they tackled important regional, economic and policy dimensions of the tourism-environment relationship. As the first edition was quite well received by the community of academics and practitioners and as all copies were already sold in 1996, the publishers requested a second edition. By that time, however, things had changed. Tourism and the environment gained importance in academic and policy circles, specialised journals and numerous books were being published, conferences were organised on this theme exclusively, and more dimensions of the relationship were being investigated, while the

[1] This publication was supported by the Saxifraga Foundation, Tilburg and the Department of Leisure Studies of Tilburg University, the Netherlands.

sustainable tourism development concept had attained top priority. Hence, there was a need to broaden the scope of the second edition to cover some of the new developments related to the topic of the book. The new title reflects, in part, these changes, by the introduction of the 'cultural' dimension. Also, the book now contains 22 instead of the original 14 contributions. Moreover, most of the old chapters have been revised to include new data and information which have been added to the body of existing knowledge in the course of the 7 years since the first edition appeared (or, even earlier). In fact, this is a new book with an old title. Linda Bokhout organised and completed the format of the book, and she fitted the chapters of the authors to this framework.

At the close of the 20th century, tourism's socio-economic profile and its role bear little resemblance to their original forms a century or 50 years ago. Tourism is recognised as a major generator of direct and indirect economic benefits – income, employment, economic activity – in both the countries of origin and the countries of destination. It is, therefore, promoted and developed not only in those places endowed with a rich and attractive cultural and/or natural environment but also in many other places, which possess other, not necessarily traditional, types of tourist resources. At the same time, tourism is no longer pursued only by the urban dwellers of the industrialised world, who wish to escape to tranquil and relaxing places and enjoy their natural and cultural beauties in order to maintain their physiological and psychological balance. Tourism is considered an obvious social necessity and right for all people in both the developed and the developing world.

But tourism's relationship with the environment – and, more importantly, people's perception of this relationship – has changed now as its development has been and is a significant cause of unwanted social and environmental disturbances. The issue is critical as the natural and manmade environments are basic ingredients of the tourist product and, naturally, their quality dramatically affects the quality of the product. The search for ways and means to maintain a balanced relationship between tourism and the environment started in the 1970s. But not until the 1980s and 1990s it became the subject of a systematic academic inquiry and research, that distinguishes itself from the broader research area of the environmental impacts of recreation and leisure activities. International organisations, such as the World Tourism Organisation, the United Nations, the OECD, the European Union and several others, have organised workshops, conducted studies, and suggested policies to preserve a healthy and attractive environment, to ensure, amongst other things, the success of tourism development (WTO, 1980; UNEP, 1987; OECD, 1980). Several international journals began to devote special issues to the tourism-environment relationship (for example, International Journal of Environmental Studies, 1985; Annals of Tourism Research, 1987; Land Use Policy, 1988). In the 1990s, journals specialising in the issue also appeared such as the Journal of Sustainable Tourism. Some of the numerous developments in this area are dealt with within this volume, which complements several other noteworthy contemporary publications. The rest of this introductory chapter elaborates on the tourism-environment relationship, briefly reviews the literature on the environmental impacts of tourism, and introduces the chapters included in the volume.

2. Tourism and the environment: the relationship

Tourism, a multifaceted economic activity, interacts with the environment in the framework of a two-way process. On the one hand, environmental resources provide basic ingredients, critical production factors, for the production of the tourist product: the natural and/or man-made setting for the tourist to enjoy, live and relax. On the other hand, tourism produces a variety of unwanted byproducts, which are intentionally and unintentionally disposed and modify the environment: negative environmental externalities. Moreover, other economic activities, besides tourism, use up, modify and affect the quantity and quality of environmental resources available for tourism purposes. Successful tourism development, on all spatial scales, depends in many important ways on the proper handling of the relationship between tourism and the environment; i.e. on integrated tourism and regional planning providing for optimal allocation of environmental resources and other production factors to tourism and other competing economic activities. A first prerequisite towards this purpose is a careful analysis of the tourism-environment relationship and of the associated principal planning and policy-making concerns in tourism development.

Tourism development depends on the availability of attractive natural and/or man-made resources in an area which tourists demand and pay for. It is questionable if tourism could exist as an economic activity, and could be distinguished from other activities, in the absence of a well-preserved and highly valued resource base. Natural, unspoiled scenery, beaches, mountains, ancient monuments, traditional, picturesque towns and villages and many more factors constitute the primary inputs to the production of the tourist product. The specific type of tourism development of an area (e.g. beach resort, ski resort, etc.) depends primarily on the nature of its environmental resources. Qualitative and quantitative differences in the distribution of these resources over space account for differences in tourism development at the regional, national and international level, with consequent differences in the spatial and temporal distribution of tourism's economic, environmental, social, and other impacts. Moreover, an important requirement for sustained tourism development is the preservation of the quality and quantity of these resources at levels acceptable to the consumers, the tourists.

It is not so simple to identify and analyse tourism's demand for environmental resources, as one may think, because tourism is not a single economic activity with a standard pattern of input requirements, and a standardised output. Instead, it should be conceptualised as a complex of interdependent and inseparable activities (travel, lodging, shopping, recreation, and services), each one with its own demand for inputs and a characteristic output. The demand for inputs and the product of tourism is, ultimately, a synthesis (not a mere addition) of these individual demands and outputs. Hence, the analysis of tourism demand for environmental inputs involves analysis of the demands made by its constituent activities as well as the interrelationships between these individual demands. This analysis is important for two reasons: (1) tourism development of an area must take into account the availability of local resources which are necessary for its growth and maintenance, and (2) tourism-related activities compete for the environmental resources of an area with each other and with other economic activities (industry, trade, transportation, etc.), and as a result, conflicts among different uses arise. These conflicts either result in deterioration of the quality and quantity of the tourist product (because of undesirable spill-over effects (externalities) from one activity to another, and, consequently, losses to the tourism industry), or in a struggle for domination of the most economically profitable activity. Changes in the physical, spatial, and socio-economic structure of a tourist area, as well as the existence of several, sometimes burdensome, environmental problems testify to the presence of these conflicts. It requests some form of conflict management (resolution or reduction) which leads to a more desirable allocation of environmental resources.

The other facet of the tourism-environment relationship concerns tourism's demand for the residual receptor services of the environment. Once an area becomes a tourist attraction pole, its resources undergo changes simply because they have been used up, directly for the production and consumption of the tourist product on the one hand, and indirectly by activities linked to the tourist-related ones on the other. The residuals generated by these activities are inevitably disposed in the environment and modify it. The extent and intensity of the modifications caused basically depend on two interrelated groups of factors: (1) the type and spatial-temporal characteristics of tourism development, and (2) the characteristics of the area (UNEP, 1982). The first group includes such factors as the type of tourist activity, the socio-economic and behavioural characteristics of tourists, the intensity and spatial-temporal distribution of use, and the strength of linkages among activities. The second group includes the natural environmental features of the area, its economic and social structure, the forms of political organisation, and the level of tourism development. The term carrying capacity is used to denote an area's maximum tolerance to tourism development before negative impacts set in (Pearce and Kirk, 1986; Lindsay,1986). Although interest is mostly centred on environmental carrying capacity, social and physical carrying capacity are also important in the ultimate determination of the maximum amount of tourism development an area can tolerate.

The results from the interaction between the two groups of factors mentioned above are the so-called direct environmental impacts. But there are also indirect impacts, caused by activities indirectly related to tourism (e.g. local handicrafts, trade, entertainment, etc.), as well as developments induced by the presence of tourism in an area (e.g. second homes, recreation and shopping facilities, transport networks, etc.). Therefore, the total impact of tourism on the environment is the result of direct, indirect, and induced impacts, of which the latter very frequently are difficult to distinguish from one another.

Overall, tourism's use of an area's environmental resources has two consequences. Firstly, the quantity of available resources diminishes and sets limits to further tourism development of the area. Physically and/or economically non-augmentative resources (e.g. beaches, sites of natural or archaeological interest) become limiting factors in this respect. For other types of resources, planning and management are necessary to maintain quantity at levels necessary for continued tourism activity. Secondly, the quality of resources deteriorates with negative effects on tourism. Firstly because the tourist product offered is of inferior quality, and secondly because the quantity of good quality products, which was the initial reason for tourism development is reduced. To avoid these negative impacts, it is imperative to plan tourism such that the tourism-environment relationship occupies a central position in providing guidelines and determining the limits to growth and development of the activities involved. The main concerns arising in this context are briefly discussed below.

Negative externalities generated by tourism, unlike those produced by other economic activities, must be controlled in the same location in which they arise and in the short-term, otherwise they have negative repercussions on the tourism industry itself. Assessment of an area's carrying capacity, at least in relative terms, is an absolute necessity in order to set some limits to growth and to avoid undesirable impacts on the economic vitality of the industry. First priority must be given to the carrying capacity of the natural environment, as defined by its major components: air, water, terrestrial, and aquatic ecosystems. However, the area's social and economic capacity must also be assessed for a comprehensive account of its tourism development potential.

Because tourism is not a single economic activity but a complex of interrelated activities, planning must encompass all these activities, their interrelationships, and their demands on environmental resources and services, some of which will be compatible with one another while others may be antagonistic. Also, despite measures that have been introduced to reduce extremes, tourism is more or less seasonal in nature and its impacts also have a characteristic seasonal pattern. For these reasons, tourism development must be embedded in a comprehensive planning framework for a given region. This region which will seek to avoid the unwanted consequences of conflicts over incompatible land use and the over-development of one activity at the expense of others and the region itself, at least in the long run. For tourism, specifically, planning should prevent extreme, seasonal negative impacts and, at the same time, should avoid investments in environmental protection which remain unused for long periods of time. Finally, this planning framework should ensure a reasonable allocation of local environmental resources and services to competing uses, directed at maximising local welfare, and at assisting the area to develop sustainability.

3. Environmental impacts of tourism: State-of-the-Art

Until the early 1980s, literature on the environmental impacts of tourism was considered a luxury, because of the paucity of relevant studies and the diversity of sources from which they had to be retrieved (Farrell and McLellan, 1987). The first efforts towards environmental impact assessment basically focused on the impacts of leisure activities and, especially, outdoor recreation (Wall and Wright, 1977). The first studies on the environmental impacts of tourism appeared after the mid-seventies (Tangi, 1977; Baud-Bovy and Lawson, 1977), and were followed by more research activity in the 1980s. Useful reviews on the subject can be found in Dunkel (1984), Pearce (1985), Farrell and McLellan (1987), and Farell and Runyan (1991). Naturally, most studies have concentrated on areas that are experiencing some form of adverse environmental impacts as a result of tourism development, for example the Caribbean islands, the Mediterranean, ski resorts. Throughout the 1990s, the tourism-environment issue has been examined, more or less, within the broader framework of sustainable tourism development. Now, at the end of the decade, and in no more than 20 years, finding literature on the subject is no longer difficult. Instead, it is a Herculean task, given the proliferation of studies and diversity of viewpoints involved, to select relevant literature. The following discussion has, thus, to be considered as a timid effort to offer a broad and by no means exhaustive account of the main topics and the most prominent issues covered in the relevant literature.

The environmental impacts of tourism have been looked at from many perspectives such as the biological and ecological, the behavioural, the planning and design, and the policy perspective. The biological and ecological impacts of tourism have been studied in specific environments – islands, coastal zones, alpine areas, national parks, etc. (Edwards, 1987; Gartner, 1987; Jackson, 1986; Lindsay, 1986; Miller, 1987; Rondriguez, 1987). The most important consideration is the assessment of the environmental carrying capacity that is necessary to plan for tourism development that is in harmony with the environment.

From a behavioural point of view, visitors' satisfaction with an area's environment as well as residents' perceptions of tourists have received scholarly attention (Liu and Var, 1987). This viewpoint relates indirectly to the notion of the social carrying capacity of an area, i.e. the amount of social disruption beyond which both visitors and the local population experience negative consequences (Pearce and Kirk, 1986).

Planning and policy-making for tourism development have been heavily concerned with the goal of attaining a balanced relationship between tourism and the environment. The most important issues in this respect are: assessment of an area's carrying capacity and the limiting factors that determine the extent of tourism growth, proper planning approaches ensuring balanced and sustainable tourism development (Baud-Bovy and Lawson, 1977; Inskeep, 1987; Inskeep 1991; Inskeep, 1994), and suitable policies for implementing the prescribed planning measures (OECD, 1980; UNEP, 1982; UNEP, 1987). The latter borrow elements from the broader class of environmental and development policies and adapt them to tourism. Their core concerns are: control of tourism growth away from environmentally sensitive areas, restrictions imposed on types, the extent and intensity of activities permitted in an area, the proper management of residuals generated by tourism, and minimisation of conflicts between tourism and competing land use.

However, in the 1990s, a re-orientation can be observed in the analysis of the environmental impacts of tourism after the broad adoption of the notion of sustainable tourism, which followed the introduction of the concept of sustainable development (Farell and Runyan, 1991; WCED, 1987). This concept is as old as humanity itself, and has been practised in the pre-industrial age. However, t became prominent after the publication of the well-known Brundtland Report by the World Commission on Environment and Development as a governmental-level response of several industrialised countries to the growing threat to the environment posed by industrial activities. The definition of sustainable development as "development which meets the needs of the present without compromising the ability of future generations to meet their own needs", although vague and subject to many interpretations, has since become the guiding principle and basis for economic and environmental policy-making in many countries in the developed and the developing world. Modern industry also became gradually convinced that polluting the environment was no longer prudent, from an economic point of view, and new concepts such as 'greening the industry', 'environmental awareness of businesses', 'ecological modernisation' are now widely used in modern management (Mol, 1995; Jänicke, 1993; Schmidheiny, 1992).

Amongst these broader developments, 'sustainable tourism' became a catchword in the 1990s. The World Tourism Organisation, in reinterpreting the definition of sustainable development, defined sustainable tourism as "tourism which meets the needs of present tourists and host regions while protecting and enhancing opportunity for the future" (Stabler, 1997). Many leading tourist enterprises and government agencies, realising the significance of a high quality natural and cultural environment as an input factor to the tourist product, adopted the idea. Changes in consumer preferences – with high value accorded to an unspoiled environment in holiday destinations – further reinforced the importance of the environment as the most critical constituent of the tourist product. This has led to forms of 'alternative tourism' – ecotourism, nature tourism, rural tourism – which are better attuned to the sustainability dictum. In sum, sustainable tourism is now the broad research setting within which the tourism-environment relationship is examined. Several noteworthy publications on the subject have appeared in the last decade which cover topics such as sustainable tourism planning, policies and management, sustainable tourism indicators, the role of national parks, ecotourism, and rural tourism development, as well as sustainable tourism development in specific geographical settings (Bramwell and Lane, 1994; Bramwell et al., 1996; Coccossis and Nijkamp, 1995; Clifford, 1995; Lindberg and

Hawkins, 1993; Nelson and Serafin, 1997; Smith and Eadington, 1994; Stabler, 1997).

The sustainable tourism movement, in conjunction with broader socio-economic and political changes, has added more dimensions to the tourism-environment research agenda. One dimension concerns the integration of tourism with other activities to achieve greater synergy in achieving the goals of environmental preservation and socio-economic vitality or, viability. Hence, the focus has shifted to e.g., agrotourism, especially in the European Union where it is currently considered to be a viable answer to a host of problems facing rural areas: declining populations, rural-urban migration, decrease in income from agriculture, landscape quality, and loss of biodiversity. Low impact tourism, traditional agriculture, and nature protection are concepts in vogue in rural development and planning (Bramwell and Lane, 1994; Schoute et al., 1995; Winter, 1996; Bramwell et al., 1996).

Another dimension is the socio-cultural dimension, which has two facets - the role of societal values and culture in mediating and structuring the tourism-environment relationship, and material and immaterial culture as an important tourism resource. The first facet springs directly from the conception of sustainability as socio-culturally defined and impossible to achieve without broad societal support of development choices. This has resulted in the focus on 'responsible tourism', 'participatory sustainable tourism', and the related discourse (Eber, 1992; D'Amore, 1993). The second facet relates to the emergence of cultural tourism, which encompasses the older forms of historic and urban tourism and the exploitation of the cultural as well as the environmental elements of host areas, in an effort to regain the dual purpose of preserving cultural heritage and boosting local economies through cultural tourism development (Dixon and Fountain, 1989; Ashworth and Larkham, 1994; Fortuna, 1997; Stebbins, 1997; Ashworth, 1993; Ashworth and Dietvorst, 1995; Turnbridge and Ashworth, 1996).

Recapitulating, the study of the environmental impacts of tourism has advanced considerably over the last two decades and has provided the grounds for corrective policy and planning actions. Starting from the traditional focus on the biological and ecological dimensions of the tourism-environment relationship, methods of analysis have proceeded to incorporate the social, cultural, political, ethical and other dimensions under the influence, in part, of the sustainable tourism model. Naturally, several issues still need to be explored as new developments add new topics to the list. The ecological impacts of tourism remain a difficult issue to tackle as it varies over space, alternative forms of tourism, including ecotourism, and across different types of environmental media. Therefore, it may be difficult to isolate impacts, which can be safely attributed to each one of the activities constituting tourism, since several activities are commonly shared with the local population (e.g. shopping, recreation, and travel). It is also difficult to distinguish the ecological impacts of tourism from those caused by natural processes or other activities occurring at the same time and in the same place (Mathieson and Wall, 1982). In addition, a lack of reliable and accurate empirical evidence for measuring and explaining the impacts observed, as well as significant variability in the factors influencing the frequency and magnitude of impacts (e.g. type of tourism activity, intensity, duration, spatial-temporal distribution, etc.), gives rise to problems of comparability among regions and makes it difficult to generalise findings from specific locations and over time.

Similarly, tourist carrying capacity assessments still need to become more precise and quantitative in nature in order to occupy their proper place, and play a more decisive role in tourism planning. This effort cannot be separated, however, from the broader research effort in the area of sustainable development which is in progress in several fields, including tourism development, to provide working definitions and guidelines. Placed in this context, tourism development must be co-ordinated and integrated with the development of the host area along lines dictated by the goal of sustainable regional development. The implications of this requirement for tourism policy and planning research are numerous. Firstly, the spatial level of analysis must not exceed the regional since sustainable development requires grass roots efforts and co-operation among tourism producers and consumers in order to succeed. At the same time, co-ordination among spatial levels is necessary to avoid conflicting actions and interventions. Secondly, the planning horizon has to be extended without losing sight of the present, and without forgetting the considerable uncertainty of the future. It seems that a process of adaptive planning (Holling, 1978) is best suited to this purpose. Thirdly, more integrated approaches must be developed to analyse tourism's environmental impacts that are capable of distinguishing direct from indirect impacts and from impacts resulting from tourism-induced development. This is an important requirement for developing suitable and effective policies directed not only to tourism-related but to other economic activities in the area. Fourthly, the proper planning tools and measures – physical, socio-economic, institutional, legislative, financial – which will put tourism and regional development on the sustainability path have to be investigated further, and their introduction and implementation must be explicitly studied within particular geographic settings. In case such tools exist but are not implemented, their effective implementation must become an important research theme. Last, but not least, proper ways to educate effectively both tourism producers and consumers have to be researched thoroughly ▢ mainly through case studies in various cultural settings. Because only a change in the mentality of the main actors can guarantee the sustainability of development in which tourism represents just one of the many interacting components.

4. Organisation of the volume

The chapters in this revised edition reflect the concerns expressed in the earlier volume and represent further efforts to analyse and understand the various dimensions of the tourism-environment relationship. The chapters are grouped into four parts, following the dimensions of the tourism-environment relationship they address and appear on the title of the book: regional, economic, cultural, and policy issues.

The first part covers broad concerns of the tourism-environment relationship in regional context. The first three chapters revolve around a common issue; namely, modelling the environmental and other impacts of tourism. In the first chapter, Helen Briassoulis focuses on the analytical aspects of the tourism-environment relationship. She first reviews the literature on the environmental impacts of tourism and identifies the main methodological issues involved in environmental impact analysis for tourism. Drawing on this analysis, she then proposes and outlines the structure of an integrated economic-environmental modelling framework based on the materials balance paradigm. Finally, she describes the use of the modelling framework, for impact analysis and evaluation for tourism.

Jeroen van den Bergh, in the second chapter, employs a dynamic simulation model to analyse the relationship between tourism development and the natural environment for an island region in Greece, the Northern Sporades. The model describes the development of the economies of the three main islands of the region and their interactions with the terrestrial and marine environment. The relationship between tourism and the environment is examined on various levels. In addition to direct tourism impacts on the environment through, for instance, pollution, noise, and disturbance, indirect, irreversible and long-term consequences are considered also. In this perspective, tourism development patterns over time, recreational attractiveness of the region, land use patterns, and the growth rate and direction of economic development dominated by tourism receive special attention. Finally, the model is used to test development scenarios in order to find out which scenario can provide an acceptable level of protection of nature and the environment. In addition, the effects can be determined of this scenario on the economic and tourism development of the region.

Patricia Kandelaars uses the same simulation model as Van den Bergh to analyse recent development in the Yucatán peninsula in Mexico, where the well-known tourist resort Cancún is located. In 1970, the village had only 117 inhabitants, in 1976 it had 18,000 inhabitants, while in 1991, the population had skyrocketed to 300,000. Tourism development was, and still is, the main reason for this tremendous growth. As Cancún is located alongside marvellous beaches with a relatively vulnerable ecosystem, environmental disruptions were, naturally, a negative side effect of this growth. The model includes five modules: economic, tourism, demographic, environmental, and governmental. Two scenarios have been developed: a base and an environmental scenario. The latter provides for complete cleaning of the water. By changing the parameters in the model, several options for policy interventions and outcomes are identified for the two scenarios.

Harry Coccossis and Apostolos Parpairis discuss the concept of carrying capacity. They argue that this concept is difficult to define, as many starting points are possible. Attention should be given to ecological as well as economic, sociological, psychological, and cultural dimensions. The concept of sustainable development is more or less related to carrying capacity, which is of particular importance as an operational tool for tourism planning.

Michael Keane discusses the role tourism can play in rural development. He argues that the mechanism of tourism can only bring about rural development in a few rural communities. A more effective strategy for rural development is to make tourism development part of a community integrated development plan. There are good a priori economic arguments as well as encouraging pieces of empirical evidence to support this view. A characteristic of rural community tourism is that it is a community product and that it is developed from local structures. A key factor in the development of community tourism is local co-ordination, linked to wider product and market structures.

The contributions in the second part examine selected economic aspects of the tourism-environment relationship. Jan van der Straaten discusses the problems, which arise when the economic valuation of nature is at stake as is often the case in a sustainable tourism approach. The central question is to what extent it is possible to assign economic value to nature and the environment. If this can be done, the decision-making process regarding the use of nature and the environment in tourism activities is facilitated as it is based on sound economic information. In this context, nature and the environment are different topics. In most cases, the environment is defined as the abiotic part of the ecosystem, which implies that water, air, and soil pollution and the use of natural resources such as iron ore, natural gas, wood, and plastics are the main issues involved. The latter are economic goods, which are priced as they are exchanged in a market. Water, air, and soil pollution do not have a price, however, although people are often economically affected. These costs for consumers and producers can be measured. With respect to nature, the context is different. Nature, the biotic part of the ecosystem, is not traded in a market, particularly in tourism, which means that it is not possible to assign a proper price to represent its economic value. Beautiful beaches, mountain scenery, and attractive Mediterranean landscapes, for example, are very important for tourists. They are input in the production process of the tourism industry and have a high economic value. However, it is not clear how to measure these economic values. In this chapter the author examines to what extent it is appropriate to 'construct' a price on using nature. In the first part of the chapter, the principal questions regarding economic values are discussed, while in the second part, various methods are discussed which have been developed to tackle this problem.

In the next chapter, Jan van der Straaten discusses the traditional welfare economic approaches of valuing nature and the environment. He expresses the opinion that these instruments are no longer appropriate for describing and analysing important environmental problems. This is demonstrated in the case of tourism in mountain areas, which are threatened considerably by erosion and acid rain. By comparing the impact on nature and the environment of a traditional ski resort with an ecotourism development in Italy, he concludes that alternative approaches, in which more attention is given to the relationship between economics and ecology, should be used, and goes on to present such an approach.

Ken Willis and Guy Garrod focus on the recreational value of inland waterways, which are both private and public goods. Commercial recreation extracts some form of payment, for example, from cruising which requires mooring licenses and fees, or from fishing which needs a permit. However, for other types of recreation, such as walking along the towpath, viewing the canal scene, or watching boats pass through locks, there is usually no charge. With the help of an Individual Travel Cost Method (ITCM), the utility or economic benefits are estimated that are associated with informal, non-priced, or 'public good' forms of recreation along selected inland waterways and canals in the United Kingdom. Given that no price is charged for access, considerable benefits from non-priced recreational activity may accrue to individuals by way of consumer surplus. The ITCM permits consumer-surplus estimates to be determined of the value of recreation over the canal system as a whole. At £62 million, this value is considerably higher than the 1989 government subsidies of £44.5 million allocated to the British Waterways Board. The authors pay special attention to typical problems, which can occur during sample collection and the processing of results. Truncation bias problems are quite significant, particularly for recreational visits.

The third part includes contributions related to the cultural dimension of the tourism-environment relationship. First, Greg Richards discusses the complex relationship between tourism and culture. In the past, studies of tourism and the environment have tended to concentrate on the 'natural' environment and did not pay attention to the role of culture in creating environments for tourism and in mediating the way in which environments are consumed by tourists. Recent studies of the tourism phenomenon have begun to redress the balance, by pointing to the way in which the production and reproduction of 'nature' is highly culturally determined. The modern production of nature is closely bound up with the growth of a culture of tourism. National parks and nature reserves which are specifically demarcating natural areas, came into being only as increasing numbers of people began to appreciate their value through tourism. This growing demand from tourists threatens the very wildness that tourists come to consume, causing the adoption of a range of management strategies to solve the conflicts between the needs of wildlife, residents, and visitors. But most areas tourists visit today are not a true wilderness but rural areas. This fact often leads to conflict between urban-based tourists and their rural hosts.

Modernisation threatens the urban landscapes as well. The rhetoric once employed against the 'rape of the countryside' is increasingly being voiced by those struggling to preserve 'unique' urban landscapes and elements of cultural heritage located in major urban centres. The process of redevelopment in urban areas is currently seen as a modern scourge. As is the case for natural environments, the creation and maintenance of conservation environments is increasingly dependent on tourism and leisure consumption. As a result, interesting parallels emerge between the processes at work in both 'natural' and 'cultural' environments. The increasing scarcity of certain types of landscape or places imbues them with a certain symbolic value, which, in turn, generates real economic value through the commodification process, creating even greater pressure to transform these places for economic purposes. Tourism is often seen as one of the least harmful ways of maximising the economic potential of these symbolic places. The chapter analyses the way in which the environment has been transformed into a cultural product for tourism consumption. The way in which culture is produced and reproduced for tourism consumption is analysed in both urban and rural contexts, with specific emphasis on recent trends in cultural tourism consumption in rural areas.

Theano Terkenli focuses on the broader question of assessing the extent to and the ways in which tourism contexts have been informed by the essential nature of the tourist phenomenon, as opposed to the social and cultural histories and geographies within which they have arisen. She employs the concept of landscape, as a cultural geographic construct and as a concrete site of human life (the landscape of Serifos, Greece), to illustrate the role and significance of geographical and, especially, cultural specificity in tourism analysis. To negotiate the larger question of the nature of tourism in the Western World in the last century or so, she analyses the cultural dimension of tourism from a geographical theoretical perspective and suggests an analytical framework for tourism research; namely, the study of landscapes of tourism. The landscape is presented as a social interface where local and global analytical perspectives, public and private life, planning priorities and consumption patterns come together in the ready construction and consumption of place identity. She presents a synthetic reading of selected modern and postmodern symbolic landscapes of the Western World tracing the relevance of cultural factors in historical tourist space construction and perception. She discusses a number of distinctive qualities that are prominent in the composition of contemporary landscapes of tourism as a social phenomenon, such as a convergence of the public with the private spheres of life and the infusion of visual and leisure elements into a growing number of local activities and preoccupations. She concludes with a case study on the island of Serifos, Greece, to assess the impact of tourism locally, in light of the proposed broader transformations wrought by tourism onto the landscape of the Western World.

Paris Tsartas analyses the problems and implications of the use and management of environmental and cultural resources in tourist areas. First, he presents the development trends in modern tourism which have affected the management of environmental and cultural resources in tourist areas: development of 'alternative' tourism, changes in organised mass tourism, organisational and economic changes in world tourism, changes and reorientation of tourist motives, the emergence of 'socially responsible' tourism, and transformation of tourism into a consumer good in modern society. He goes on to examine the factors which affect the development and management of tourist, cultural, and environmental resources: degree of growth of organised mass tourism, special characteristics of local resources, the massive development of alternative tourism, tourist marketing and advertising, tourism policy in sustainable development programs, the definition of tourist resource, the tourism perspective on legislation for the protection and management of culture and the environment, organisational development of a country's tourist sector, and its position in the international division of leisure and local actors engaging in tourist development. He concludes that currently (1) the notion of tourist resource is differentiated and includes a wide range of elements, infrastructure, and activities related to the culture and the environment of tourist areas, (2) planning and policy-making for the management of cultural and environmental resources in tourist areas is becoming all the more complex because of the special characteristics of these resources and their significance for modern tourists, and (3) culture and the environment will continue to form the basic fields in which tourist resources are produced and will constitute the poles of attraction for more and more specialised tourist trips.

Joseph Stefanou suggests that experimental iconology can be used for the analysis of landscape quality, when improvements are necessary to promote a place to tourists. This approach can help to identify those landscape elements, which make the greatest contribution to the appeal an area might or might not hold for tourists. The results obtained in this way are used in synthesis, i.e. in the creation of tourist images and places which will ensure the sustained appeal of an area, as well as the design of policies to make this attractiveness pay.

In the following chapter, Joseph Stefanou elaborates on the use of the image of a place, its landscape, as a principal means for the development and promotion of tourism in this place. He utilises the postcard, which is simultaneously a communication medium, a consumer product and mass art, as a tool for the analysis of the landscape of tourist places. This technique is based on the presumption that the aesthetic and semantic interpretation of the landscape depends on the way and mode by which it is formed and perceived as well as on the degree of mental, psychological and practical appropriation of the landscape by an individual.

The last part of this volume is devoted to policy issues that have been raised in the context of the tourism-environment relationship. Cees van Woerkum, Noelle Aarts and Cees Leeuwis discuss the problems and challenges of communication in the potential conflict between the various users of nature and the environment. The authors argue that it does not make sense to implement rules without communication with the potential users of natural assets. They describe two approaches to communication for sustainable tourism. In the first instrumental approach communication has the function of influencing behaviour directly or, perhaps even more importantly, supporting other instruments, by informing people or by gaining acceptance. In the second approach, communication in the process of policy-making itself, is regarded as a more viable policy and learning process regarding nature and tourism. These approaches do not exclude each other. Interactive policy-making seldom achieves all we wanted. Instrumental campaigns are still necessary. But the interactive approach and the broad communication it provokes can provide the basis for more public awareness and understanding. Unless the whole communication network around tourism is activated in the right way, the instrumental approach is difficult to pursue. The tourists will still get contradictory messages from different sources. It is unclear which messages appeal to them most.

Helen Briassoulis presents an ex ante critical examination of the implementation potential of theoretically prescribed sustainable tourism development policies within current tourism policy implementation environments. After briefly describing the tourism system, sustainable tourism development, and the tourism policy implementation environment, the main policy prescriptions for sustainable tourism development are screened on the basis of three evaluation criteria: tractability of the policy problem, compliance, and influence of non-statutory factors. Conceptual, scientific-technical, economic, socio-cultural, institutional, legislative, administrative, political, physical, and communication implementation issues are also identified in this chapter. Three general courses of action are proposed to, at least partially, address these issues: (1) retain and improve the positive aspects of current implementation environments, (2) introduce new forms of tourism policy implementation and (3) feed back into tourism policy to design policies that can be implemented more easily.

George Ashworth focuses on cities as specific environments, even though the natural attributes of site, vegetational cover, building material, and the like have been restructured by deliberate intervention and design. The distinction between the so-called natural and the built environment is determined by the degree of human intervention rather than its existence. Which planning strategies can be used to develop cities from a tourist perspective? Although some examples are offered, no general lessons for the development of a planning strategy can be drawn. The difficulty is that the very selectivity of these cases stresses their particular characteristics. There is no clear-cut blueprint. Neither the size, antiquity, dominant political ideology, type of commodified heritage environment, nor a particular mix of functions seems to offer clear guidelines. A high quality urban environment and urban heritage tourism can be incompatible or mutually supporting opportunities; the choice between these alternatives is not predetermined by any particular set of conditions and, thus, remains open to deliberate decision.

Jan van der Borg argues that, when analysing the impact of tourism on the environment, reference is usually made to the devastating effects mass tourism has on the natural environment. It is only recently, that the question has arisen whether cities, originally designed to host people, might have similar problems with tourism. An affirmative answer to this question implies that a city's policy for tourism development has to account for the city's limits for absorbing visitor flows. In other words, urban tourism development strategy has to be compatible with the urban environment. The aim of his contribution is to discuss the principal characteristics of such a strategy, as far as environmental issues are concerned. The intention is to give a comprehensive answer to the question whether, and under what circumstances, urban tourism may be worth developing, a crucial question for many cities that are currently considering the promotion of tourism development.

Nikos Konsolas and Gerasimos Zacharatos discuss the problem of regionalisation of tourism activity and related policies in Greece. Compared with the autonomous character of policies for regional development during the 1974-1989 period, the exclusive focus on the monetary aspects of international tourism has gradually been abandoned, and the regionalisation of tourism development has come to be recognised as the second most basic aspect (after currency) of this development in Greece. In this perspective, tourism development is now being promoted as one of the basic instruments of regional policy, especially in problematic and socio-economically depressed areas.

George Chiotis and Harry Coccossis highlight some of the basic policy issues relating to the role of tourism in national and regional development, with a particular focus on the strong interrelationships between tourism policies and the environment. The basic question in their chapter revolves around the role of tourism in Mediterranean countries and particularly in certain regions, which are sensitive to tourism and to the preservation of their natural resources and their environmental quality. To illustrate the issues involved in the context of tourist development and environmental protection, the experience of Greece and some of its regions are used as an example. Special reference is made to the role of the European Community and international co-operation.

Frank Convery and Sheila Flanagan deal with the relationship between tourism and the environment in Ireland. The environment – natural and man-made – is of vital importance to tourism in Ireland. It represents the backdrop to many other activities and comprises a major attraction in its own right. The purpose of the authors is to examine the development of environment-based tourism in Ireland, and its possible impact on the landscape. They discuss rural and urban threats to the environment in relation to tourism, and examine and compare the tourism management strategies that are available for environmental protection in Northern Ireland.

Finally, Theodosia Anthopoulou addresses the question of whether agrotourism can contribute to the preservation of the rural environment. She focuses on the mountainous and less-favoured zones of the Mediterranean which, even after the reform of the Common Agricultural Policy of the European Union in 1992, are still penalised. As they cannot compete with the high productivity regions, they are socio-economically marginalised, and face significant environmental problems. Diversification of farming into activities other than the production of raw materials, such as agrotourism, is presented as an alternative for sustainable development in these areas. The author describes the basic characteristics of agrotourism, describes its features and assets in the less-favoured areas of the Mediterranean, and identifies three main types of agrotourism units which can be distinguished in these areas, and especially in Greece. She describes the introduction and organisation of agrotourism in Greece and summarises a case study of a women's agrotourism co-operative in Petra, on the island of Lesvos. Based on this experience, she identifies the problems which the development of agrotourism in Greece faces, and evaluates its contribution to the preservation of the rural environment. She suggests that an agrotourism development policy for the Mediterranean mountainous and less-favoured areas is necessary as it would take into account the need to retain both the population and the agricultural activities which are gradually being abandoned in order to preserve the fragile Mediterranean environment.

References

Ashworth, G.J. (1993) Culture and tourism: conflict of symbiosis in Europe. In W. Pomple and P. Lavery (eds), *Tourism in Europe: Structures and Developments.* CAB International, Wallingford.

Ashworth, G.J. and A.G.J. Dietvorst (eds) (1995) *Tourism and Spatial Transformation: Implications for Policy and Planning.* CAB International, Wallingford.

Ashworth, G.J. and P.J. Larkham (eds) (1994) *Building a New Heritage: Tourism, Culture and Identity in the New Europe.* London, Routledge.

Baud-Bovy, M. and F. Lawson (1977) *Tourism and Recreation Development.* The Architectural Press, London.

Bramwell, B. and B. Lane (eds) (1994) *Rural Tourism and Sustainable Rural Development.* Channel View Publications, Clevedon.

Bramwell, B., I. Henry, G. Jackson, A. Goytia Prat, G. Richards and J. van der Straaten (eds) (1996) *Sustainable Tourism Management: Principles and Practice.* Tilburg University Press, Tilburg.

Clifford, G. (1995) *What Tourist Managers Need to Know.* Paper presented at the World Conference on Sustainable Tourism, April, 1995.

Coccossis, H. and P. Nijkamp (1995) *Sustainable Tourism Development.* Aldershot, Avebury.

D'Amore, L.O. (1993) A code of ethics and guidelines for socially and environmentally responsible tourism. *Journal of Travel Research* **31**, 64-66.

Dixon, M. and K. Fountain (1989) *Contribution to the Drafting of a Charter for Cultural Tourism.* ECTARC - EEC, Wales.

Dunkel, D.R. (1984) Tourism and the environment: A review of the literature and issues. *Environmental Sociology* **37**, 5–18.

Eber, S. (ed) (1992) *Beyond the Green Horizon.* A discussion paper on Principles for Sustainable Tourism, WWF, Surrey, UK.

Edwards, J.R. (1987) The UK heritage coasts: An assessment of the ecological impacts of tourism. *Annals of Tourism Research* **14**:1, 71–87.

Farrell, B.H. and R.W. McLellan (1987) Tourism and physical environment research. *Annals of Tourism Research* **14**:1, 1–16.

Farrell, B.H. and D. Runyan (1991) Ecology and tourism. *Annals of Tourism Research* **18**:1, 26–40.

Fortuna, C. (1997) Cultural tourism in Portugal. *Annals of Tourism Research* **24**:2, 455–457.

Gartner, W.C. (1987) Environmental impacts of recreational home developments. *Annals of Tourism Research* **14**:1, 38–57.

Holling, C.S. (ed.) (1978) *Adaptive Environmental Assessment and Management.* John Wiley, New York.

Inskeep, E. (1987) Environmental planning for tourism. *Annals of Tourism Research* **14**:1, 118–135.

Inskeep, E. (1991) *Tourism Planning: An Integrated and Sustainable Development Approach.* Van Nostrand Reinhold, New York.

Inskeep, E. (1994) *National and Regional Tourism Planning: Methodologies and* Case *Studies.* Routledge, London.

International Journal of Environmental Studies (1985), Tourism and the Environment, Special Issue.

Jackson, I. (1986) Carrying capacity for tourism in small tropical Caribbean islands. *Industry and Environment,* January/February/March, 7–10.

Jänicke, M. (1993) Über ökologische und politische Modernisierungen. *Zeitschrift für Umweltpolitik und Umweltrecht* **2**, 159–175.

Lindberg, K., and D.E. Hawkins (eds) (1993) *Ecotourism a Guide for Planners and Managers.* The Ecotourism Society, North Bennington.

Lindsay, J.J. (1986) Carrying capacity for tourism development in national parks of the United States. *Industry and Environment,* January/ February/March, 17–20.

Liu, J.C. and T. Var (1987) Resident perception of the environmental impacts of tourism. *Annals of Tourism Research* **14**:1, 17–37.

Mathieson, A. and G. Wall (1982) *Tourism: Economic, Physical, and Social Impact.* Longman, London.

Miller, M.L. (1987) Tourism in Washington's coastal zone. *Annals of Tourism Research* **14**:1, 58–70.

Mol, A.P.J. (1995) *The Refinement of Production. Ecological Modernization Theory and the Chemical Industry.* International Books, Utrecht.

Nelson, J.G. and R. Serafin (eds) (1997) *National Parks and Protected Areas.* Springer Verlag, Berlin.

OECD (1980) *The Impact of Tourism on the Environment.* OECD, Paris.

Pearce, D.G. (1985) Tourism and environmental research: A review. *International Journal of Environmental Studies* **25**, 247–255.

Pearce, D.G. and R.M. Kirk (1986) Carrying capacities for coastal tourism. *Industry and Environment,* January/February/March, 3–7.

Rondriguez, S. (1987) Impact of the ski industry on the Rio Hondo watershed. *Annals of Tourism Research* **14**:1, 88–103.

Schmidheiny, S. (1992) *Changing Course. A Global Business Perspective on Development and the Environment.* MIT Press, Cambridge.

Schoute, J.F.Th., P.A. Finke, F.R. Veeneklaas and H.P. Wolfert (eds) (1995) *Scenario Studies for the Rural Development.* Kluwer Academic Publishers, Dordrecht.

Smith, V.L., and W.R. Eadington (eds), (1994) *Tourism Alternatives.* John Wiley, Chichester.

Stabler M.J. (ed.), (1997) *Tourism Sustainability: Principles and Practices.* CAB International, Oxon.

Stebbins, R T. (1997) Identity and cultural tourism. *Annals of Tourism Research* **24**:2, 450–452.

Tangi, M. (1977) Tourism and the environment. *Ambio* **6**:6, 336–341.

Turnbridge, J.E. and G.J. Ashworth (1996) *Dissonant Heritage: the Management of the Past as a Resource in Conflict.* Wiley, London.

UNEP (1982) Tourism. In: *The World Environment 1972–1982.* Ch.14, Tycooly International Publishing Co, Dublin.

UNEP (1987) *Report of the Seminar on the Development of Mediterranean Tourism Harmonized with the Environment.* Priority Actions Programme, Regional Activity Centre, Split, Croatia.

Wall, G. and C. Wright (1977) *The Environmental Impact of Outdoor Recreation.* Publication Series No.11, University of Waterloo, Department of Geography, Waterloo.

WCED (1987) *Our Common Future* (Brundtland-Report). Oxford University Press, New York/Oxford.

Winter, M. (1996) *Rural Politics.* Routledge, London.

WTO (1980) *The Manila Declaration on World Tourism.* World Tourism Organization, Madrid.

ENVIRONMENTAL IMPACTS OF TOURISM: A FRAMEWORK FOR ANALYSIS AND EVALUATION

HELEN BRIASSOULIS
Department of Geography
University of the Aegean
Karantoni 17, Mytilini, Lesvos 81100
Greece

1. Introduction

Tourism grew rapidly during the 1960s and 1970s, but it was soon realised that this growth was not without costs. The social, cultural, economic, and environmental impacts of tourist growth became subjects of serious study and research, and entered in the policy agendas of national and international organisations. Tourism is no longer considered a 'clean industry' as opposed, say, to heavy manufacturing. Tourism planning is advocated as a tool for controlling the negative impacts of tourism development and for protecting the very same resources upon which the profitability of the industry depends. However, many tourism development studies and plans as well proposed policies have not been (and still are not) based on rigorous quantitative and integrated analyses of the several dimensions of tourism, neither have they placed the impacts of tourism development within the broader spatio-temporal context of their occurrence. There are many reasons for this omission, the analytical complexity of the task and the lack of proper data being among the most important of them. This chapter focuses on the environmental impacts of tourism and proposes an integrated framework for their analysis and evaluation. It is a revised version of the original chapter which appeared seven years ago in the first edition of this volume. It keeps the essential methodological orientation and gist but attempts to refine the integrated modelling framework, which was proposed originally. Moreover, it addresses explicitly the question of sustainability, which, in the meantime, became a central concern in all discussions about tourism development. The first section of the chapter reviews briefly the literature on tourism and its environmental impacts and identifies the main methodological issues in their analysis. The second section details the proposed methodological framework while the last section discusses its advantages and drawbacks.

2. Tourism and its environmental impacts

The study of tourism in general occupied a rather minor position in the academic literature until the late 1960s since most studies were concerned with the broader activity of recreation. But, while recreation refers to any pursuits taken up during leisure time (other than those to which people are normally 'highly committed'), tourism refers to longer journeys and temporary stays (more than 24 hours) of people at a place other than their own (Baud-Bovy and Lawson, 1977; WTO, 1981). Tourism is not a single activity but a complex of activities involving, mainly, travel, accommodation, sightseeing, shopping and entertainment (Baud-Bovy and Lawson, 1977; Komilis, 1986). Tourism has a strong international as well as a domestic dimension. The tourist product is a multidimensional entity, which includes: resources at a destination (inherent attractions); facilities at a destination (accommodation, catering, recreation, transport, information and assistance); and transport to the destination (Baud-Bovy and Lawson, 1977). Hence, tourism cannot be defined as a separate economic sector, in the traditional sense, because most functions performed during tourism, although inseparable from one another, are accounted for by the conventional economic sectors and secondly because its product has both tangible and intangible components. Tourist activities can be classified according to three broad dimensions:
1. the purpose and motives of tourists (leisure and vacation, business, health)
2. the organisation/participation forms and socio-economic status of tourists (mass versus individual)
3. its space-time characteristics (winter/summer, coasts/mountains, agro-tourism) (Komilis, 1986)

The above conceptualisation of tourism has important implications for the analysis of its environmental (as well as its social, economic, etc.) impacts. Firstly, tourism's environmental impacts originate in various economic sectors (Fletcher, 1989) as well as in activities which are not directly recorded by the economy (e.g. hiking, snorkelling, etc.). In addition, certain tourism-related activities are also activities of the host area's population (e.g. shopping, entertainment, services). Secondly, impacts originating in a given tourism-related sector may cause other impacts in other sectors. Thirdly, the type and intensity of the environmental impacts of tourism depend on the interaction between the type of tourism development, the socio-economic and other characteristics of tourists and the natural, socio-economic and institutional characteristics of the host area (Mathieson and Wall, 1981). An analysis, which disregards these considerations cannot be but an incomplete account of tourism's impacts. On the other hand, however, the practical difficulties of making all relevant factors operational and obtaining the proper data cannot be discounted easily and may be one of the main reasons for the narrowness of current approaches to the analysis of the environmental impacts of tourism.

The study of the environmental impacts of tourism started basically after 1970. The topics studied have ranged from perceptual, ecological, economic-environmental impacts to questions of planning and policy measures for tourism harmonised with the environment (UNEP, 1987; WTO, 1983). Most studies have concentrated mainly on heavily developed tourist regions which experienced earlier than others the negative impacts of tourism development, such as coastal, mountainous (ski resorts) and wilderness areas (national parks). The analysis of the environmental impacts of tourism has been predominantly qualitative and mostly descriptive. Usually, only a few aspects of the impact of tourism on the environment of a region are examined, as well as the difficulties of quantifying the ecological impacts and the related carrying capacity and sustainability concerns. Among the reasons accounting for these difficulties are the lack of generally accepted environmental indicators for a variety of impacts and the rapid changes in environmental problems caused by technological and industrial development.

Similarly, the various policy measures proposed (some of which are already effective) – in order to secure the long-term viability of the environment and the economy of host tourist regions (EEC, 1992; OECD, 1980; UNEP, 1987; UNEP, 1982) – have a mostly qualitative basis and their design does not draw from some broader theoretical and methodological framework. An OECD study (1980) attempted a more comprehensive categorisation of impacts and asked from member countries to provide quantitative estimates of the reduction in tourist flows due to environmental degradation (caused by tourism development, accidents, etc.). The study recognises the difficulty to operationalise the tourism-environment relationship and to compare the experiences of geographically, culturally, and economically different countries.

In a UNEP report (1982), the direct and indirect environmental impacts of tourism were given particular attention and attributed to the nature of the host region, the type of tourism promoted, the planning and management system in effect, and the ideology and behaviour of tourists.

The most thoroughly studied and emphasised issue in the study of the environmental impacts of tourism is the carrying capacity of a region, the physical, social, biological, and psychological capacity of the environment to support tourist activity without diminishing environmental quality or visitor satisfaction (Lindsay, 1986; Coccossis and Parpairis, this volume). This concept, borrowed from the natural sciences and employed, originally, in recreation and development studies, although of instrumental importance in guiding environmental management and policy making, has met several difficulties in becoming operational in the case of tourism. First because it is multi-dimensional (Pearce and Kirk, 1986), secondly is it difficult to identify the critical 'limiting' resource to be used as a basis for estimating the carrying capacity of a region, thirdly the number of individuals a region can accept without undergoing significant environmental deterioration depends on the type and intensity of tourist activity (Lindsay, 1986; UNEP, 1982) as well as on technological factors, and finally there are usually more than one competing explanatory factors of environmental degradation one of which may be tourism. It has been suggested that assessment of the carrying capacity of a region must be made on a case by case and activity by activity basis (UNEP, 1982).

The omnipresent concept, however, of the 1990s is 'sustainable tourism development'. Despite its wide use in academic, policy and business circles, sustainability of tourism development remains loosely defined, mostly associated with environmental protection but never treated holistically as its original definition in the Brundtland Report (WECD, 1987) dictates. Nevertheless, it provides a sound reference point for directing tourism away from destructive and towards environmentally benign modes of development.

Finally, most studies make a plea for holistic and integrated approaches to the study of tourism and its impacts. Systems analysis is the most widely advocated analytical tool (Sessa, 1988; UNEP, 1982) but very few operational models have appeared in the literature yet thus far (see, for example, Coccossis and Nijkamp, 1995; Nijkamp and Verdonkschot, 1995). Lack of adequate quantitative data restricts most system analytic applications to abstract and very general formulations, which address general policy problems.

From the literature reviewed the following methodological issues in environmental impact analysis of tourism development can be identified:

1. Because tourism is essentially an activity complex, its impacts are spatially and temporally diffuse, difficult to categorise, identify, measure exactly, and distinguish from the impacts of other contemporaneous activities or natural processes. Because many natural and man-made resources are shared in common by tourists and non-tourists, the critical question is how to isolate impacts due to tourism from those due to other activities. Moreover, these impacts are rather cumulative arising out of many, small, individual actors (the equivalent of non-point sources of pollution) rather than out of a few, large actors (the equivalent of point sources of pollution).

2. The environmental resources, which constitute part of the touristic attractions of a region are simultaneously receptors of the undesirable by-products of tourism (solid and liquid wastes, air pollutants, congestion, visual and noise pollution); i.e., the external costs of tourism are internal to its production. The conflicts created thereof cannot be ignored, as it frequently happens when source regions are spatially separated from receptors, but must be resolved locally and in the short term for the region to avoid long-term reduction in tourism due to environmental degradation.

3. Because the conduct and form of tourism activities change over time due to changes in the economy, technology, social habits and norms, etc., the spatial and temporal pattern, degree of concentration, intensity, etc. of its environmental impacts change as well, making their assessment difficult and

requiring continuous monitoring and methods sensitive to the variable nature of the phenomenon.
4. Tourism 'induces' further development of the host region caused either by speculative land development to tap agglomeration economies or by provision of infrastructure (roads, water supply, sewage) which attracts activities which would not otherwise locate in the particular region.

Summarised, both the nature of the phenomenon and the relevant literature reveal a need for an integrated framework for the analysis of the environmental impact of tourism which can address the methodological issues raised above. This is the topic of the next section.

3. A framework for the integrated analysis of the environmental impacts of tourism

The aim of the proposed framework is to provide a tool for the comprehensive assessment of the environmental impacts of tourism, a necessary prerequisite for making sound tourism planning and policy decisions. The conceptual framework and an impact categorisation scheme are presented first, the proposed modelling framework is described next, the general procedure for applying this framework for environmental impact analysis of tourism development is outlined and, finally, the question of assessing the sustainability of current or proposed tourism development is discussed.

3.1. CONCEPTUAL FRAMEWORK AND IMPACT CATEGORISATION

In order to model the interface between tourism, an economic activity complex, and the environment, the materials balance principle is used as a broad, conceptual frame of reference. The Materials Balance Model (Figure 1) has long been proposed as a suitable conceptual, at least, tool for the analysis of economic-environmental problems (Kneese *et al.,* 1970; Victor, 1972). The version suggested for the case of tourism is shown in Figure 2. The model postulates mainly that the economic system interacts with the environment in two principal ways: first, the economy uses environmental resources as inputs to production and consumption of economic goods and services and, second, it disposes residuals (usually, harmful by-products) to environmental media. According to the first law of thermodynamics, all material and energy inputs to the economic system must equal the material and energy outputs of this system. Despite limited information, which makes its complete operationalisation and application difficult, the materials balance model provides a useful organising scheme for studying the environmental impacts of one or more economic activities.

Based on the above framework, four major classes of environmental impacts of tourism can be distinguished: (1) production-related impacts on resources; (2)

consumption-related impacts on resources; (3) production-related residuals impacts; (4) consumption-related residuals impacts. The resources impacts concern depletion and competition for resources between tourism and other economic activities, while the residuals impacts refer to pollution and environmental degradation of resources and receptors.

Within the above four broad classes, environmental impacts can be categorised further on the basis of other criteria, such as: (1) directness of impact B direct, indirect, and induced; (2) spatial scale and extent B local, regional, national as well as point and non-point; (3) temporal scale B short-, medium-, and long-term impacts. Direct impacts are directly emanating from tourism activities such as accommodation, travel, and sightseeing. Indirect impacts are caused by activities linked to tourism such as retail trade, entertainment, etc. Induced impacts are due to non-tourist-related activities which are induced by tourism development of a region and would not have occurred otherwise (e.g. first and second home development, recreation, etc.). The other categories of impacts are more or less self-explanatory.

At a given spatial scale, the total environmental impact of tourism can be considered as the resultant of production and consumption impacts on resources (direct/indirect/induced) and production and consumption residuals (direct/indirect/induced) impacts, properly distinguished according to their temporal dimension. Interaction effects between resources-related and residuals-related impacts have to be taken into account, also since competition for resources among tourism-related activities is not uncommon and resources used as inputs to the 'production' of tourism serve as receptors of its residuals as well.

3.2. AN INTEGRATED MODELLING FRAMEWORK FOR THE ANALYSIS OF THE ENVIRONMENTAL IMPACTS OF TOURISM

Drawing from the conceptual framework of the materials balance model, an integrated economic/environmental modelling framework (Figure 3) is proposed for the analysis of the environmental impacts of tourism at the regional level. It consists of four modules: a societal, an economic, an environmental, and an interfacing module. Exogenous information, such as national socio-economic and environmental forecasts and technical information is also fed to the modules when necessary.

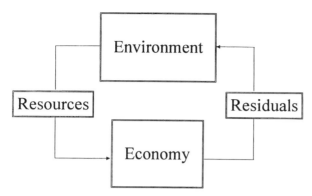

Figure 1. A schematic representation of the Materials Balance Model

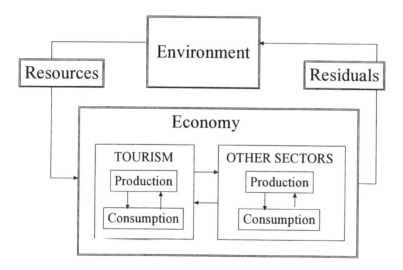

Figure 2. The Materials Balance Model version for tourism

The Societal Module represents, on the one hand, the demographic and social structure of the study region (the regional profile part) and, on the other, the characteristics of tourism demand (the tourism profile part) related to this region. Representative variables, which can be used, are shown in Table 1. This module provides input to the Economic Module in the form of final demand for the products of the economic sectors of the region, which originates both in the local population and in tourists flowing to the region. Naturally, in a dynamic analysis context, the Societal Module can receive feedback from the Economic Module as changes in the welfare of the region (due, among other things, to tourism) may cause demographic and social changes. The Interfacing Module translates the characteristics of the Societal Module into demand for resources and residuals generated due to activities which are not accounted for by the economic accounts of the study region (e.g. open air recreation activities of both locals and tourists).

The Economic Module offers a description of the present economic structure of the region and tourism's position in it. It comprises of the profiles of the economic sectors of the study region and the structural relationships between them. Relevant variables are shown in Table 1. For an explicit treatment of tourism activities a distinction is drawn between tourism-related sectors and other sectors (based, e.g., on the estimated share of tourism in the total receipts of a sector). Moreover, that part of economic activity, which can be safely considered as induced development, triggered by tourism, has to be treated separately. Table 2 shows representative tourism-related economic sectors. The structural relationships among the sectors distinguished before may be modeled by an inter-industry model B industry by industry (Fletcher, 1989) or industry by commodity input-output or regional econometric model (Solomon, 1985) B capable of capturing the direct, indirect, and household consumption induced effects of regional growth. If resources do not permit, simpler models representing the relationships among the most closely associated sectors can be used.

The Economic Module receives input from the Societal Module as mentioned before. Moreover, it may provide input to the Societal Module in two respects. Firstly, changes in regional economic activity are translated into changes in the demographic and social characteristics of the region. Secondly, changes in the economic conditions of the host area may imply changes in the associated profile of tourism (e.g. influx of wealthier or poorer tourists). The Economic Module is connected to the Interfacing Module which translates the output of the economic system into demand for resources and generation of residuals. Finally, Exogenous Information (e.g. population and tourism projections, changes in prices and wages, changes in technology, etc.) are provided as input to the Economic Module.

The Interfacing Module's function is to translate the output of the Economic Module into: (1) demand for resources of various types, by means of a resources utilisation model, and (2) quantities of residuals generated, by means of a residuals generation model. Operationally, these models can range from simple linear relationships – resources utilisation rates (e.g. acres of land/1000 tourists) and residuals generation coefficients (e.g. mg SO_2/currency unit$ output) – to complex multivariate forms. The resources utilisation model should be able to depict also the competition for resources between tourism-related and the other sectors in the study region; for example, by assessing each sector's economic return per unit of resource used. The output of the Interfacing Module is fed into the Environmental Module. Conversely, the Environmental Module can provide input to the Interfacing Module to be translated into environmental constraints (e.g. upper limits on available resources) imposed on the economic system.

Table 1. Representative Variables of the Societal and the Economic Modules

SOCIETAL MODULE	
Regional population profile	*Tourism profile*
Local Population – volume	Number of tourists
Age/sex structure of population	Nationalities – distribution
Occupational structure	Age/sex structure of tourists (average)
Active population by main categories	Types of tourists (various classifications)
Educational level	Tourist income distribution
	Consumption expenditures by product type
	Seasonality of tourist demand
	Length of stay; duration of tourist period
Policy profile	
Supranational policies	
Agricultural policies	
Industrial policies	
ECONOMIC MODULE	
Sectoral composition	Sectoral imports
Sectoral output	Sectoral exports
Sectoral employment	Sectoral final demand
Sectoral value added	Number of firms
Product prices	Wages by sector
Interindustry (technical) coefficients	Consumption coefficients

The Environmental Module consists of: (1) an inventory of regional environmental and physical resources (Table 3) and estimates of their carrying capacity; (2) a resources impact model; and (3) a residuals impact model. The resources impact model receives input from the resources utilisation model of the Interfacing Module and shows how the demand for resources by the economy (including tourism) impacts on the quantity of available regional resources (of a given quality class). Similarly, the residuals impact model receives input from the residuals generation model of the Interfacing Module and assesses how the residuals generated by regional economic activity (including tourism) affect the quality of regional resources and whether they overload the carrying capacity of the regional environmental and physical receptors. The models can be either quantitative or qualitative depending on the type of resource and the state-of-the-art in environmental model building. Both the resources and residuals impact models are linked to the resources inventory to assess changes in the quality and quantity of available resources and check if the carrying capacity of some of them has been exceeded. Although the latter cannot be measured unambiguously, and any measure can be revised due to changes in technology, society, the environment, etc., for the foreseeable future, the most limiting regional resource may be identified and used for figuring out the maximum number of human users (locals and tourists) of regional resources.

Table 2. Representative tourism-related economic sectors

Accommodation (all categories)
Recreation and entertainment facilities
Specialised facilities (marinas, ski lifts, etc.)
Transportation B local, supralocal
Utilities (electricity, telephone, etc.)
Water supply and sewage disposal
Solid wastes disposal
Retail services (food, clothing, etc.)
Specialised tourist services
Car rentals
Banking services

Input from the regional resources inventory can be translated – via the Interfacing Module – into environmental and physical limits on the economy imposed by the carrying capacity of regional resources; in other words, maximum allowable amounts of resources inputs and residuals levels necessary for the normal functioning of the economic (and, of course, of the environmental) system. The Environmental Module accepts also environmental information concerning exogenous conditions, which may affect the regional environment (e.g. climate change).

Table 3. Regional Resources Inventory

Natural	Cultural
Abiotic	Traditional settlements
Land (by class)	Monuments
Landscape/landforms	Historic buildings
Soil (by class)	Historic districts
Surface waters (quantity/quality)	Archaeological sites
Groundwater (quantity/quality)	Historic sites
Air	Scenic areas
Biotic	Landmarks
Forests	National parks/ forests
Ecosystems (terrestrial, marine, coastal, littoral, alpine, etc.)	Traditional feasts
Agro-ecosystems	Festivals
Wetlands	Expositions
Wildlife – endangered species	

3.3. IMPACT ANALYSIS PROCEDURE USING THE INTEGRATED MODELLING FRAMEWORK

The analysis of the environmental impacts of tourism aims at assessing the magnitude of the basic categories of impacts identified in Section 2.1. The integrated modelling framework proposed before can be used either for ex post or ex ante impact assessment depending on whether a region is already developed touristically or if it is considering some form of new (or additional) tourism development. In this section, the basic steps followed in the application of the modelling framework are presented first followed by a discussion of certain methodological differences between ex post and ex ante analysis.

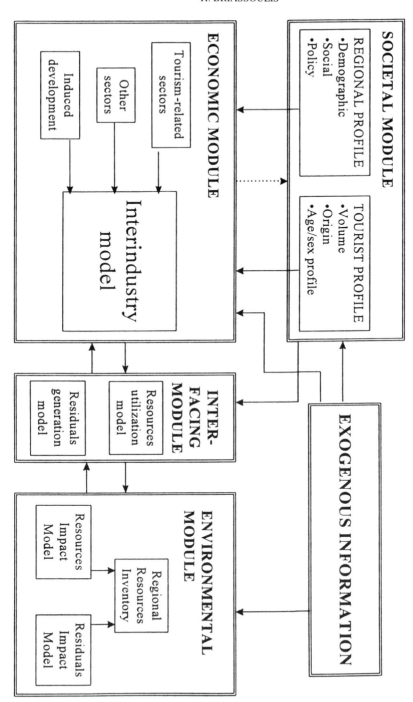

Figure 3. The integrated impact analysis model

The procedure starts from the Societal Module. Given an adequate description of the regional demographic and social structure as well as of the pertinent tourist demand, the related final demand for the output of the economic sectors of the region placed by the resident and the tourist population is either estimated or it is projected (in an ex ante impact analysis context). Exogenous projections of the future resident and tourist population as well as of anticipated changes in technology, tastes, etc. can be used also to assess more closely future final demand. This is fed then into the Economic Module.

The Economic Module accepts the estimated or projected final demand from the Societal Module to assess the output of the economic sectors necessary to meet this demand. These assessments incorporate the direct, indirect and induced effects of quantitative and qualitative changes in tourist and/or resident population. Exogenous information regarding changes in the interindustry and primary inputs (raw materials, labour, capital) coefficients, projected changes in imports and/or exports, etc. can be fed also into the economic module to be taken into account when assessing the sectoral outputs, employment, income, etc.

The output of the Economic Module can be fed back to the Societal Module to relate changes in sectoral output (as well as income, employment, value added) to changes in the demographic and tourism demand structure of the region and vice versa. In this way, the assessment is not static but dynamic and may stop whenever successive rounds of economic-societal interactions do not add significantly to changes in the demographic, tourist and economic structure of the region.

The outputs of the Economic and the Societal Modules – sectoral outputs, population changes and changes in tourist demand – are fed into the Interfacing Module. The Resources Utilisation Model(s) assess the demand for regional resources necessary to meet the production (economic module) and consumption (societal module) requirements of tourism and other economic activities. The Residuals Generation Model(s), similarly, assess the types and amounts of residuals generated during the production and consumption activities of the resident and tourist population and the economy of the region. In this manner, tourism's direct and indirect contribution to demand for regional resources and generation of residuals can be assessed, with a distinction made between their production- and consumption-related aspects. It is noted that these assessments reflect, or should reflect, the degree of environmental protection exercised by the various economic sectors. Changes in their practices or in current policies will, naturally, lead to changes in the demand for resources and the residuals of the societal and the economic system.

Finally, the output of the Interfacing Module is fed into the Environmental Module. The Resources Impact Model(s) translate the regional demand for resources (output of the Resources Utilisation Model) into impacts on regional resources of given quality classes – whether they are shortages, carrying capacity overload, etc. Similarly, the Residuals Generation Model(s) translate the output of the residuals generation model into impacts on environmental receptors and environmental quality, in general. The outputs of both the Resources Impact and the Residuals Impact Models are tested against the regional resources inventory to identify the overall changes in the amounts of resources in given quality classes which result from tourist activity as well as from the other economic activities in the region. If sustainability indicators are available (e.g. measures of deviations of current from desirable environmental conditions) they can be assessed to provide estimates of the impacts of tourism on the sustainability of regional development.

The output of the Environmental Module is essentially the final output of the whole process. More specifically, application of the modelling framework will make it possible, to the extent the available data permit, to assess the contribution of tourism in the observed or expected environmental modification: (1) directly, during the production and consumption of the tourist product, (2) indirectly, through changes in economic, demographic, physical, and other characteristics of the region; and (3) through induced development in the host region.

Although the basic modelling framework applies in principle to all cases of impact assessment, certain methodological and practical differences between ex post and ex ante assessment should be mentioned. Ex post impact analysis attempts to assess the net environmental impacts of existing tourism development; i.e. to isolate them from those which autonomous regional growth has caused. This requires longitudinal analysis, which controls for the effects of economic activities and socio-economic events besides tourism on regional social and economic structure and the environment over time. In this case, a statistical assessment design can be used if adequate data are available to control for all factors of interest. Alternatively, a quasi-experimental assessment design might be helpful if it is possible to find a 'control' region similar to the study region which has not experienced tourism development in order to make a valid comparison between a 'without tourism' and a 'with tourism' case. If this is not possible, a plausible base case scenario may be set up referring to regional conditions 'without tourism' development, of which the results are then compared to the 'with tourism' current situation.

Ex ante impact assessment compares alternative tourism development scenarios to a base case 'autonomous growth' scenario. The impacts due to the latter can be assessed by means of longitudinal analysis, the purpose being to extrapolate past trends into the future. The impacts associated with the alternative tourism scenarios are then compared to those expected under the base case scenario. The differences in impacts are estimated to make a final decision on the most desirable tourism development scenario.

3.4. ASSESSING THE SUSTAINABILITY OF TOURISM DEVELOPMENT WITH THE INTEGRATED MODELLING FRAMEWORK

Although a complete treatment of the sustainability question is beyond the scope of this chapter, it is worth indicating how the proposed integrated assessment framework can be used to address this issue. For the present purposes, sustainability of tourism development assumes the balanced promotion of economic welfare, environmental protection and social equity in tourism areas. This is a multicriterion assessment problem (see, for example, Nijkamp and Ouwersloot, 1997) whose general objective is the minimisation of the negative and the maximisation of the positive impacts expected under alternative development scenarios. Its solution requires, on the one hand, the specification of particular sustainability criteria and, on the other hand, information on the economic, environmental and social conditions in the host area within a given time period and for a given level of technology. This information can be provided by the outputs of the economic module (sectoral estimates of employment, income and output levels), and of the environmental module (levels of resource use by type of resource and levels of environmental quality by type of receptor). In addition, information from the regional resources inventory can be used to assess the degree to which the carrying capacity of certain resources has been exceeded by their current use in the study area. In this way, the proposed assessment framework can provide support for making broader decisions on the ultimate sustainability of current or proposed tourism development patterns.

4. Conclusions

Integrated approaches to the analysis of economy-environment interactions have long been and are still advocated in the literature and applied in several circumstances (see, for example, Bergh, 1996; Van den Bergh in this volume; Braat and Van Lierop, 1987; Brouwer et al.,1983; Lakshmanan, 1983; Nijkamp, 1980; OECD,1980). The proposed framework for the analysis of the environmental impacts of tourism belongs to this strand of approaches and, consequently, possesses several of their advantages and drawbacks. Its most important advantages are that:

- it enables the systematic assessment and evaluation of the relative importance of different types of production and consumption impacts of tourism
- it accounts for the direct and indirect as well as the induced impacts of tourism, the latter two being sometimes equally important as the former (Fletcher, 1989)

- it sets tourism within the broader economic and environmental context of the study region and, thus, makes possible a reliable assessment of its relative contribution to the observed or expected environmental change
- it facilitates the comparative assessment and evaluation of tourism's regional economic and environmental impacts (the outputs of the economic and the environmental modules)
- it provides valuable input to the process of assessing the sustainability of tourism development
- its modular form facilitates the continuous improvement of its components without affecting the overall model structure

The proposed framework is not without drawbacks, however. The environmental module is usually the less well developed because of the difficulties encountered in ecological modelling, on the one hand, and in assessing environmental carrying capacity and resources stocks unambiguously, on the other. The identification and treatment of the environmental impacts of tourism is less straightforward than it is the case with conventional economic sectors, as tourism is not a pure economic sector but an activity complex. Things are complicated by the fact that several facilities are used both by tourists and the local population. Also, several of the environmental impacts caused by visitors are not recorded by the economy and have to be treated separately, a fact that may create compatibility problems among the different types of impacts assessed. Considerable judgement and discretion is, thus, required, on the part of the analyst, to isolate tourism's share in the economy and the environment. Use of the proposed modelling framework in a static analysis mode runs the risk of concealing or mistreating the particular traits of the environmental impacts of tourism; mainly, their seasonal and cumulative nature as well as the fact that several of them are internal to tourism itself and erode its long-run profitability. Finally, the framework presented is not immune to the perennial data problem of all integrated models, a factor that affects seriously its operationalisation, applicability and cost.

Despite these problems, it is believed that a valid integrated assessment of tourism's environmental impacts requires application of the proposed framework. Its application to a variety of tourist areas and environments will test its ultimate usefulness and contribution in making enlightened tourism development decisions.

References

Baud-Bovy, M. and F. Lawson (1977) *Tourism and Recreation Development*. The Architectural Press, London.

Bergh, J.C.J.M. van den (1996) *Ecological Economics and Sustainable Development: Theory, Methods and Applications*. Edward Elgar, Cheltenham.

Braat, L.C. and W.F.J. van Lierop (1987) *Economic-Ecological Modeling.* North-Holland, Amsterdam.

Brouwer, F., J.P. Hettelingh and L. Hordijk (1983) An Integrated Regional Model for economic-ecological-demographic-facility interactions. *Papers of the Regional Science Association* **52**, 87–104.

Coccossis, H. and P. Nijkamp (eds) (1995) *Sustainable Tourism Development.* Avebury, Aldershot.

European Community Council (EEC 421) (1992) A community action plan to assist tourism. *Official Journal of the European Community*, No.L231/16.

Fletcher, J. (1989) Input-Output analysis and tourism impact studies. *Annals of Tourism Research* **16**, 514–529.

Kneese,A.V., R.U. Ayres and R.C. d'Arge (1970) *Economics and the Environment.* Johns Hopkins Press, Baltimore.

Komilis, P. (1986) *Tourism Activities.* KEPE, Athens (in Greek).

Lakshmanan, T.R. (1983) A multiregional model of the economy, environment, and energy demand in the United States. *Economic Geography* **59**, 296–320.

Lindsay,J. (1986) Carrying capacity for tourism development in national parks of the United States. *Industry and Environment* **9**, 17–20.

Mathieson, A and G. Wall (1981) *Tourism: Economic, Physical and Environmental Impacts.* Longman, London.

Nijkamp, P. (1980) *Environmental Policy Analysis.* John Wiley & Sons, New York.

Nijkamp, P. and S. Verdonskschot (1995) Sustainable tourism development: A case study of Lesvos. In H. Coccossis and P. Nijkamp (eds), *Sustainable Tourism Development.* Avebury, Aldershot.

Nijkamp, P. and H. Ouwersloot (1997) *A Decision Support System for Regional Development: The Flag Model.* Research Paper, Free University of Amsterdam.

OECD (1980) *The Impact of Tourism on the Environment.* Organisation for Economic Cooperation and Development, Paris.

Pearce, D.G. and R.M. Kirk (1986) Carrying capacities for coastal tourism. *Industry and Environment* **9**, 3–6.

Sessa, A. (1988) The science of systems for tourism development. *Annals of Tourism Research* **15**, 219–235.

Solomon, B.D. (1985) Regional econometric models for environmental impact assessment. *Progress in Human Geography* **9**, 379–399.

UNEP *(1987) Report of the Seminar on the Development of Mediterranean Tourism Harmonized with the Environment. Priority Actions Programme.* Split, Croatia.

UNEP (1982) *Tourism in the World Environment 1972–1982.* Ch.14. Tycooly International Publishing, Dublin.

Victor, P.A. (1972) *Pollution: Economy and the Environment.* University of Toronto Press, Toronto.

WCED (World Commission on Environment and Development) (1987) *Our Common Future.* Oxford University Press, Oxford.

WTO (1981) *Technical Handbook on the Collection and Presentation of Domestic and International Tourism Statistics.* World Tourism Organization, Madrid.

WTO (1983) Workshop on Environmental Aspects of Tourism. Joint UNEP and WTO Meeting, Madrid, 5–8 July, 1983.

TOURISM DEVELOPMENT AND THE NATURAL ENVIRONMENT: AN ECONOMIC-ECOLOGICAL MODEL FOR THE SPORADES ISLANDS

JEROEN C.J.M. VAN DEN BERGH
Department of Spatial Economics
Free University
Amsterdam
The Netherlands

1. Introduction

This chapter presents a study of the relationship between tourism development and the natural environment for an island region in Greece, the Northern Sporades. The analysis is based on a dynamic model, which describes the development of the economies of the three main islands of the region and their interactions with the terrestrial and marine environment. The relationship between tourism and the environment is taking place on various levels. In addition to direct tourism impacts on the environment through, e.g., pollution, noise and disturbance, indirect, irreversible and long term consequences must be considered also. In this perspective, tourist patterns over time, recreational attractiveness of the region, land use patterns and the growth rate and direction of economic development dominated by tourism must be taken into account. The long-term relationships between tourism, economy and the environment are receiving special attention in the present study.

The next section of the chapter discusses the modelling of economy-environment interactions in the context of long term development analysis. A description of the case study that considers both the regional characteristics and the formulation of the problem is given in the third section. This is followed by an explanation of the structure of the case study model in the fourth section. Scenarios, indicators and simulation results obtained with the model are discussed in the fifth section while general conclusions are presented in the last section of the paper.

2. Models for tourism, economy, environment and development

The relationship between tourism and the environment can be regarded in the broader context of economy-environment relationships. This study places special emphasis on two important dimensions of this relationship, namely that of long-term change (development and growth) and the regional scale of analysis. With regard to the latter, transboundary flows of materials and influences outside the study region (external determinants) deserve special attention. In a general regional framework, the following types of relationships between development, economy and the natural environment can be distinguished:

1. direct impacts of the economy on the environment, through extractive, polluting and disturbing activities;
2. direct positive and negative impacts of the environment on the economy, through the provision of resources and assimilative capacity, necessary conditions for production and consumption (e.g., recreation), and damages occurring from ambient pollution and degradation of natural systems and quality of extracted materials; natural disasters unrelated to anthropogenic effects (e.g. volcano eruptions, earthquakes) can be mentioned as a special category of direct environmental impacts on human systems;
3. impacts of development on the economy, through capital accumulation, sectored changes, demographic changes, technical change, and the above mentioned external determinants (e.g. tourism)
4. impacts of natural, environmental conditions and changes therein on decisions and actions that form part of development processes; examples of environmental causes are natural resources scarcity, deterioration of environmental quality, overshooting of carrying capacity limits (by population, tourism, or economic growth), which may give rise to migration, introduction of new technologies, certain economic and environmental policies, and changing patterns of economic (sectional) development.

The impact sequences that go from (3) to (1) and from (4) to (3) can be regarded as *indirect* impacts from the economy to the environment, and from the environment to the economy, respectively. It is important to recognise these various relationships because of the focus of this study on long-term relationships. For a more extensive treatment of this framework see Van den Bergh (1991).

In particular, the integration of economics and ecology is important for a study of the relationships mentioned above. One approach for realising this is the use of formal models in which processes of both fields are described in relation to each other (see Van den Bergh and Nijkamp, 1996; or Van den Bergh, 1996). This may be a fruitful direction for three reasons. First, there is some similarity in approach between the sub-disciplines dealing with the aggregate level, namely regional and macroeconomics and synecology (see Van der Ploeg, 1976). Second, both fields have a tradition of theoretical and applied modelling. Third, methodological differences are circumvented to a certain extent in this way. The specific accumulated disciplinary knowledge is used to establish the structure of the model, select its elements and specify their relationships (see Braat and Van Lierop, 1986). Furthermore, a systems approach can be adopted that focuses on elements, their relationships, causality, and feedback mechanisms. Looking at the economy and the environment in terms of systems yields a "common denominator" and allows representation of interactions in a systems-theoretic framework. Thereby, the economic system can be considered within the environmental system, or the two can be perceived as two coupled sub-systems, or, finally, they can be regarded as one total system. These systems can be examined at various levels, taking notice of energy, matter and information. Process descriptions may involve biological, chemical, physical, monetary, behavioural, institutional and organisational relationships (see Bennet and Chorley, 1979). For a wider discussion and survey of integrated modelling the reader is referred to Van den Bergh (1996).

Integrated models with tourism are quite rare. Kandelaars (1997), somewhat following the approach adopted in the present study, has developed a model of tourism, development and environment for the Yucatán peninsula in Mexico, which is included in this volume, as part of a wider IIASA study on modelling the interaction between population, development and environment (Lutz, 1994). Many regional studies have focused on islands, which are of course well-bounded areas with some features of isolated economies and development (e.g., Cole, 1997). However, especially where trade and external influences occur, the stability of the region's economy and environment is threatened. In this context, also referred to as spatial sustainability, the role of carrying capacity and (un)sustainable trade are examined (Van den Bergh and Nijkamp, 1994a, 1995; Gowdy, 1997). Although the present study is focusing on internal, regional systems dynamics issues, these wider spatial sustainability elements are also relevant. Clearly, the one crucial element of openness and potential spatial unsustainability regarded here is that of tourist flows to the region (see also Coccossis and Nijkamp, 1995; and Hunter and Green, 1995).

In this study, we use a model that is compatible with the purpose of integration of economics and ecology as discussed above. Integration in this approach may include one or more of the following four levels (see Van den Bergh, 1991): (i) material flows between economy and the natural environment based on materials balance conditions; (ii) effects of human systems (economy and population) on environmental quality through immaterial or less-tangible categories of impacts such as land use, noise and soil exploitation; (iii) effects of environmental conditions on economic production, consumption and health, including for instance negative pollution effects; and, (iv) production functions with a mix of economic and natural factors of production, for instance for resource extraction, recreation and agriculture.

3. Region and problem description

The Northern Sporades is a complex of islands located in the Aegean Sea close to central Greece. They constitute an area where socio-economic development objectives conflict with environmental conservation objectives. This area embraces the inhabited islands Skiathos, Skopelos and Allonisos, as well as the uninhabited islands Kyra Panagia, Gioura, Piperi, Psathura and Ikantzoura. For the purpose of the present study a nearby, low-developed area on the mainland, Pillion, has been included in the analysis as well, since the aim (set by the Greek Ministry of Economic Affairs) is that Pillion will benefit through spill-over effects from economic development on the islands. The population levels in 1985 of Skiathos, Skopelos, Allonissos and Pillion are 5,064, 4,226, 1,621 and 600, respectively. Table 1 shows land cover data for the three main islands.

Table 1. Land cover data for the three main islands (1985); beaches not included (source: Frantzi and Despotakis, 1991)

x 1000 m2	Skiathos	Skopelos	Allonissos
Natural vegetation	17,400	45,800	72,060
Pasture	15,800	26,300	5,660
Cultivation	14,075	20,480	1,920
Urban-infrastructure	3,400	3,900	2,250
Total area	50,675	96,480	81,890

The islands have a rich vegetation including pine forests and different types of Mediterranean scrubland (maquis and garrique). Both flora and fauna include many interesting, and some unique, species. Rare birds like Eleonora's Falcon and Audouin's Gull are regular visitors of the area, and wild goats on Gioura are regarded as a valuable species. There is a natural process of transformation between different types of vegetation like forest and scrubland. Maquis is naturally transformed into forest vegetation. Occasionally, the rate at which the process in the other direction occurs – from forest vegetation to maquis and from maquis to garrique – is sped up by fires, which are frequently caused by tourists. Bare land constitutes an important part of the total land area of the islands due to particular natural conditions such as high elevations and soil erosion. The land use patterns of agriculture (pastures for grazing by goats and cultivations for crop and fruit production), urban areas and road development dominate the changes in land available for natural vegetation. Expansion of agricultural land occurs into and transforms areas with low-growing scrubs (garrique) or bare land. Urban land expands around villages, usually into formerly cultivated or agricultural areas. Degradation of the natural vegetation caused by agricultural and urban expansion, tourism, and natural processes induces soil erosion by wind and water. This makes natural recovery of transformed areas rather unlikely. Disturbance of quietness by tourists is one extra element that may harm especially the fauna (for instance motorbikes are very popular nowadays).

The marine environment is rich in species, such as corals, different types of fish, dolphins and the Mediterranean Monk Seal (Monachus monachus). This latter species is very valuable from a conservation point of view: it is unique, rare and valued highly by tourists and experts (biologists)[1] and a critical part of the marine ecosystem, occupying the top of the food chain. However, seawater pollution, fishing and tourism in the area threaten the existence of this species. The number of seals has gone down and stabilised in recent years to a present level of approximately 40. In addition, three fish and a lobster species, located at the same level in the food chain, are distinguished for their economic value: the first fish species includes fish like red mullet; the second species includes sword fish and melanuria; and the third one includes gopes and tuna. The first fish species and the lobster are the most expensive. Catches of the other two fish species, however, are much higher so that all of them may be regarded as equally important for realising economic objectives in fisheries. Other marine life such as phytoplankton, zooplankton, benthic algae, molluscs and crustacea (except lobster) are not considered (see Scholte, 1989).

A marine park was established in September 1986 to protect the marine system of the region[2]. Visits of tourists are, thus, being limited to certain parts of the marine area and fishing is controlled by rules, fishing zones, fishing periods and type of nets. Besides tourism and fishing, flows of pollution and waste generated by the island populations, tourists and economic activities impact negatively on the natural environment of the region. The present analysis will concentrate on nitrogen and phosphorus derived from tourist and population numbers, and certain economic

---

[1] The Monk seal was once found along all shores of the Mediterranean Sea (and a few other places around the world). The World Wide Fund for Nature (WWF) ranks it among the ten most endangered species in the world, with only 400 to 600 individual animals presently surviving. The greatest number is found along Greek and Turkish coasts, and among the Aegean Islands. They are the largest of seals, weigh about 250 kg, measure 2.5 to 3 metre in length, and have an average life span of 10 years (probably when subject to environmental stress). They prefer coastal areas. The main threats to their existence are habitat destruction, marine pollution, competition with fishery for food, and hunting (see for instance The World Bank/The European Investment Bank, 1990).

[2] This was part of recovery plans in line with a resolution on the protection of the Monk Seal which passed the European Parliament in 1984.

activities. Additional, derived indicators for the consequences on living conditions and organic materials in the marine environment are BOD (Biological Oxygen Demand) and chlorophyll.

The region's terrestrial and marine ecosystems provide several services and goods to the socio-economic system. The marine system provides fish populations for fisheries catch. The presence of seals and dolphins as well the nice beaches and aquatic facilities add to the attractiveness of the islands for tourists. The terrestrial system supplies firewood and resin (for Greek wine), while goats herded on the scrubby and rocky areas, provide meat and milk used for making cheese.

The past directions of development of the three main islands – Skiathos, Skopelos and Allonissos – have been greatly influenced by tourism. However, the development stages of the three islands differ significantly. Skiathos has a long history as a holiday resort for Greek citizens, mainly those coming from the two largest Greek cities Athens and Thessaloniki. Though originally characterised by a reputation of richness and exclusiveness, it is currently functioning as an attractor for various categories of Greek and foreign tourists. Tourism developed somewhat later in Skopelos, and it is known to be more peaceful. It may be considered at a stage in between those characterising Skiathos and Allonissos. This latter island is especially quiet, due to its relatively large natural area and the rather recent development of commercial tourism activities (since 1980). At first sight, it is not entirely clear what will be the economic and environmental consequences of a development driven by tourism. For this reason, special attention is devoted in the present study to the consequences for Allonissos. This means more detailed attention to its economic activities and endogenous development mechanisms in tourism activities and concern for their interactions with the terrestrial system. Table 2 gives relevant tourism data for the three islands. From these data and the respective population levels and land areas we can derive indicators of tourism pressure. Skiathos has then the highest density of tourist-nights per capita (± 23.7) as well as per square metre land (± 2.4). Allonissos comes second with regard to the first of these two-pressure indicators (± 12.3), but last in terms of the other (± 0.24) (for Skopelos these indicators have values 9.5 and 0.42, respectively). More details on the structure of the economic system will be given in the presentation of the model.

The size of the islands' economies is rather small. The following economic sectors can be distinguished: local services, fisheries, agriculture, public sector, construction and two tourism-based sectors (accommodation and services). Trade and transport are arranged by mainland firms, leaving no special benefits to the local population except via the local services. This sectoral disaggregation is based on a combination of the following considerations: operation of the labour market, interdependencies between activities, and economic-environmental relationships. In Table 3 estimated figures of value added per sector and island are listed. Especially the labour market is crucial for the operation of the economy, since it provides the opportunities and constraints for specific sectoral developments. Competition for labour exists within certain economic activities, for instance agriculture and fisheries. Labour available for these two activities is determined, to a large extent, after the labour requirements of other economic activities have been fulfilled. A strong pull force is generated especially by tourism activities. This is analysed in the next section.

Table 2. Estimated tourist data for the three main islands (1985; on the basis of data from questionnaires, and Frantzi and Despotakis, 1991); employment in man years; price level of tourism related to Allonissos

	Skiathos	Skopelos	Allonissos
Accommodation (number of beds)	1,921	1,006	535
Tourists-nights	120,000	40,000	20,000
Employment	325	117	65
Relative price level of services in tourism	131	115	100

The Northern Sporades provide an interesting case for testing integrated economic-ecological modelling on a regional scale and performing scenario analysis of tourism and economic development. The possible conflict between one-sided economic development based on tourism, giving rise to undesirable long term environmental and economic consequences, and natural conservation directed at both land and marine ecosystems (Nijkamp and Giaoutzi, 1991) is one of the important scenarios to be tested via the model presented below.

Table 3. Estimated sectoral value added figures per island (1985; source: Frantzi and Despotakis, 1991)

x million Gr. Dr.	Skiathos	Skopelos	Allonissos
Accommodation	392	115	62.5
Tourism services	3,760	987	400
Construction	84	42	131.5
Local services	287	240	92
Agriculture	3.1	3.6	57
Fisheries	8.8	10.5	83
Total	4,534.9	1,397.1	1,026

4. Model structure

The main output variables are tourism indicators (nights spent, land use by type of accommodation, indices for disamenities), economic indicators (total and sectoral value added, production per capita, unemployment rate), indicators for terrestrial disturbance and pollution (congestion index, particulates, sewage), indicators for conservation of natural vegetation (area of terrestrial vegetation and a land diversity index) and indicators for marine conservation (adult seals, stocks of fish, quality of sea water). The main input (control) variables are: investment in accommodation, investment in local fisheries, dummy variables for marine park and agricultural policies, conditions for fishing and external impacts on the number of tourists. A full account of the model equations, assumptions and data can be found in Van den Bergh (1991).[1]

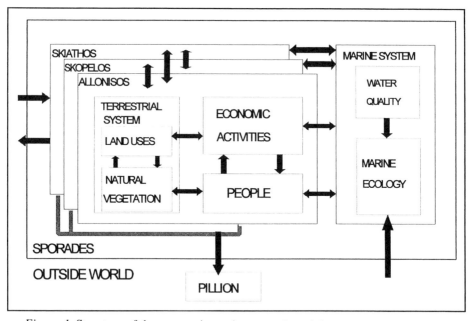

Figure 1. Structure of the economic-environmental model

---

[1] The ecological components of the model were set up by ecologists (L.C. Braat and A.J. Gilbert). The author has designed the economic modules and is responsible for the integration of the various modules. To that end, some of the ecological modules were transformed or slightly changed. A full report is available on a broader study, of which the design and analysis of the model discussed here is a part (see Giaoutzi and Nijkamp, 1993; and Van den Bergh, 1996, Ch. 11). A technical account of model equations is found in Van den Bergh and Nijkamp (1994b).

Figure 1 depicts the overall structure of the model – the modular components and their interrelationships. The three main islands in the region are treated separately since their development levels differ, and because only the Allonissos module includes a terrestrial submodule with a description of land use and natural vegetation cover. The economy and people modules are set up for each one of the islands. The socio-economic relationships with the outside world, the influence on the nearby area Pillion, and the impact of the mainland on the marine system are represented also.

The "economic activities" module includes the different economic sectors, classified on the basis of economic and integrative (economic-environmental interactions) considerations. The sectoral structure, income composition, and important determinants of sectoral development and production are set up in a similar way for each island but with more details for Allonissos. The following sectors are distinguished: accommodation sector including mainly hotels, but also bungalows and pensions; tourist services sector including restaurants, bars, rental services and souvenir shops; private service sector for the local population; construction sector for investments in private and public facilities; two resource-based sectors, namely agriculture and fisheries; public services such as education, health care, administration, and police; and utilities – especially water pumping – and infrastructure. The activity levels in the tourist sectors are determined by the number of days/nights spent by tourists, determined in the "people" module. The activity levels in local services is determined by the population level and in public services by both tourist and population levels, given by the "people" module. Construction activity depends on the expansion of hotels, and the increased demand for housing, given by the "people" module. The remaining labour supply, determined in the "people" module, an agricultural incentive policy, and competition for labour by a mainland fishery industry, determine agricultural and fisheries activity levels. For some of the activities, physical capacity (e.g., number of beds) and environmental or biological limits (e.g., land and water for agriculture, fish stocks for fisheries) are taken into account as well. These are generated by the terrestrial and marine system modules.

The description of socio-economic relationships with Pillion is kept very simple. It refers only to agricultural and general tourist sector activities. These are assumed to experience spill-over effects of development in the three main islands, given by the respective "economic activities" modules. In addition, they may be subjected to sectoral incentive policies.

The "people" module includes sub-models for demography, housing development, tourist flows and the labour market. Labour supply is determined by population size (indirectly including effects of migration and seasonal commuting), exogenous trends in participation rates and a combination of productivity and working hours. As mentioned above, labour demand by some sectors, generated by the "economic activities" module, determines labour allocation between those and the remaining sectors. The stock of houses depends on the population and tourist levels (based on the demand for rented rooms and second homes).

The main driving force behind economic development is tourism. The structure of the tourist model is shown in Figure 2. The number of tourists is determined by a regional trend, which, to some extent, follows a national pattern (with implicit assumptions regarding international developments in tourism), and a local attraction mechanism, based on economic and environmental amenities and several disamenities. This local attraction depends on regional development and provides, through the impact of tourism on regional development, a feedback mechanism from economic and environmental conditions, generated by the "economic activities", the "terrestrial system", and the "marine system" modules, to tourist growth. A curbing effect on initially high growth in the number of tourists may be the result of such a mechanism.

Prices of goods and services are assumed constant and in real terms. Only for the price level of tourist services and accommodation in Allonissos and Skopelos a change over time is assumed, so that prices converge within a specific time period to the respective price levels of Skiathos.

The determinants of development and growth include: number of tourist visits (for the tourist sectors), profit levels (tourist sectors), economic policies (e.g. agricultural development), environmental management practices (e.g. sustainable fishing), and ecological limits (e.g. carrying capacity). Since the model will be used to analyse different long term development scenarios, the level of detail in the description of cost and profit structures for each sector has been kept to a minimum.

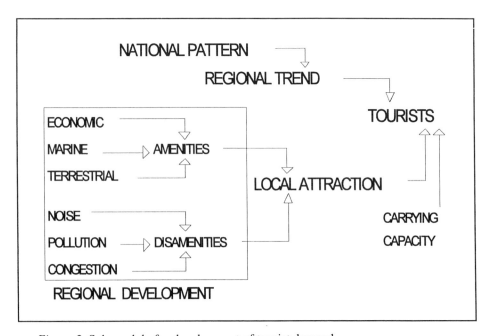

Figure 2. Sub-module for development of tourist demand

A terrestrial module represents structural, long-term economic-environmental interactions that occur through land transformation (see Section 3). Dependent variables are land areas by type of vegetation or land use. Their driving forces, generated by the "economic activities" and "people" modules, are: land use by agriculture, which transforms areas with natural vegetation into cultivated land; grazing by cattle, encouraging, for instance, the transformation of one type of scrubland into another; and urban growth (via new houses, tourist sectoral investment, infrastructure development), mainly transforming agricultural land around existing urban areas. Natural processes of transformation are described between different types of vegetation (forest, maquis, garrigue). Tourism, generated by the "people" module, has a direct effect on these processes through increased probabilities of fires. The processes and exogenous factors that affect the composition of natural areas (four categories) and the two main land uses were already explained in Section 3. A physical constraint based on data on altitudes and ground structure sets limits to changes in land uses. A land diversity index is generated to serve as an input to the tourism module.

The marine system is influenced by variables of the "people" and "economic activities" modules. The "marine system" module consists of a water quality component and an ecological system description. Nitrogen and phosphorus flows are determined by population and tourist levels, generated by the "people" module, and impact indirectly upon BOD and chlorophyll in the water along the coastline. This has then a slight effect upon the marine ecology. The ecosystem is modelled as a food chain with the Monk Seal predating on two species of fish (see Section 3). Lobster and two species of fish are fished. Vessels from the three islands and the mainland generate a certain amount of fishing effort (determined in the "economic activities" module), which is one, but important, determinant of catch. There are two other determinants. One is the sum of the sizes of the four stocks of fish and lobster species being subjected to fishing. The other is the constraint on catch to realise sustainable fishing patterns. This may be operationalised by restrictions on types of nets, fishing periods, and zones.

5. Scenario analysis

Two possibly conflicting objectives, which have to be resolved for the Sporades region, are regional economic development dominated by tourism and environmental protection of the terrestrial and marine ecosystems (especially conservation of the Monk Seal colony). A constraint to the first objective could be the economy's high dependency on one sector, namely tourism. Therefore, anything with an adverse impact on tourism provides a threat to future regional income. Economic and human activities cause a range of environmental impacts: water pollution, loss of landscape values, loss of traditional values (culture), transformation of old villages, congestion (loss of recreational values), litter, noise and overfishing. These effects could adversely impact future tourism and threaten economic development (see Figure 2). They contradict also directly the second objective. A constraint to the second objective is partly overlapping with that to the first one: growth of certain activities and the size of human impacts may not be sustainable from a natural conservation point of view.

Scenarios are designed to represent various policies and exogenous patterns of change or events. Several types of policy actions can be considered: environmental, conservation, economic, infrastructural, and social programs and even changes in institutional arrangements. Three types of exogenous patterns of change or events can be distinguished: internal socio-economic, external economic, and natural events. The future development of the Sporades islands may range from an extrapolation of present trends to strong sectoral deviations from these trends due to environmental or economic opportunities, constraints and policies. The following eight scenarios are used to represent the most realistic and important policy and development options for the Sporades islands:

1. *Steady growth*: all sectoral growth rates are in line with present trends; tourist numbers continue to grow; tourists are allowed to visit the marine park area.
2. *Steady growth and marine park policy*: like scenario one, but now the tourist flows to the marine park are controlled.
3. *Strong growth*: like scenario two, but with a higher potential growth rate of tourists; tourists control as in scenario two.
4. *Limiting tourism growth*: like scenario two, but tourist growth is zero.
5. *Sustainable fishing*: like scenario two, but fishing effort is confined to levels for which the stocks of fish are not reduced; tourist control as in scenario two.
6. *Tourist limits and sustainable fishing*: a combination of scenarios four and five.
7. *Agricultural incentive*: like scenario two, with strong emphasis on land cultivation.
8. *Very strong growth*: like scenario two, but tourist numbers follow a very high growth rate.

Performance indicators generated by the "economic activities" module will be used to test for the performance of the scenarios in terms of economic efficiency, equity, and conservation. Value added and unemployment can be used to test for efficiency, be it on sectoral, island or total region's level. Testing for equity may pertain to equity between sectors or islands.

The number of tourist-nights spent ("people" module), the land use of accommodation industry ("economic activities" module), the percentage of drinking water that is imported ("economic activities" module) may serve as indicators of regional tourism development. Import of water depends on the local extraction rate and the demand created by tourists and locals. Finally, many natural and economic aspects related to tourism are summarised in indices for amenities and disamenities based on a set of natural, economic and congestion variables (tourist model in "people module"). These latter indices can be regarded as attractiveness indicators for tourism.

Congestion (based on total number of tourists and local people, generated by the "people" module), solid wastes (generated in the "economic activities" module and dependent on certain activity levels such as in construction), sewage (determined by population and tourist levels in "people module") serve as indicators of terrestrial disturbance and pollution.

For land vegetation conservation, two indicators have been selected: the total area of forest and maquis and an index for land diversity. These are determined on the basis of the land areas under human control or with different vegetation types, given by stock variables in the "terrestrial system" module.

Finally, three indicators, generated in the "marine system" module, were chosen for marine conservation, namely number of adult seals, the total stock of fish, and the quality of marine water. The latter is based on the concentration of nitrogen, phosphorus, and on BOD and chlorophyll.

In the remainder of this section we will discuss results of scenario analysis based on the model of the last section, the eight scenarios mentioned above, and a choice out of the possible indicators mentioned. The performance of economic and tourism indicators for Allonissos will be compared with those for the other islands. The performance of the attractiveness and environmental indicators for tourism on Allonissos will be discussed separately from the environmental indicators for the marine system.

5.1. TOURISM AND ECONOMIC DEVELOPMENT IN ALLONISSOS AND THE OTHER ISLANDS

Figure 3 shows the patterns of total value added for Allonisos, for Scenarios 1, 6, 7 and 9 (see above) and, for the other two islands under the steady growth scenario.

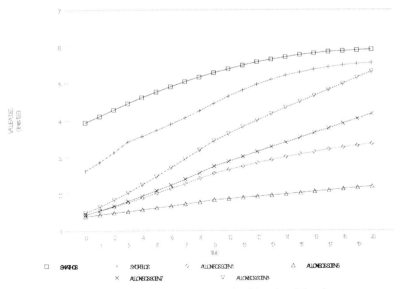

Figure 3. Allonisos' economy compared with other islands

The slowing down observed on some curves is explained by tourism development under stationary growth (Figure 5). Both the national pattern and the environmental feedback through amenities and disamenities are responsible for this result (see Section 4). Under the strong growth scenario, the size of the economy of Allonissos will approach that of Skopelos, while both will approach that of Skiathos. Scenario 6, which can be considered as reflecting the strictest environmental conservation measures, leads to the slowest pace of growth. The agricultural incentive (Scenario 7) leads to a higher growth rate than under stationary growth.

Figure 4 presents the unemployment rates for Allonissos under the various scenarios, which show trends consistent with those in Figure 3. It is clear that the environmental Scenario 6 gives rise to the highest and (increasing) unemployment level over the entire period. Unemployment is also increasing under the stationary growth scenario, especially because supply of labour increases faster than demand, as a result of increased participation (stimulated by clean, simple work in tourism for which little education is necessary), population growth, migration and seasonal commuting. The agricultural incentive and high growth Scenarios 7 and 8 lead to the best performance from the viewpoint of employment. Although growth in production per capita is higher for Scenario 3 than for the agricultural incentive Scenario 7, unemployment over time is always higher. This can be explained by noticing that the growth in Scenario 3 is dominated by the tourism sectors, where the receipts per unit of labour effort are higher than in agriculture.

Tourist numbers (indicated by total nights spent per year) are shown for Allonissos and Skopelos in Figure 5. For most scenarios, the pattern of growth in the number of tourists is curbed as a result of feedback from attractiveness indicators and changes in national trends. The number of tourists is not allowed to increase under the very strict environmental conservation Scenario 6.

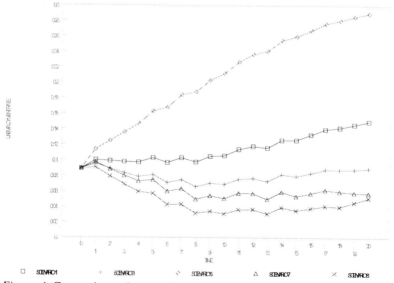

Figure 4. Comparison of unemployment rate on Allonisos under six scenarios

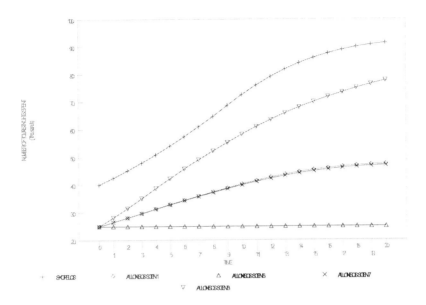

Figure 5. Tourist numbers for different scenarios of Allonisos, compared with Skopelos

5.2. ENVIRONMENTAL AND ATTRACTIVENESS INDICATORS FOR ALLONISSOS

A complete characterisation of the regional situation at the final time horizon includes, in addition to the economic performance levels of the islands, information on the state of the environment, summarised in the amenities and disamenities, a land diversity index, and the marine ecosystem species. Initially, amenities for Allonissos decrease under all scenarios, as indicated by an index over time in Figures 6 and 7.

After some time, under Scenarios 3, 4, 5, 7, and especially 8, amenities increase. The reason is that economic amenities are included in the index, and they are generated by economic growth in the tourism sectors. This is not the case under Scenario 7 with agricultural development and relatively strong growth. Only for Scenarios 2, 4 and 6, which have some form of tourism regulation in common, the trend keeps being negative. This is explained by the direct negative attraction effect of the environmental protection measures on tourism, and as a derived effect by the negative impact upon economic amenities (e.g. fewer hotels, less shops, less services).

The patterns of a disamenities index, that summarises congestion indicators, are not shown. All patterns have positive trends, though very modest ones for the environmental conservation Scenarios 4 and 6. Growth Scenarios 3 and 8 show the strongest increases. This means that the positive final parts of the curves denoting amenities for these scenarios are counterbalanced by increasing levels of disamenities. The ultimate combined effect is reflected in the tourists' trends. Essentially, this can be regarded as conflicting development patterns: positive trends of economic amenities and negative trends of natural and congestion amenities. Land diversity indicator patterns over time for Allonissos are all following a downward trend. The strong growth Scenario 8 gives rise to the steepest fall of this indicator. Only for Scenarios 4 and 6 stabilised levels are attained by restricting negative impacts of tourists (e.g. fires), tourism activities, and derived activities (e.g. construction).

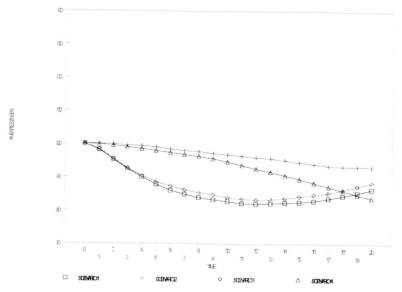

Figure 6. Development of amenities index over time under Scenarios 1 to 4

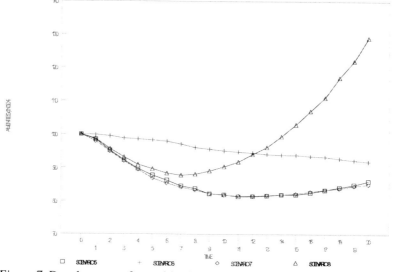

Figure 7. Development of amenities index over time under Scenarios 5 to 8

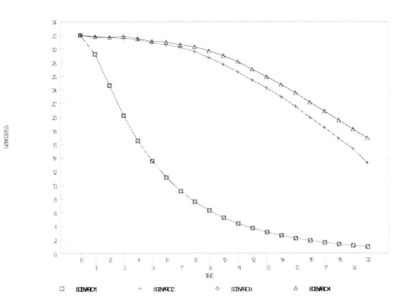

Figure 8. Monk seal population over time for Scenarios 1 to 4

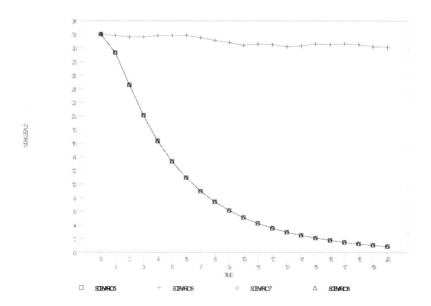

Figure 9. Monk seal population over time under Scenarios 5 to 8

5.3 ENVIRONMENTAL INDICATORS FOR THE MARINE SYSTEM

The patterns of Monk Seals are given in Figures 8 and 9. Scenarios 2, 4 and 6 perform best, though only in Scenario 6 the level of Monk Seals remains stable over time. This means that the negative effects of fisheries and tourists has to be minimised, i.e. restricting tourist entrance in certain marine areas and keeping fish catches below sustainable yield. Fish stocks are depleted in most cases (not shown), except for Scenarios 5 and 6, i.e. when sustainable fishing is applied. Scenario 2, which includes the marine park policy of restricting tourist visits to a sensitive marine area, gives better results than the steady growth scenario. Scenario 4, which also limits tourist growth, gives rise to a slowing down of the depletion of fish stocks via a derived effect on the fisheries industry, and, consequently, to a slowing down of the decrease in the population of Monk Seals.

6. Conclusions

The results in Section 5 indicate that compared with the tourism growth Scenarios 3 and 8, the environmental conservation Scenarios 4 and 6 give rise to worse performances of indicators of island production and employment and tourism amenities. However, the results for tourism disamenities, as well as for terrestrial and marine system indicators point in the other direction. The agricultural scenario performs very well in terms of island production, and especially employment, but exhibits identical behaviour as in the steady growth scenario for the other indicators. From the point of view of environmental conservation, Scenarios 2 and 5 can be regarded as behaving worse than Scenario 4 and 6.

The case study shows that formal, dynamic models can provide insight into the functioning of long-term processes of tourism and economic development and on economic-environmental interactions at the regional scale. It is too early to draw definite conclusions on the conditions of the Northern Sporades. The model indicates that if the Monk Seal is to be protected, policy measures should be directed at restricting both negative environmental impacts of tourism and fisheries. This means restricting the size of the fisheries fleet or effort to sustainable fishing capacities and pursuing a marine park policy of limiting tourist visits to sensitive areas. Even limits on tourism growth may by required. High unemployment is, however, likely to result from such policy measures. In order to prevent this from happening, tourism growth may be allowed but should go along with a set of very restrictive measures to prevent negative impacts on fisheries and Monk Seals. Alternatively, the agricultural incentive scenario can be combined with a low level of tourism growth to achieve positive employment effects.

The most important beneficial effects of tourism development are clearly an increase in local (economic) welfare. Several negative impacts are possible, however, which were not dealt with in this study (see, e.g., Mathieson and Wall, 1982). First, most of the receipts in the tourist sectors may be going to non-local investors, (seasonal) commuters, or foreign organisations. Secondly, the domination of tourism results in a one-sided economic development of the island economies, which may have negative consequences as regards the system's capacity to adjust to sudden and drastic changes in the pattern of tourism demand. Last, tourism does not only have an important impact on the economic but also on the social and cultural structure of the islands. This is especially relevant for small communities such as those of the Sporades islands.

Some of the results of the model simulations are straightforward since simplifying assumptions had to be made. Furthermore, exogenous patterns of certain variables had to replace complicated long term, dynamic processes. In order to perform a deeper analysis of long-term relationships between the environment and the economy, it is necessary that more attention is given to endogenous growth and development processes. This requires the study of sector-specific decision processes and determinants of change. The results here are very much dependent on the structure of the "economic activities" and "people" modules. The labour market as it is being modelled at present has a large impact on the opportunities for certain activities to expand and on the way labour is allocated to the remaining sectors. An improvement could be obtained by more insight into the operation of the labour market, thereby including seasonal patterns and moonlighting. A drawback of the indicators, such as in the case of the amenity and disamenity indices, is that they are sensitive to the aggregation procedure used. Finally, both the environmental impact of economic activities, resource extraction and pollution, as well as the impact upon economic activities require more study. Although there is much information on a lot of these issues separately, there is still a long way to go in integrating these.

Acknowledgement

This study was supported by the Foundation for Advancement of Economic Research (Ecozoek), which resorted under the Dutch Organization for Scientific Research (NWO), Project No. 450-230-007. It has benefited from the support of Leon Braat, Alison Gilbert, Peter Nijkamp and Hans Opschoor, as members of the SPIDER team at the Free University of Amsterdam, and Vasilis Despotakis, Maria Frantzi, and Maria Giaoutzi from the National Technical University of Athens. I am grateful to Frits Soeteman for reflections in an early stage of this study on assumptions underlying the model that is presented here, and to Jan van der Straaten for helpful comments.

References

Bergh, J.C.J.M. van den (1991) *Dynamic Models for Sustainable Development.* Thesis Publishers, Amsterdam.

Bergh, J.C.J.M. van den (1996) *Ecological Economics and Sustainable Development: Theory, Methods and Applications.* Edward Elgar, Cheltenham.

Bergh, J.C.J.M. van den, and P. Nijkamp (1991) Operationalizing sustainable development: dynamic ecological economic models. *Ecological Economics* **4**, 11–33.

Bergh, J.C.J.M. van den, and P. Nijkamp (1994a) Sustainability, resources and region. *The Annals of Regional Science* **28**, 1–5.

Bergh, J.C.J.M. van den, and P. Nijkamp (1994b) An integrated model for economic development and natural environment: An application to the Greek Sporades Islands. *The Annals of Operations Research* **54**, 143–174.

Bergh, J.C.J.M. van den, and P. Nijkamp (1995) Growth, trade and sustainability in the spatial economy. *Studies in Regional Science* **25**, 67–87.

Braat, L.C. and W.F.J. van Lierop (1987) *Economic-Ecological Modelling.* North-Holland, Amsterdam.

Coccossis, H., and P. Nijkamp (eds) (1995) *Sustainable Tourism Development.* Avebury, Aldershot.

Cole, S. (1997) Economic cultures and ecology in a small Caribbean island. In J.C.J.M. van den Bergh and J. van der Straaten (eds) *Economy and Ecosystems in Change: Analytical and Historical Approaches.* Edward Elgar, Cheltenham, pp. 231–269.

Giaoutzi, M. and P. Nijkamp (1993) *Decision Support Models for Regional Sustainable Development.* Avebury, Aldershot.

Gowdy, J. (1997) Trade, equity, and regional environmental sustainability. In J.C.J.M. van den Bergh and J. van der Straaten (eds). *Economy and Ecosystems in Change: Analytical and Historical Approaches.* Edward Elgar, Cheltenham, pp.166–184

Hunter, C. and H. Green, H. (1995) *Tourism and the Environment: A Sustainable Relationship?* Routledge, London.

Kandelaars, P.P.A.A.H. (1997) A dynamic simulation model of tourism and environment in the Yucatán Peninsula. Interim report IR-97-18/April, IIASA, Laxenburg, Austria.

Lutz, W. (ed.) (1994) *Population–Development–Environment: Understanding their interactions in Mauritius.* Springer Verlag, Berlin.

Mathieson, A., and G. Wall (1982) *Tourism: Economic, Physical and Social Impacts,* Longman Group, London.

Ploeg, S.W.F. van der (1976) Ecology and economics: synthesis or antithesis. In P. Nijkamp (ed.), *Environmental Economics, Vol. 1: Theories.* Martinus Nijhoff, Leiden.

Richmond, B. et al., (1987) *STELLA Userguide*, Dartmouth College, Lyme, New Hampshire.

Scholte, J. (1989) *Allonissos: A Conceptual Ecological-Economic Model.* Master Thesis, Faculty of Economics, Free University, Amsterdam.

World Bank and The European Investment Bank (1990) *The Environmental Program for the Mediterranean: Preserving a Shared Heritage and Managing a Common Resource.* World Bank/European Investment Bank, Washington and Luxembourg.

A DYNAMIC SIMULATION STUDY OF TOURISM AND ENVIRONMENT IN THE YUCATÁN PENINSULA IN MEXICO

PATRICIA KANDELAARS[1]
Free University
De Boelelaan 1105
1081 HV Amsterdam
The Netherlands

1. Introduction

This chapter presents results generated by an integrated dynamic model used to study the interactions between population, environment, and economy. The model is based on the Population-Development-Environment (PDE) approach developed by Lutz (1994). Recent PDE models have dealt with the small islands of Mauritius (Lutz, 1994) and Cape Verde (Wils, 1996). After studying islands the next step is to develop a more elaborate PDE model for a peninsula. For the present study, the Yucatán peninsula in Mexico was chosen for several reasons. First, it is a region that mostly borders the sea, which poses constraints to trade and human migration. Only recently, the peninsula has been connected by land infrastructure to the rest of Mexico. Second, it has a specific historical culture – the Mayan civilisation whose population suddenly decreased – of which many archaeological sites remain (Lutz *et al.*, 1996). Third, drastic economic and demographic changes have taken place in the last twenty years, such as a large population increase, resulting mostly from the growth of the tourism industry. Fourth, conflicts exist between the goals of economic growth, the population and the environment which are, to a large extent, due to tourism growth. In the PDE modelling of the peninsula special attention is given to the tourism sector for three reasons: (i) it is the most important economic sector in the peninsula; (ii) the tourism sector has developed rapidly in the last decades and is expected to

---

[1] Department of Spatial Economics, Free University, De Boelelaan 1105, 1081 HV Amsterdam and Tinbergen Institute, Amsterdam. The present research was performed at the International Institute of Applied Systems Analysis (IIASA), Laxenburg, Austria. This research was financially supported by a grant from the Dutch Organisation for Scientific Research (NWO).

continue growing in the coming years; and (iii) there are many interesting interactions between tourism, other economic sectors, population and environment worth analysing.

The purpose of the following application of the model is to represent the dynamics of the PDE interactions and to explain the observed relationships and the behaviour of the system. It does not intend to forecast future developments, but to show how interesting questions may be answered with the use of dynamic modelling and simulation. It should be emphasised that the model is not a black box in which the reader or user has no insight. It is an interactive model in which variables, relationships, and even modules may be changed or added. This allows the model to be updated easily. The model may provide also a basis for researchers or firms to forecast future development. The model, as presented in this chapter, may be useful for teaching and policy analysis. However, it cannot offer guidelines for policy recommendations, because it does not include all relevant relationships needed for this purpose. The equations of the model in its present form are not calibrated on historical data and, also, for many variables and interactions the correct data are missing. The model is meant to give indications on how dynamic modelling may be applied in a transparent way as it is user-friendly and the graphical interface allows one to see the interactions in the system. Thus, the model with its graphical and mathematical interface may serve as a basis for further exploration of PDE and tourism interactions.

The remainder of this chapter is structured as follows. Section 2 describes the tourism industry and the environment in the Yucatán peninsula. The modelling method is presented in Section 3. Section 4 describes the dynamic model comprehensively. Section 5 contains the model's results of testing alternative scenarios and Section 6 the model's results of the sensitivity analysis. Conclusions are drawn in the last section and suggestions for future research are offered.

2. The Yucatán peninsula, tourism and the environment

A short description of the peninsula is given first, followed by a presentation of tourism and the environment in the peninsula. The Yucatán peninsula (Figure 1) is a region in the Southeast of Mexico, surrounded by the Gulf of Mexico, the Caribbean Sea (Atlantic Ocean), Guatemala and Belize.

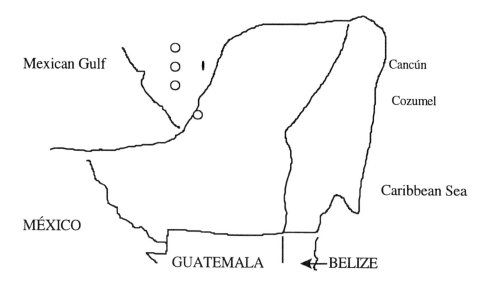

Figure 1. The Yucatán Peninsula

The southern border of the peninsula consists mainly of tropical forests that restricted the access to this area until very recently when roads were built to mainland Mexico and Belize. No road to Guatemala exists yet. Figure 1 shows that the Yucatán peninsula is not actually a peninsula, but it is considered as such due to limited access to the region. Centuries ago this region was a centre of the Mayan civilisation of which many archaeological and historical sites remain. The history of the peninsula has been different from that of the rest of Mexico, mainly because of the lack of communication between the peninsula and the mainland. The Yucatán Peninsula consists of three states: Yucatán, Campeche and Quintana Roo. Although these three states form one peninsula, they are different administrative regions without common regional policies.

Until the end of the 1960s, the economy of the peninsula was based on agriculture for the local population. Nowadays, tourism is the main economic sector in which the population is occupied. Table 1 shows the population in the three states and the importance of tourism for the regional GNP and employment. Other economic sectors which are important in the Yucatán Peninsula are fisheries, agriculture, food and clothes production, forestry and wood manufacturing (for fisheries, see Hale, 1996; for agriculture, see de la Cuanalo et al., 1996). These sectors are related to tourism as, for example, in hotel construction. The decrease in the number of persons occupied in agriculture has been very fast, just like in other regions which were stimulated by the government (for example, Long, 1991) for Santa Cruz in the Huatulco area).

Table 1. GNP, employment in tourism and population in the three states of the Yucatán Peninsula

	Yucatán	Campeche	Quintana Roo	Total
% of GNP in tourism (1993)	22%	27%	58%	38%
Employment in tourism (1990)	69690	20836	45904	136430
(% of total employment)	(17%)	(15%)	(30%)	
				(20%)
Population (1990)	1362940	535185	493277	2391402

Source: INEGI (1992).

2.1. TOURISM

Mexico has had many beach resorts for domestic and, since the late 1940s, also for international tourism. International tourism can be divided into three types: urban tourism (Mexico City, Guadalajara and other cities); border tourism in the north; and resort tourism (for example, Cancún). Urban and border tourism are highly dependent on the domestic economy. Resort tourism is less dependent on the domestic economy because many tourists come from abroad. These international tourists make that the growth path of tourism differs from that of the whole economy (Inskeep and Kallenberger, 1992).

In the late 1960s, the Mexican government made a plan to stimulate the development of tourism in economically disadvantaged areas. The direct goal was to develop tourism in rural areas that have tourist attractions (beach or historical sites) and where other sources of employment or economic development are scarce. Indirect goals were to stimulate other sectors of these underdeveloped areas, to stimulate tourism in Mexico, and to generate income in foreign currencies. The long-term program focused on developing five resort areas: Cancún, Ixtapa-Zihuatanejo, Loreto, Huatulco and Los Cabos (Inskeep and Kallenberger, 1992).

Cancún was chosen to become a tourist resort because of its geographical features: a strip of land and beaches, which enclose a large lagoon. In the beginning of the 1970s, four main elements were planned and developed: beach hotels; an international airport; a new urban zone; and, conservation areas. A new town was needed for the increasing population of Cancún, which developed from a small village of only 117 inhabitants in 1970 to a city of 18 thousand in 1976 and to a major resort with 300 thousand in 1991 (Inskeep and

Kallenberger, 1992). The Cancún area has still potential to grow. The conservation areas were mainly designed to protect the lagoons.

The expansion of tourism in the peninsula, which was stimulated by local (regional) policy makers to develop the region has been very fast. Tourism has increased from 1.3 million tourists in 1970 to 6.3 million in 1989, mostly due to the increase in beach tourism (Inskeep and Kallenberger, 1992). The share of tourism in the gross national product (GNP) was 2.8% in 1989. The number of people employed in tourism was estimated to be 1.3 million in 1989.

The per capita income in the peninsula is higher than in most other regions of Mexico, which is mainly due to the earnings from tourism. As a result of these income differences and the economic development of the region, there has been migration from other parts of Mexico to the Yucatán peninsula. The development of the Cancún area for tourism has had a great impact on the economy and the society of the Yucatán Peninsula. In the past, the population of Yucatán mostly lived in rural areas while now most of the people live in urban areas (Aguilar and Rodriguez, 1995). Together with the natural growth rate, immigration is the cause of population growth. The infrastructure has improved with the addition of the international airport, regional highways, water supply, electric power and telecommunications (Inskeep and Kallenberger, 1992).

The tourism sector has evolved during the last twenty-five years. Especially in the last ten years the increase in the number of tourists has been very fast, mainly because of the growth of the number of international tourists. Only Hurricane Gilbert caused damage and a decline in tourism in 1988. The percentage of international tourists has increased much more than that of national tourists (INEGI, 1994; World Tourism Organization, 1994). Figure 2 shows the total number of tourists in Quintana Roo in the 1981–1994 period.

The development of Cancún as a tourist resort has had a direct impact on the region around Cancún, for example on the island of Cozumel. Table 2 presents the increase in the number of tourists in these places and also in Mérida, a city in which tourism development did not get special attention.

Table.2. The number of tourists in the three main tourist destinations

	Cancún	Cozumel	Mérida
Tourists (in 1000s) (1981)	540.8	174.3	558.7
Tourists (in 1000s) (1994)	1958.1	321.0	459.2
% growth between 1981 and 1994	262 %	84 %	- 18 %

Sources: INEGI (1994); World Tourism Organization (1994).

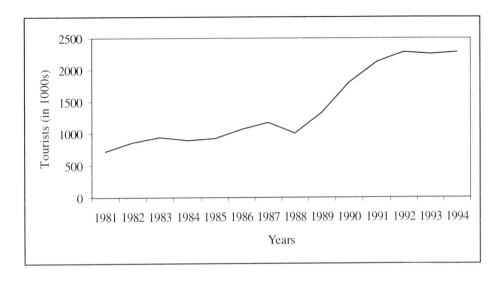

Figure 2. The total number of tourists (in thousands) visiting Quintana Roo from 1981 to 1994. (Sources: INEGI (1994); World Tourism Organization (1994))

The increase in the number of tourists has a direct effect on services needed like, for example, accommodation. Table 3 shows the number of rooms in 1981 and 1994 in the three main tourist places: Cancún, Cozumel and Mérida.

Table 3. The number of rooms in the three main tourist destinations

	Cancún	Cozumel	Mérida
Rooms (1981)	5225	1725	3138
Rooms (1994)	18859	3350	3331
% growth between 1981 and 1994	261 %	94 %	6 %

Sources: INEGI (1994); World Tourism Organization (1994).

2.2. ENVIRONMENT

The environment provides resources to the economic system. The marine ecosystem provides fish (shrimp, lobster and red grouper, see Hale, 1996) and the terrestrial ecosystem is used for agriculture. Besides fisheries and agriculture, the natural environment supplies 'services' for tourists. For example, the beaches and lagoons, which offer the opportunity to do several types of water sports, such as diving, make the area very attractive for tourists. Tourists come to Yucatán for the beaches and to 'discover' the ancient Maya culture of which many historical and archaeological sites remain (e.g., Chichen Itza, Tulum and Coba), and for the natural attractions (e.g., the Sian Ka'an Biosphere Reserve and the marine reserve with a protected reef ecosystem at Cozumel).

In the early 1980s, the rapid growth of the region caused environmental problems; the demand for the disposal of sewage waste was too high and caused an algal bloom in a lagoon (Inskeep and Kallenberger, 1992). For environmental reasons a marine transportation system through the lagoon was not implemented and only a few buildings have been constructed around the lagoon. Those few that were built were damaged by hurricane Gilbert in 1988. This hurricane also damaged the beaches by erosion, but natural processes are recovering that erosion.

The lagoons in Cancún are very sensitive ecosystems, which need to be protected. Therefore, boating in the lagoons is restricted and because of the algae problems water quality of the lagoons is being continuously monitored. According to Inskeep and Kallenberger (1992) 'it appears likely that the natural beauty and ecological functions of the lagoons will be retained.'

3. Modelling the Yucatán Peninsula

Various interesting questions may be raised for the tourism and PDE interactions. How can tourism growth affect the migration rate into the tourism region? What may be the impact of tourism policies on tourism itself, and also on foreign investments in the region? What may be the effect of tourism or population growth on water use and on water quality? Will investments in other sectors in the region decrease or increase due to investments in tourism? May a policy to clean up the water have an effect on the tourism sector? How will the tourism industry affect the environment? To address this sort of questions a dynamic model will be employed. First, an overview of studies on tourism will be given and then the dynamic model will be discussed.

The interactions between tourism, the economy and the environment have been studied for various regions and countries. Most of these studies are (historical) descriptions. For example, Ramsamy (1994) describes the development of tourism and the environmental problems caused by it for the island of Mauritius. Sinclair (1991) studies the distribution of earnings from safari and beach tourism in Kenya. Shah (1995) analyses the demand and supply for wildlife viewing by different methods. An exception to these descriptive studies is a model for tourism development of the Northern Sporades, a group of islands in Greece. This study indicates that tourists visiting certain fragile areas should be limited to protect the marine system and the species living there (Van den Bergh, this volume). The modelling approach adopted in the present chapter is comparable to that employed by Van den Bergh. A model is described in which the interactions between tourism, the rest of the economy, population, and the environment are considered with the goal to obtain insight into their dynamic interactions.

Programming of, and dynamic simulation with, the model that is presented in Section 4 is performed in the graphical modelling language Stella/ithink[2]. This model allows for integrating separate modules (i.e. partial models) into one model, allowing one to obtain a view on direct or indirect relationships between various modules (see Van den Bergh, this volume). Furthermore, the dynamic model allows for identifying delayed effects, accumulation over time, and changes over time.

With quantitative modelling, additional insights may be obtained, compared to descriptive studies, as regards, for example, population changes and changes of the age-structure of the population due to tourism development and growth. The government and the tourism industry, i.e. the hotel owners, may be interested in being able to estimate the impacts resulting from policies of cleaning the water on the number of tourists visiting the Cancún area. A question that may be asked is to which level water should be cleaned and who will pay for it: the government, the tourism industry, the local population, the tourists, or a combination of these? If the government policy will be that the tourism industry needs to pay for a part of water clean up that will upset the hotel owners, unless there is a positive impact due to a higher tourist demand with higher prices. As regards population, development, environment and tourism, many kinds of interesting questions may be asked, especially with respect to their interactions. For instance, the impacts of changes in investments or policies on (foreign) investments on tourism and the rest of the economy, and the relationship between the environment and tourism.

Several scenarios may be set up to simulate policies, future developments, and radical changes. The government may impose policies, for example, to clean up the water, to limit investments in hotels, or to curb international investments. Future developments or changes may be simulated in all modules, variables, and parameters. Examples of possible developments are: a lower birth rate, a change in the popularity of Yucatán as a tourist destination, less migration because other parts of Mexico become more attractive, changes in investments, reduced attractiveness of the beaches or of the archaeological sites. Radical changes that may affect the Yucatán peninsula are, for instance, a hurricane like Gilbert in 1988 where the tourism industry experienced a substantial setback.

---

[2] Stella/ithink is a software package for developing dynamic models and performing dynamic simulation (see Peterson and Richmond, 1994 and 1996). Hannon and Ruth (1994) and Grant *et al.* (1997) provide good introductions to dynamic modelling and systems analysis using Stella/ithink.

4. Model structure

This section describes the structure and the main ideas, questions and relationships of the model that is designed to obtain insight in the dynamic development of tourism, the other economic sectors of the region, the population and the environment. The model is a descriptive model with causal relationships and differential equations, and consists of five modules: the economy, the tourists and rooms, the population, the environment, and the government module. These modules are the building blocks of the model and interact with one another. Often these modules are studied separately.

The model has 1994 as a base year, a time horizon of 20 years (1995-2015) and it is run for every year. The initial conditions and the relationships are based on data from various sources: on tourist and room numbers (Inskeep and Kallenberger, 1992; World Tourism Organization, 1994; INEGI, 1994); on economics of Mexico (World Bank, 1994; OECD, 1995); on water use (Gelting, 1995); on economics of the Yucatán peninsula (Cinvestav, 1996); and on demographics (INEGI, 1992 and 1994; CONAPO, 1995). The initial conditions and equations of the model are given and explained in the Appendix.

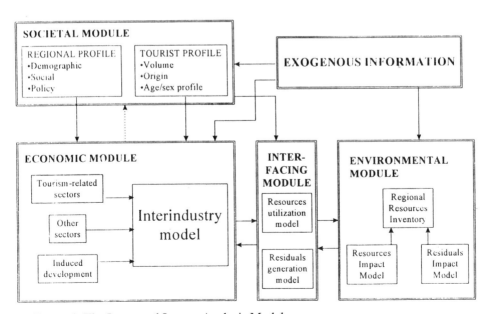

Figure 3. The Integrated Impact Analysis Model

Figure 3 shows the modules, the main variables and the direct linkages between the modules. The modules and the interactions between them are described shortly. There are various indirect linkages between the modules, for instance, between tourism and the population via the economy. The more tourists come in, the more people are needed to work in the tourism industry, which may cause migration into the tourist region.

4.1. THE ECONOMIC MODULE

The economy is modelled as consisting of two sectors: the tourism sector and the rest of the economy which is an aggregate of all other sectors. Ideally, the economic module would deal with various economic sectors separately to give a more accurate description of the economic system. For practical reasons, however, only the tourism sector is dealt with in detail.

The gross output of the tourism sector depends on the price per night, the number of nights and the number of tourists visiting the area. Part of the gross output goes to wages. The wage per worker depends on the occupancy rate, the intermediate consumption of the tourism sector, the gross output and the number of people employed in tourism. The profit of the tourism sector depends on the gross output and the occupancy rate of the hotel rooms.

Total investments are the sum of three types of investments: international investments that are done only in the tourism industry and depend on the profit per room; national investments that depend on profit per room; and regional investments that depend on wages and profits in both sectors.

The distribution of regional investments in the two sectors depends on the change in the profit per room in the previous year. For the aggregate of the other sectors, gross output depends on the labour force, the capital stock and prices.

The gross output of this sector is also divided into intermediate consumption, wages and profits. The wage per worker is determined by the part of the gross output dedicated to labour and the number of people working in the sector.

4.2. THE TOURISM MODULE

Tourism demand depends on the price per room, water quality, beaches and archaeological sites, and exogenous factors (e.g., the general popularity of Mexico that may depend on marketing strategies or even the political situation in various countries). The exogenous factors are based on the historical growth rate of tourism in the region (INEGI, 1994). Price is the equilibrium price between the demand and supply of rooms. The supply of rooms depends on the total availability of rooms and price[3]. When the price is very low, the hotel owners

---

[3] The supply and demand interactions that result in an equilibrium price may be interpreted for the tourism sector as follows. Tourist agents base the price on the expected demand and the

will not supply all rooms, because the costs of opening are higher than the revenues. When the price is very high, the supply of rooms will be equal to the total number of rooms available. The number of rooms available (the capital stock of the tourism sector) depends on investments in tourism and on the depreciation rate of rooms.

The quality of the beaches depends on the occupancy rate of the hotels and not directly on the number of tourists, because if more hotels are being built, more kilometres of beaches are available. This is the case in the Yucatán peninsula because there are still new beach areas to develop (Inskeep and Kallenberger, 1992). Naturally, in the long term, the availability of new beaches to develop will be limited. The quality of the archaeological sites depends directly on the number of people that visit these places.

Tourism has many different impacts of which the most important will be described here. First is the impact of tourism on employment. If tourism grows, jobs will be mainly created in the three tourist resorts of the peninsula (Cancún, Cozumel and Mérida). This increase in job opportunities may attract people from the rural areas of the peninsula and from other areas of Mexico. This internal migration will cause population increase in addition to the natural growth rate. Therefore, tourism growth has an impact on the population magnitude of the Yucatán peninsula. The impact of migration from rural to urban (resort) areas on the rural areas is not obvious. On the one hand, it may have a negative impact because it can leave the countryside without a sufficient labour force and, on the other, it may be beneficial because people in rural areas will have access to off-farm work which is badly needed given the poor conditions of many rural areas. Many people choose to work part of the year in other sectors when the harvest or the economic conditions in the rural area are stringent.

4.3. THE DEMOGRAPHIC MODULE

The natural growth rate of the population depends on the birth and death rates (natality and mortality) which are exogenous to the model. Migration to Yucatán from other parts of Mexico depends on an exogenous migration rate and the difference in the wage rate in the tourism sector and the general wage rate in Mexico in the previous year. The underlying assumption is that potential migrants decide on the existing wage rate differential and will enter the peninsula one year later.

The population variable does not include an age structure and therefore the labour force is assumed to be a fixed proportion of the whole population. The total labour force is either working in the tourism sector or in the other economic sectors, e.g., fisheries, agriculture, construction and semi-formal sectors.

---

availability of rooms. If the actual demand is lower than the expected demand the price may lower, i.e. discounts are given. If the prices are very low, hotel owners may decide to close their hotel, for example, during low season.

Although the unemployment rate is officially around 6%, but effectively much higher, it is not taken explicitly into account because many unemployed people are working in the semi-formal sector (INEGI, 1994).

4.4. THE ENVIRONMENTAL MODULE

This module consists of three broad environmental quality indicators: water quality, quality of beaches and quality of archaeological sites. Tourism cannot grow uncontrolled due to the environmental impacts it generates. To include the attractiveness of the beaches and the archaeological sites, the environment is defined broader than 'nature' or the living environment. The impact on the environment and the damage caused by tourists is hard to measure, but the impact of tourists on water quality, the beaches and the archaeological sites cannot be neglected. The effects on culture are difficult to describe objectively, let alone to quantify them. Therefore, cultural impacts are not included in the model. For practical reasons, all these issues will not and cannot be included in the model.

Water quality depends on the amount of water used, the natural clean up and possibly on government policies to clean up the water. The initial value of water quality (1994) is set at 100. The impacts on water use per day caused by the local population and the tourists are differentiated for a geographical and a recreational reason. The geographical reason is that tourists stay mostly at the coast line, on a strip of 25 kilometres along the coast, where the use of water has more impact on water quality. The recreational reason is that the average use of water by tourists is higher than that of the local population (for instance, due to the swimming pools of hotels). A drop in water quality has a negative impact on the demand for tourist nights.

The quality of the archaeological sites depends on the number of tourists visiting these sites. The quality of beaches depends on the occupancy rate of the hotels; that is, the number of tourists divided by the number of hotel rooms. When a new hotel is being built, a new strip of beach will be developed and thus becomes available to tourists. It is assumed that the average tourist likes a moderately crowded beach and not a totally empty or an overcrowded beach (see Casti, 1996, for the same phenomenon for visits to a bar).

4.5. THE GOVERNMENTAL MODULE

The government is modelled in a limited way. The government may only impose two policies: the (percentage of) cleaning of the water used by the tourists and by the population. The government only pays for the cleaning of the water used by the population. This payment from the government to the peninsula may be seen as a subsidy. The tourism sectors, i.e. the hotel owners, pay for the cleaning of the water used by the tourists.

4.6. INTERACTIONS BETWEEN THE MODULES

An important interaction between the economic module and the tourist module occurs through the tourists. The demand for tourist nights and the supply of rooms determines the equilibrium price per tourist per night that is an important variable in the economic module.

The price per tourist night, the occupancy rate and the profit per room partly determine the level of investments done in both sectors (and regions). The labour force, a proportion of the population, is important for both sectors in the economy. The labour force needed in the tourism industry depends on the number of tourists coming in every year. Internal migration is flexible. If labour is not needed any more in the tourist region, people out-migrate. The proportion of the labour force not working in the tourism industry is assumed to work in the rest of the economy. Thus, unemployment is included as concerning the rest of the economy. The labour force in the rest of the economy directly influences the output of this sector.

Water quality diminishes because of water use by tourists and the population. The government may decide to impose policies to clean up the water used by tourists and/or the population. The government may impose a policy for a certain percentage of water cleaning. The tourism sector pays for the cleaning of the water used by tourists, while the government pays for the cleaning of the water used by the population.

The demand for tourist nights depends on the quality of the beaches and the archaeological sites, on water quality and on other (exogenous) factors. The quality of the beaches and of the archaeological sites has a delayed effect on the number of tourists. Water quality has a direct impact on the number of tourists. Water quality is an accumulated variable; it equals the quality of the previous year plus the changes that have occurred (positive or negative).

5. Scenarios and model results

In order to address some of the questions raised in Section 3, this section outlines two scenarios: a base scenario in which no policies are employed and a policy scenario in which the government imposes policies to clean up the water used by the tourists and the population. A time horizon of 30 years is chosen for the simulations, limited by the forecasts of CONAPO (1995). The main variables of each module, presented in the results, are indicated with an asterisk (*).

1. Environmental indicators: water quality*, quality of archaeological sites and of beaches.
2. Demographic indicators: population*, labour force, migration rate.
3. Economic indicators: (a) tourism sector: gross output*, price per tourist night*, investments*, wage rate, labour*; (b) other economic sectors: gross output*, aggregate price, investments*, wage rate, labour*.

4. Tourism indicators: tourist nights*, number of rooms*, occupancy rate, profit per room*, exogenous demand factor.
5. Governmental indicators: policies, subsidy.

5.1 BASE SCENARIO

In the base scenario, the government does not impose policies to clean up the water. This means that water quality improves only due to natural clean up. In this scenario, the exogenous demand factors increase over time, as expected by the World Tourism Organization (1996). The results of the base scenario show that the number of tourists is increasing until 2005, after which it decreases until 2009 (see Figure 4). Then, there is another cycle of increase and decrease with its lowest point in 2018. Figure 5 presents the price per tourist night, which is cyclical with a decreasing average. These cyclical patterns can be explained by the following mechanism. The price per tourist night is the equilibrium price between the demand and the supply of rooms. The supply of rooms depends on the price per room and the total availability of rooms. The higher the availability of rooms, the higher the supply of rooms at a certain price. When price increases, the profit per room will increase which affects the investments in rooms positively. Investments are made in rooms, which increases the availability of rooms. But with price being constant the supply of rooms will be higher, so the price should decrease to correspond to the same number of rooms. Thus price decreases. The demand factor also plays a role in this cyclical pattern. If price is low, many tourists will visit the peninsula which will decrease the quality of the archaeological sites and may decrease the attractiveness of the beaches. In the following years, this will influence the demand for tourist nights negatively. These two effects on the supply and demand side cause the cyclical pattern.

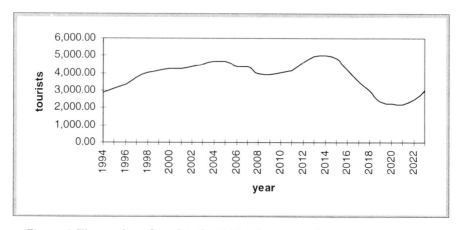

Figure 4. The number of tourists (in 1000s), base scenario

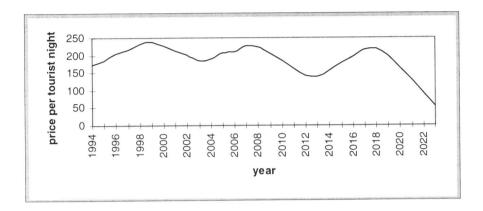

Figure 5. The price per tourist night (in 1994 pesos), base scenario

The wage rate and profits exhibit cyclical patterns as they are dependent on the demand for tourist nights and the price per night. For those years in which wages in the tourism sector in Yucatán are higher than in Mexico, immigration from other parts of Mexico is higher than when wages are lower than in the rest of Mexico. It might even be possible that the migration rate becomes negative when the economy in Yucatán enters a deep recession.

The number of rooms is increasing over time until 2019 after which investments are lower than the depreciation rate (from 2019 to 2023). This is due to a low profit per room, which implies that national and international investments are not made in Yucatán. Total regional investments depend on wages and profits in both sectors. The allocation of these regional investments to the two sectors depends on the change in profit per room. If the profit per room increases, then more investments will go to the tourism sector, otherwise they will go to the rest of the economy (see Tables 4 and 5). In some years (1994-2000, 2005-2006, 2008 and 2014-2016) it is more attractive to invest in tourism than in the rest of the economy due to the increasing profit per room.

The output of sector 2 (the rest of the economy) depends on the labour force and the capital stock. As it was mentioned before, the capital stock is slightly increasing, while the labour force in sector 2 strongly increases. Therefore, the output, in physical terms, increases too. Table 5 shows that the gross output, which is measured in monetary units, also increases over time, although the price paid for the output is slightly decreasing. Wages in sector 2 are decreasing which means that the increase in the gross output in sector 2 is outweighed by the increase in the labour force in sector 2. Profits in sector 2 are increasing due to the increasing gross output.

Table 4. Number of tourists, price per room, number of rooms, profit per room, water quality and population (base scenario)

Years	Tourists	Price per room	Rooms	Profit per room	Water quality	Population
1994	2876	175	9487	224	98	2534
1995	3092	186	9408	264	96	2682
2000	4249	228	10230	430	88	3505
2005	4686	207	10818	410	79	4446
2010	4039	184	11854	294	70	5477
2015	4822	178	12512	310	58	6584
2020	2269	164	12775	83	44	7751
2023	4976	53	11199	58	34	8463

Table 4 shows that water quality is decreasing over time. This decrease in water quality is caused by water pollution due to the use of water by tourists and the population. The government does not impose any policies to improve water quality. The quality of the archaeological sites is directly related to the number of tourists visiting those sites which is assumed to be a fixed percentage of the total number of tourists. Therefore, the quality of the sites also exhibits a cyclical pattern. The attractiveness of the beaches is a function of the occupancy rate of the rooms. This can be interpreted also as the number of people on a strip of beach, considering that new rooms, i.e. new hotels, are built on new developed strips of beach.

The natural increase of the population is the ratio between the exogenous birth and death rates. The migration rate depends on an exogenous factor, the base migration rate, and on an endogenous factor which depends on the difference between the wage rate in Mexico and the wage rate for tourism in Yucat□n. Migrants will come to work in the tourism sector, although not all of them will be employed there, so they will work in other economic sectors.

Population will increase as can be seen in Table 4 due to migration and natural population growth.

The conclusions that can be drawn from this scenario are that, although there is a potential of tourism growth because of the popularity of Yucat☐n, this potential is not optimally used. The main reasons lie on the supply and demand side of tourist nights. The supply side factor is that the more rooms available, the more rooms are supplied at a certain price. The demand side factors include the drop in water quality, the quality of the archaeological sites and the attractiveness of the beaches. To avoid the cyclical patterns and to make the tourism industry profitable and sustainable, water quality may be improved by cleaning it up and the quality of the archaeological sites may be improved by making them more attractive (for example, by making them more accessible by improving the infrastructure). On the supply side, the interactions are more complex: the more tourists come, the higher the price, the higher the profit per room, more investments will be made, more rooms are available, the lower the price. According to the model, the best policy of the government is to limit investments in rooms which makes the existing rooms more profitable and maintains the potential growth in tourism in the peninsula.

Table 5. Labour force, investments and gross output in the two sectors (base scenario)

Years	Labour in tourism	Labour in other sectors	Invest-ments in tourism	Invest-ments in other sectors	Gross output in tourism	Gross output in other sectors
1994	144	1630	3958	4630	5622	38160
1995	155	1723	5346	3880	6443	39178
2000	212	2241	7420	4852	10829	43705
2005	234	2878	8279	5633	10867	52813
2010	202	3632	6704	7697	8303	62134
2015	241	4368	9600	7134	9615	72562
2020	113	5312	860	7737	4156	83654
2023	149	5775	870	7830	1768	88027

5.2. TOTAL WATER CLEAN UP POLICY

In this policy scenario, water, which is used by tourists and the population. is cleaned. Table 6 shows that with this policy water quality will remain at the same level as in 1994. Figure 6 shows the number of tourists in this scenario and the base scenario. Compared to the base scenario, the number of tourists is lower only in 2007. The pattern of the number of tourists is not cyclical as in the base scenario. This is due to the fact that water quality does not drop, which makes the demand for tourists nights dependent on the beaches and the sites which are fluctuating per year, and the increasing exogenous demand factors. Until 2002, the cleaning of the water does not influence the number of tourists more than in the base scenario, but after that the number of tourists is higher than in the base scenario. The fluctuating pattern is due to the changing quality of the beaches and the archaeological sites and the supply of rooms. The quality of beaches and sites fluctuates per year, because it is not an accumulated variable like water quality. Therefore, the number of tourists fluctuates more in the short term than in the base scenario. In the long term, the number of tourists fluctuate less.

Table 6. Number of tourists, price per room, number of rooms, profit per room, water quality and population (policy scenario)

Years	Tourists	Price per night	Rooms	Profit per room	Water quality	Population
1994	2876	176	9485	223	100	2534
1995	3138	188	9406	268	100	2682
2000	4348	262	10388	495	100	3505
2005	5942	184	11218	460	100	4450
2010	5684	239	12206	515	100	5478
2015	5931	176	13596	339	100	6593
2020	4578	200	14729	261	100	7756
2023	5977	206	15604	352	100	8466

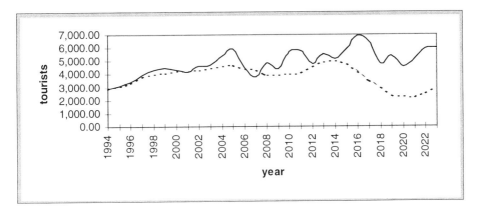

Figure 6. Number of tourists (x 1000), policy scenario (solid line) compared with base scenario (broken line)

The supply of rooms is determined by the level of investments in tourism which follow a cyclical pattern (see Table 7). Table 6 shows that the number of rooms is increasing over time and faster than in the base scenario due to a higher profit per room (compare Table 6 with Table 4). The attractiveness of the beaches changes more due to the changing occupancy rate of the rooms. The quality of the archaeological sites depends only on the number of tourists, but has a delayed effect on it. The exogenous demand factors are the same as in the base scenario; they are increasing over time.

The price per tourist night is generally higher than in the base scenario (see Figure 6 and compare Tables 4 and 6). The profit per room increases in the first 15 years after which a decrease in the profit per room is visible. This profit is always higher than the (inter)national profit rate which implies that (inter)national investments are made in Yucatán. Due to these investments, the number of rooms increases over time. The number of rooms in this policy scenario is higher than in the base scenario, because in the base scenario the profit per room is sometimes lower than the (inter)national profit rate.

Gross output first increases until 2010, then decreases for some years and at the end of the simulation period it increases again (see Table 7). The gross output depends on the number of tourists and the price per tourist night. Wages in the tourism sector follow the same pattern as gross output and because wages in the tourism sector are always higher than the wage rate in Mexico migration into Yucatán is stimulated.

Table 7. Labour force, investments and gross output in the two sectors (policy scenario)

Years	Labour in tourism	Labour in other sectors	Invest-ments in tourism	Invest-ments in other sectors	Gross output in tourism	Gross output in other sectors
1994	144	1630	3953	4659	5671	38158
1995	157	1721	5379	3888	6592	39149
2000	247	2236	7973	5005	12738	43842
2005	297	2818	8673	5732	12215	52075
2010	284	3550	10249	6742	15219	61292
2015	297	4319	8096	9074	11396	71876
2020	229	5201	10478	8026	10232	84363
2023	299	5627	11985	8813	13760	90148

The higher migration rate in this scenario causes a higher population growth, because the exogenous birth and death rates remain the same (compare Table 6 with Table 4). Wages in the rest of the economy, sector 2, are decreasing over time, but less than in the base scenario. This is a result of the higher gross output in sector 2 (compare Table 7 with Table 5).

For the other economic sectors, sector 2, water clean up is beneficial, too. Gross output increases more than in the base scenario which is mainly due to the higher capital stock in sector 2 resulting from higher investments (compare Table 7 with Table 5). Although the labour force is higher than in the base scenario due to a higher migration rate, the labour force in sector 2 is lower than in the base scenario because more labour is needed in the tourism sector.

An interesting conclusion of this scenario is that the profit per room increases although the hotel owners have to pay for the cleaning up of the water used by tourists. For hotel owners it is not attractive to pay a tax or charge per tourist (for their use of water), but because this policy, together with the government paying to clean the water used by the population, is beneficial for the profit per room, they might be interested in co-operating for cleaning up the water.

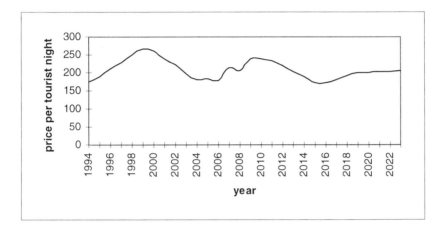

Figure 7. Price per tourist night (in 1994 pesos), policy scenario (solid line) compared with the base scenario (broken line)

6. Sensitivity analysis

The parameters, functional relationships and assumptions in the basic model as described in Section 4 were those considered the most probable to occur, but many other values may be possible also. With a sensitivity analysis, each of the parameters, relationships and assumptions may be examined to see how responsive the model is to their changes. This section presents the sensitivity analysis to the exogenous factors determining the demand for tourist nights.

In this sensitivity run, the exogenous factors that influence the demand for tourism are assumed to increase over time as in the base scenario, but the reaction of demand to other factors will be weaker. Figure 7 shows that demand still exhibits a cyclical pattern but this pattern is generally smoother and lower than in the base scenario. The demand for tourist nights is lower than in the base scenario until 2017 after which the number of tourists is higher. The price per tourist night shows a cyclical pattern as in the base scenario but price is generally lower than in the base scenario (see Figure 9). This is due to the lower impact of the exogenous demand factors.

The number of tourists is generally lower and the price per tourist night is also lower than in the base scenario which implies that the gross output of the tourism sector is lower, too. This makes the profit per room often lower than the (inter)national investment rate. The number of rooms depends on investments in tourism and because these are lower than in the base scenario, the number of rooms is lower, too. In the long run (30 years), the number of rooms has decreased by more than 2000 since the base year 1994 (see Table 8).

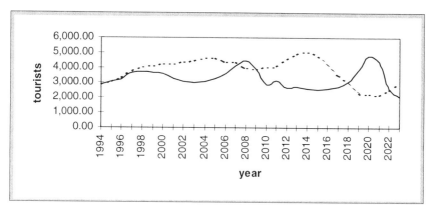

Figure 8. The number of tourists (x 1000), sensitivity run (solid line) compared with base scenario (broken line)

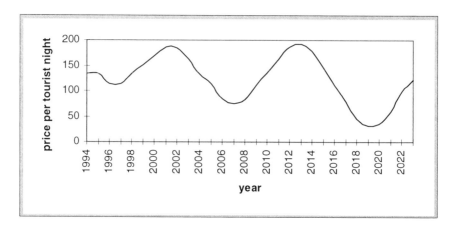

Figure 9. The price per tourist night (in 1994 pesos), sensitivity run

Water quality is only slightly better than in the base scenario after thirty years (compare Table 8 with Table 4). This is due to a smaller population and fewer tourists than in the base scenario. The migration rate is lower than in the base scenario because the wage rate in tourism in Yucat□n is not often higher than the wage rate in Mexico.

For sector 2, the results are not as good as in the base scenario. Investments are lower which implies lower capital stock and output. Together with the lower prices, the gross output is also lower than in the base scenario.

The conclusion of this sensitivity run is that a lower, but positive, impact of the exogenous factors on the demand has a negative impact on tourism and the rest of the economy. This run shows that the dependency of the tourism sector and the rest of the economy on the exogenous factors is very important. Therefore, a signal may be given to the hotel owners and the government to take this dependency into account. A recommendation for the tourism sector of the Yucatán peninsula might be to emphasise the attractiveness of the peninsula for tourism by stimulating and promoting, for example, ecotourism.

Table 8. Number of tourists, price per room, number of rooms, profit per room, water quality and population (sensitivity run)

Years	Tourists	Price per night	Rooms	Profit per room	Water quality	Population
1994	2876	133	9477	171	98	2534
1995	3084	133	9362	189	96	2682
2000	3567	169	9460	288	88	3501
2005	3293	113	9810	165	80	4437
2010	2839	135	9110	179	70	5459
2015	2518	145	10298	143	59	6563
2020	4794	35	8298	97	45	7719
2023	2053	121	7340	141	35	8419

Table 9. Labour force, investments and gross output in the two sectors (sensitivity run).

Years	Labour in tourism	Labour in other sectors	Invest- ments in tourism	Invest- ments in other sectors	Gross output in tourism	Gross output in other sectors
1994	144	1630	3988	4473	4286	38143
1995	154	1723	3829	4656	4596	39101
2000	178	2272	6226	4531	6754	43999
2005	165	2941	574	5168	4169	51826
2010	142	3679	7069	5740	4294	58739
2015	126	4468	728	6555	4090	68585
2020	240	5163	758	6820	1882	75629
2023	103	5791	831	7475	2783	81812

7. Conclusions and further research

This study aimed to analyse the interactions between tourism, population and the environment, and the rest of the economy under different policy and development path scenarios. Therefore, a dynamic model has been developed to simulate those interactions and to analyse the impacts of various policies and development paths. This model is used to show the impacts over time including delayed and accumulated effects. The model has a mathematical and a graphic interface in which variables, interactions and scenarios can be easily added and analysed. It is important to underline that the model does not provide predictions. It is intended to explain the structure and the dynamic behaviour of the whole system instead of separate parts of the peninsula. Like all models, it is a simplification of reality and therefore not all interactions and variables have been included.

In comparison to many other models (optimisation, equilibrium or static models), the dynamic descriptive model is very flexible. It allows one to change interactions, variables, or parameters. Therefore, the model can be updated when new insights, data or relationships are acquired which makes it a useful tool for future studies. The model is useful for policy makers because it gives some insights into the impacts of policies. The graphic interface, which is user-friendly and easy-to-learn, allows users to visualise the interactions that are part of the model and change them according to their own insights.

The model is based on two economic sectors, the tourism sector and the rest of the economy. The environmental factors influencing the demand for tourist nights are water quality, the quality of the archaeological sites and the quality of the beaches. The government may impose policies to affect the demand for tourist nights.

It is important to note that the results of the scenarios, these are illustrative and explanatory, instead of forecasts. The results of the base scenario show that the number of tourists, which are visiting the Yucatán Peninsula follows a cyclical pattern. This pattern results from the interactions between the demand (water quality, quality of archaeological sites and beaches and exogenous demand factors) and supply (number of rooms available) sides of the tourism industry that shape the price per tourist night. The policy scenario shows that water-related policies may be beneficial for tourism and the other economic sectors. The profits of the tourism sector are higher although they have to pay for the clean up. For the other economic sectors, the policies are also beneficial because more investments will be made in the Yucatán Peninsula which is good for the output of this sector. A sensitivity run shows that the tourism sector is highly dependent on the exogenous demand factors: the cycle is smoothened, but the number of tourists and the profit per room are generally lower than in the base scenario.

The environmental part of the model can be further specified to produce a more accurate model. The module for the environment is basic and more aspects could be included to make the model more realistic, for example, the inclusion of land use and infrastructure. The population module can be refined and updated when population projections become available. This may include, for example, the age structure of the population. The economy may be divided into more than two aggregate economic sectors. This means that the model serves as a basic tool in which many components may be changed and added to make it more appropriate for a more detailed scenario analysis.

The model integrates the interactions between the economy, the environment and the population to analyse the possible direct and indirect effects of policies and development paths. In the future, updating, refining and improving the model is needed to make it more useful for policy makers and other agents in various economic sectors.

Appendix: The equations of the model

The equations of the model are presented with a short explanation. The variable names that are not directly clear are listed first. In the equations, sector 1 is the tourism sector and sector 2 are the other economic sectors. The explanations of the equations of the model are preceded by '#'. The monetary unit used is the Mexican peso (1$ = 3.23 peso = 1.82 Dfl in 1994) .

cap2(t) = capital stock of sector 2 at time t (analogous: cap2_in, cap2_out)
p2(t) = price of sector 2 at time t
wage1 = wage in sector 1 (analogous for sector 2)
inv1 = investments in sector 1 (analogous for sector 2)
invint = international investments
invyuc1 = regional investments in sector 1 (analogous for sector 2)
invyuc = total regional investments
invmex1 = national investments in sector 1 (analogous for sector 2)
intprofitrate = international profit rate
nat_profitrate = national profit rate
gdpmex = GDP of Mexico
Si1 = rate of intermediate consumption of sector 1 (analogous for sector 2)
Sk1 = rate of capital costs of sector 1 (analogous for sector 2)
Sl1 = rate of labour costs of sector 1 (analogous for sector 2)
va1 = value added of sector 1 (analogous for sector 2)
va = total value added
out2 = output (production) in sector 2
wagemex = the wage in Mexico
rooms(t) = availability of rooms
rooms_out = the depreciated rooms
occrate = occupancy rate
tourproom = the number of tourists per room.
beach_attr = quality of beaches
sites_quality = quality of archaeological sites.
other = exogenous demand factors
demvar = demand variable
dem = demand
sup = supply
pop = population
lab1 = labour force (employment) in sector 1 (analogous for sector 2)
basemig = base migration rate
water_qual = water quality
cleaning = cleaning of the water
nat_cleaning = natural cleaning
pop_clean (tour_clean) = cleaning of water used by the population (analogous for tourists)
perc_pop_clean (perc_tour_clean)= percentage of the water cleaned that is used by the local population (analogous for tourists)

The economic module
cap2(t) = cap2(t - dt) + (cap2_in - cap2_out) * dt
INIT cap2 = 4000
cap2_in = inv2/12
cap2_out = 0.1*cap2
# The capital stock of sector 2 increases with investments and decreases with a fixed depreciation rate of 10%. The initial capital stock (1994) is 4000.

p2(t) = p2(t - dt) + (p2_in - p2_out) * dt
INIT p2 = 1500
p2_in = if dummyd>0 then dummyd*50 else 0
p2_out = if dummyd<0 then -dummyd*50 else 0
dummyd = ((delay(wage2,1)-delay(wage2,2))/delay(wage2,2))
# The price of the output of sector 2 depends on the growth rate of wages in sector 2.
inv1 = invint+invyuc1+invmex1
# Investments in tourism are the sum of the international, national and regional investments in tourism. Note that all international investments go to tourism.
inv2 = invmex2+invyuc2
# Investments in sector 2 are the sum of the national and regional investments in this sector.
invint = if profit_perroom<intprofitrate then 0 else 5*(profit_perroom-intprofitrate)
intprofitrate = 170
# International investments are zero if the profit per room is lower than the international profit rate.
invmex = if profit_perroom<nat_profitrate then 0 else .005*gdpmex
invmex1 = 0.5*invmex
invmex2 = invmex-invmex1
nat_profitrate = 170
gdpmex = 686406*(1+0.05*time)
# National investments are zero if the national profit rate is higher than the profit per room, otherwise they are a proportion of the GDP of Mexico. National investments are in the base scenario equally divided amongst sector 1 and 2. The GDP of Mexico increases over time.
invyuc = profit1*0.4+0.1*wage1*lab1+.4*profit2+.05*wage2*lab2
invyuc1 = if profit_perroom<nat_profitrate then 0.1*invyuc else dummyb
dummyb= if delay(profit_perroom,1)-delay(profit_perroom,2)>0 then 0.6*invyuc else 0.4*invyuc
invyuc2 = invyuc-invyuc1
# Regional investments depend on profits and wages in both sectors. A small part of those investments will go to tourism if the profit per room is smaller than the national profit rate. Otherwise this part depends on the changes in the profit per room.
gross_output1 = p1*tourists*11.2/1000
Si1 = 0.2
Sl1 = 1-Si1-Sk1
va1 = gross_output1*(1-Si1)
profit1 = gross_output1*Sk1-costs_to_clean
costs_to_clean = tour_clean*10/(2876*11.2*1.5/1000000)
wage1 = Sl1*gross_output1/lab1
Sk1 = GRAPH(occrate)
(0.00, 0.11), (0.1, 0.12), (0.2, 0.16), (0.3, 0.213), (0.4, 0.345), (0.5, 0.37), (0.6, 0.383), (0.7, 0.403), (0.8, 0.408), (0.9, 0.42), (1, 0.428)
# The gross output of sector 1 depends on the price and the number of tourists. The gross output is divided amongst intermediate consumption (Si1), capital (Sk1), labour (Sl1), costs to clean up water and profits. The part which goes to capital depends on the occupancy rate of the hotels. The value added is the gross output minus intermediate consumption. The part of the gross output dedicated to labour is divided amongst the number of workers.
out2 = 0.01*SQRT(lab2)*sqrt(cap2)
gross_output2 = out2*p2
profit2 = va2-wage2*lab2
Si2 = 0.5
Sl2 = 0.3
va2 = gross_output2*(1-Si2)
wage2 = Sl2*gross_output2/lab2

# The output of sector 2 depends on labour (lab2) and the capital stock in sector 2 (cap2). The gross output is the output times the price. The gross output is divided amongst intermediate consumption (Si2), labour cost (Sl2) and profits. Value added is the gross output minus intermediate consumption. Wages in sector 2 depend on the part of the gross output dedicated to labour costs and the number of people working in sector 2 (lab 2).
va = va1+va2
# Value added is the sum of value added in both sectors
wagemex = 15
# the wage in Mexico

The tourism module
rooms(t) = rooms(t - dt) + (new_rooms - rooms_out) * dt
INIT rooms = (18859+3350+747+3331)*365/1000
new_rooms = inv1/10
rooms_out = .05*rooms
# The number of rooms depends on investments in tourism (inv1) and on the depreciation rate of 5%.
tourists(t) = tourists(t - dt) + (tour_in - tourist_out) * dt
INIT tourists = 1958+321+138+459
tour_in = dem
tourist_out = if time>0 then dummye else 2876
dummye = delay(tour_in,1)
# The number of tourists depends on the demand and supply of rooms and the price.
occrate = min((tourists*11.2/(rooms*tourproom)),10)
tourproom = 6
# The occupancy rate depends on the number of tourists, the number of rooms and the number of tourists per room.
beach_attr = GRAPH(occrate)
(0.00, 0.63), (0.1, 0.69), (0.2, 0.735), (0.3, 0.765), (0.4, 0.843), (0.5, 0.9), (0.6, 1), (0.7, 1.03), (0.8, 1.08), (0.9, 0.945), (1, 0.84)
# The higher the attractiveness of beaches the higher will be the demand for tourist nights. The attractiveness of beaches depends on the occupancy rate of the rooms. If the occupancy rate is very low tourists do not like to come and also when the occupancy rate is very low tourists do not like to come either.
sites_quality = max(100-5*(tourists/1000),10)
# The quality of the sites depends on the number of tourists. The fewer tourists come in the higher the quality of the sites.
other = GRAPH(time)
(1994, 1.66), (1998, 2.04), (2001, 2.25), (2005, 2.38), (2008, 2.52), (2012, 2.58), (2016, 2.68), (2019, 2.74), (2023, 2.79)
# The impact of factors which are not included in the model changes over time. In the base scenario the factor 'other' increases which means that Yucat□n becomes more popular.
demvar = other*delay(beach_attr,1)*sqrt(water_qual/100)
*delay(sites_quality,3)/100
# The demand for tourist nights depends on the other factors, the attractiveness of beaches, water quality and the quality of the sites in the previous period.
dem = 1/(sqrt(pr))*demvar*demand*20000
sup = 2*rooms*((exp(0.0055*(pr-50))/(1+exp(0.0055*(pr-50))))-0.5)
dem = sup
# Price makes the demand equal to the supply. The supply of rooms depends on the price and the number of rooms. When the number of rooms increases, the supply of rooms at a certain price increases too. When the price is zero than the supply of rooms will be zero and when the price is

very high, the supply of rooms will be equal to the number of rooms. The demand for rooms depends on the price and the various variables (demvar, see tourists and rooms).
Note: Stella II cannot solve simultaneous equations which means that demand, supply and price have to be solved by an iterative process within each year.
profit_perroom = profit1*1000/rooms
# The profit per room is the profit divided by the number of rooms.

The demographic module
pop(t) = pop(t - dt) + (pop_in - pop_out) * dt
INIT pop = 535.185+493.277+1362.940
pop_in = pop*(birth_rate+migration)/100
pop_out = pop*death_rate/100
birth_rate = GRAPH(time)
(1994, 6.16), (2009, 4.87), (2024, 4.29)
death_rate = GRAPH(time)
(1994, 3.65), (2009, 2.90), (2024, 2.84)
The initial population is exogenous. The population increases (pop_in) by the birth and migration rates times the population and the population decreases (pop_out) by the death rate times the population. The birth and death rates change over time.
labour_force(t) = labour_force(t - dt) + (labour_in) * dt
INIT labour_force = 0.7*(535.185+493.277+1362.940)
labour_in = labour_force*(birth_rate+migration-death_rate)/100
# The labour force is a proportion of the population. It increases with the birth and the migration rates and it decreases with the death rate times the labour force.
lab1 = tourists/20
lab2 = labour_force-lab1
# The labour needed in tourism depends on the number of tourists. Labour in sector 2 equals the entire labour force minus the persons working in tourism. This means that there is no unemployment.
migration = if dummyc>0 then basemig+dummyc*.1/15 else basemig
dummyc = delay(wage1,1)-delay(wagemex,1)
basemig = GRAPH(time)
(1994, 3.46), (2009, 2.13), (2024, 1.31)
#Migration depends on a base migration rate (basemig) which depends on time and the difference in the wage rate between tourism in the peninsula and the wage rate in Mexico. If the wage rate in tourism is higher than the wage rate in Mexico then migration will be higher than the base migration rate.

The environmental module
water_qual(t) = water_qual(t - dt) + (cleaning - water_use) * dt
INIT water_qual = 100
water_use = (tourists*11.2*4+pop*365*2)/1000000
# Water quality depends on water use and the water clean up.
cleaning = nat__cleaning+tour_clean+pop_clean
nat_cleaning = (1-(water_qual/100))*(water_qual/100)*10
pop_clean = perc_pop_clean*pop*365*2/1000000
tour_clean = perc_tour_clean*tourists*11.2*4/1000000
# Water cleaning is the sum of natural cleaning and cleaning of the water used by the tourists and the population. Natural cleaning depends on water quality and the population cleaning depends on the percentage the government wants to clean (see government) and the number of people. The tourist clean up depends on the percentage the government decides to clean (see government) and the number of tourists.

The governmental module
perc_pop_clean = 0
perc_tour_clean = 0
# The government can impose a policy to clean up the water used by the population (perc_pop_clean) or the tourists (perc_tour_clean).
subsidy = pop_clean*10*(87/5)/((535.185+493.277+1362.940)*365/1000000)
# When the government imposes a policy to clean up the water used by the population it has to pay for it. This can be seen as a subsidy, which depends on the amount of water that has to be cleaned.

References

Aguilar, A.G. and F. Rodriguez (1995) The dispersal of urban growth in Mexico, 1970-90. *Regional Development Studies* **1**, 1–26, Winter 1994/95.

Bergh, J.C.J.M. van den (1998), this volume.

Casti, J. (1996) "What if…". *New Scientist* **13**: July 1996.

Cinvestav (Centro de investigaciones y estudios avanzados) (1996) Reporte de los avances del modelo de economia del proyecto poblacion y medio ambiente para un modelo de desarollo sustentable (Progress report for the project on population and environment on the economic module in the model of sustainable development), IIASA-Cinvestav, Unidad Mérida, Mexico.

CONAPO (Consejo Nacional de Poblacion) (1995) Estimaciones y proyecciones demograficos de Mexico (Demographic estimations and projections), Mexico.

Cuanalo, C.H. de la, R. Navarrete and A. Gomez (1996) La agricultura del estado de Yucatán (Agriculture of the provincie of Yucatán), Cinvestav-Mérida, Mexico.

Gelting, R. J. (1995) Water and population in the Yucatán Peninsula, WP-95-87, IIASA, Laxenburg, Austria.

INEGI (Instituto Nacional de Estadistica, Geografia e Informatica) (1992) XI censo general de poblacion y vivienda (General census on population and housing), Aguascalientes, Mexico.

Grant, W.E., E.K. Pederson and S.L. Marín (1997) *Ecology and Natural Resource Management – Systems Analysis and Simulation.* John Wiley, New York.

Hale, L. (1996) Population, development and environment in the Yucatán Peninsula: the fisheries module. IIASA, Laxenburg, Austria.

Hannon, B. and M. Ruth (1994) *Dynamic Modeling.* Springer Verlag, New York.

INEGI (Instituto Nacional de Estadistica, Geografia e Informatica) (1994) Anuario estadistico de los Estados Unidos Mexicanos (Annual statistics of Mexico), Aguascalientes, Mexico.

Inskeep, E. and M. Kallenberger (1992) An integrated approach to resort development: six case studies. World Tourism Organization, Madrid.

Long, V.H. (1991) Government–industry–community interaction in tourism development in Mexico. In M.T. Sinclair and M.J. Stabler (eds) *The Tourism Industry: An International Analysis.* C.A.B. International, London.

Lutz, W. (ed.) (1994) *Population, Development and Environment: Understanding their Interactions in Mauritius.* Springer Verlag and IIASA, Berlin.

Lutz, W., W. Folan, J. Gunn and B. Faust (1996) Possible effects of climate change on the Classic Maya collapse, paper presented at the Population Association of America (PAA) annual meeting, 9–11 May 1996, New Orleans.

OECD (1995) OECD economic surveys, Mexico 1994-1995. Paris.

Ramsamy, M. S. (1994) Sustainable tourism. In W. Lutz (ed.) *Population, Development and Environment: Understanding their Interactions in Mauritius.* Springer Verlag and IIASA, Berlin.

Shah, A. (1995) *The Economics of Third World National Parks: Issues of Tourism and Environmental Management.* Edward Elgar, Cheltenham.

Sinclair, M.T. (1991) The tourism industry and foreign exchange leakages in a developing country: The distribution of earnings from safari and beach tourism in Kenya. In M.T. Sinclair and M.J. Stabler (eds) *The Tourism Industry: An International Analysis.* C.A.B. International, London.

Wils, A. (1996) PDE-Cape Verde: A systems study of population, development and environment. WP-96-9, IIASA, Laxenburg, Austria.

World Bank (1994) Mexico, country economic memorandum. Report 11823-ME, Washington.

World Tourist Organization (1994) El Turismo en Mexico 1994 (Tourism in Mexico). Madrid.

World Tourism Organization (1996) http://www.world-tourism.org/ trends95.htm.

TOURISM AND THE ENVIRONMENT: SOME OBSERVATIONS ON THE CONCEPT OF CARRYING CAPACITY

HARRY COCCOSSIS
and
APOSTOLOS PARPAIRIS
Department of Environmental Studies
University of the Aegean
Mytilini, Lesvos 81100
Greece

1. Introduction

The relationship of tourism and the main components of the man-nature system are subject to much discussion but relatively little investigation. This chapter is based on the hypothesis that tourism generates environmental externalities and is often affected by them. When the negative effects of such externalities exceed certain levels of disturbance of the environment, significant and irreversible changes occur which alter the basic processes and characteristics of the environment. The carrying capacity (C.C.) concept has often been used to identify the limits of a system to absorb changes. The concept of carrying capacity as a planning tool is investigated in a systematic and comprehensive manner, by analysing its major dimensions. The aim of this analysis is to increase our understanding of this important and intrinsically too extensive and complex multifaceted concept, to discuss its applicability in the studies of the growing tourist industry and demonstrate how composite perspectives of the concept can provide valuable insights into some phenomena of interest to academic observers and researchers in allied disciplines such as resource management, planning, economics, sociology, anthropology, geography, business administration, as well as to those involved in the development and management of the tourist industry at various levels. There is in addition a new opportunity for detailed consideration of the resource base tourism in a period where the industry shows many signs of approaching a point of crisis caused by considerable weaknesses in understanding the relationship of increasing pressure – demand to natural, cultural and heritage resources (Parpairis, 1992).

This chapter is designed to provide a theoretical base as well as an operational framework in an effort to establish carrying capacity as a useful tool for tourism planning and management. It will review a variety of different scientific approaches and methodological issues, which focus on tourism and different aspects of carrying capacity. It will evaluate and compare techniques and methods suitable for the treatment of the issue, and it will indicate some critical factors affecting capacity and the related concept of saturation. The development of carrying capacity as an operational analytical tool could allow the reversal of damaging mistakes of the past, the anticipation of potential problems created by tourism development on the environment and the negative feedback effects on tourism itself in areas under development. Carrying capacity could also be applied at the planning stage for the imaginative design of projects and programs, which are simultaneously environmentally sound and supportive of rising economic and social prosperity – the two major goals of sustainable development. As a management tool, the concept of carrying capacity could also be applied to mitigate negative effects in areas, which have already experienced intensive tourism development.

2. Factors determining tourist flows to an area

Tourism and recreational behaviour are increasingly becoming important subjects of research in the environmental and social sciences as the result of many socio-economic and environmental factors which influence the movements of persons (Nijkamp, 1974). Research on tourism is generally less developed than on other human activities from which tourism differs both in nature and in character (Pearce, 1987). Leisure and tourism have only recently attracted the interest of planners, psychologists, sociologists and others, although the majority of studies, during tourism's short history, almost always saw tourism as an economic activity and indeed concerned with only one zone of the tourist system – tourist destinations (the other two being tourist origins and linkages between origins and destinations).

The spatial interaction arising out of tourist movements from origin to destination has not been examined comprehensively in much of the existing tourism literature, although a sense of this interaction emerges from several studies (Miossec, 1977; Defert, 1966). One of the most interesting problems, from a planning perspective, is saturation, which arises within each of the above mentioned three zones of the tourism system (WTO, 1983a):

- origin (demand), where saturation is caused by failure to stagger departures (due to a number of limitations like climate, school holidays, employment conditions, etc.) or time concentration of demand on certain types of tourism
- linkage (transit), where saturation takes the form of bottlenecks in travel

- destination (supply), where saturation produces a wide range of negative impacts by overloading facilities and, thus, damages the environment and the image of the host area both in physical and human terms

Although this chapter focuses on the zone of supply, the significance of the other two zones must be recognised and, indeed, an investigation of the total tourist system is needed because common issues underlie the' problem of saturation, stemming from the dynamic interrelationship between and within zones. However, the fundamental question, which underlies the phenomenon of saturation in tourism, is why people leave their home area to visit other places? Which are the main factors determining tourist flows to a certain area? Indeed, the issues of motivation and demand directly affect the phenomenon of saturation, particularly in the zone of destination.

Briefly, the basic factors influencing the ability of a person to travel are: the rise in welfare (Chali, 1977) and the availability of more leisure time (Coccossis, 1987), the decline in the quality of life (Nijkamp, 1977), increased accessibility, improved telecommunication services and information systems (Vernicos, 1987), the increase in the supply of tourist services, social progress, the disappearance of old superstitions and prejudices, the response to pull factors (Leiper, 1984), socio-psychological motives and cultural motives (Nielsen, 1977), the development of more sophisticated advertising campaigns, the desire for change and a number of other minor but interrelated factors. The above mentioned motivation factors have in turn rapidly changed the conditions influencing the demand for travel and tourism on such a scale that within a short period of time international tourist demand has risen from 284.8 million arrivals in 1980 to 567.0 million in 1995 and the tourist receipts from 102.372 million US $ in 1980 to 371.682 million in 1995 (see Figure 1), an unquestionably rapid increase compared to the past and also one of the most economically successful stories of the last 40 years (Coccossis and Parpairis, 1995).

The rapid growth of demand expected to become even stronger in the next century, suggests increasing pressure on tourism destination areas, many have already occurred many difficulties in the above mentioned three-zone tourist system, one of which is saturation (Parpairis, 1992). Two reasons are responsible for this particular problem. One relates to the high concentration of demand in time and space and the other to the time lag involved in the creation of tourist centres (supply) and the seasonal fluctuations of tourism activities. The increase in tourist demand has been more rapid than the creation of new tourist centres for absorbing tourists, equipped with the minimum appropriate installations for optimum tourist development. This situation has not been brought about solely by the massive concentration of demand in the majority of traditional tourist areas.

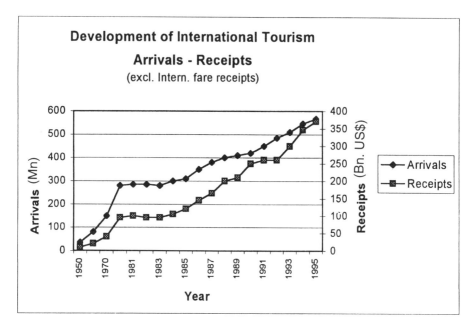

Figure 1: Development of international tourism arrivals - receipts (exc. intern. fare receipts). Source: World Tourism Organisation (WTO).

In most cases, such concentration takes place only during one or two months of the year, the capacity for reception being left virtually unused for the rest of the year. Saturation may also come about through an excessive growth of supply. Accordingly, we should acknowledge that there are two distinct and interrelated kinds of saturation, supply-related and demand-related, which take place within the origin and destination zones. Each kind has its own characteristics and distinctive features, which in turn determine distinctive types of saturation.

3. Definition of carrying capacity

The concept of carrying capacity is well-defined in ecology in terms of population (Ricci, 1976; Lursen and Wolft, 1981) and describes the upper asymptote in the rate of population growth. An example that typifies the measurement of carrying capacity and gives a practical interpretation of the concept is provided by solving the logistic equation, under the assumption, for example, that fresh water is the limiting factor to population growth. Ecologists have developed also more complex approaches which range from the early work of Lotka (1932) and the population model of Volterra (1931), which couple two differential equations that describe different growth rates to represent the predator-prey interaction, to the work by May (1973) studying the effects of deterministic and stochastic environments on biological population growth and stability (Conrad, 1983).

Additional insight into the concept, with the inherent preoccupation of how much and how far a population can expand, is provided by anthropology and geography. Thus, carrying capacity is defined as the theoretical limit up to which a population can grow and still be supported permanently by the environment (Vernicos, 1985). Here it involves calculations of the maximum number of individuals forming a stable population given a set of environmental resources. The capacity is also defined in terms of ecosystem characteristics. An ecosystem's carrying capacity is often estimated by dividing the estimate of the total available strategic resource or mix of resources, considered to be the limiting factor, by the individual resource need over a certain time period. The denominator (individual need) is normally weighed by a coefficient that reflects effectiveness of resource use and possible energy maximisation (e.g. the feeding strategy in which organisms maximise food yields). Obviously, it makes a lot of difference whether the resources are renewable or not.

Cybernetics has contributed also to the definition of the concept, starting from the idea that every species has a number of interlinked variables which form a mechanism of control and which are closely related to 'survival'. Carrying capacity is derived from the idea that an organism can exist only within a limited range of physical conditions.

A related concept which is broad enough to embrace most of the viewpoints regarding the ability of the natural environment to sustain itself is that of sustainability (WCED, 1987) or sustainable development (De Vries, 1989; Nijkamp, 1988). In other chapters of this volume, special attention is given to the concept of sustainable development in tourism. The definition of sustainable development originates in ecology and biology and has been transferred to economics where it has gradually found wider and richer, although less well-defined, applications (Conrad, 1983). In most definitions the carrying capacity of a certain system is related to its potential for sustaining certain population species and it denotes a living system (ecosystem) which is capable of sustaining itself indefinitely and without help from another such system.

4. Definition of tourist carrying capacity

Concerning the application of the concept of carrying capacity to tourism and recreation, most of the research has approached it mainly through the ecological and man-made environment traditions (Burton, 1979). It was pointed out, many years ago, that carrying capacity must be considered as a means to an end and not as an absolutely definite limit that is unalterable for each type of environment under discussion. In that respect, one of the most convenient definitions is the number of users that a recreation or tourist area can provide each year without permanent biological or physical deterioration of the area's ability to support recreation and without appreciable impairment of the recreational experience. From an ecological point of view, the definition of carrying capacity in this context is: The maximum level of recreation use, in terms of visitor numbers and activities, that can be accommodated before a decline in ecological value sets in.

Theoretically, the carrying capacity of a tourist area could be defined as the point where the minimum infrastructure/superstructure requirements, as well as the natural resource assets (beaches, etc.) which create demand, become insufficient to meet the needs of both the resident population and the visiting tourists, whereupon the threat of environmental hazards appears. The problem, therefore, refers to the quantitative levels of change of the environment, which can be permitted in the area under consideration. However, the above ecological definition of carrying capacity does not seem to take into account the ways in which recreational activities interact with natural ecosystems. The length of stay, the time and level of use, the way it is distributed over time and space and the desires of management, should be considered in any comprehensive definition in ecological terms.

From a sociological point of view, the definition of carrying capacity incorporates a relationship between 'amount of use and user satisfaction' (Nielsen, 1977; Stankey, 1982). Some geographers have used the term of carrying capacity, defined as the maximum theoretical population that can be supported by a given resource base:

$$\left[\frac{\text{available land}}{\text{utilized land}} * \frac{\text{theoretical population}}{\text{actual population}} \right]$$

From a physical planning point of view, the concept of carrying capacity has to do with determining the spatial ability of an area to accommodate visitors (physical capacity is concerned with the size and number of places suitable to this activity). The existence of three different types of carrying capacity – environmental, physical and perceptual or psychological – of an area has been suggested by Pearce (1989), relating broadly to the degradation of the environment, the saturation of facilities and the enjoyment of visitors.

However, continuing progress will undoubtedly require combined research efforts from a variety of disciplines since the dynamic and complex phenomenon of tourism requires a comprehensive approach (Coccossis and Parpairis, 1995). In that respect, one suitable definition of the concept could be as follows: The number of user unit use periods that a tourist area (island etc.) can provide each year, without permanent biological/physical determination of the site's ability to support recreation and tourism and without appreciable impairment of the recreational experience. A similar definition of the concept under discussion could also be one referring to life cycle products (Parpairis, 1992) where the product can be quickly adapted or replaced and completely removed from the market by a new technological solution. This cannot be so easily done with the tourist site as a market product if the phase of its saturation is not objectively recognised – particularly in the case of exaggerated commercialisation.

The above definition of the concept of carrying capacity is only a sample of those already developed from a wide range of perspectives which obviously vary depending on the type of an area, type of ecosystem and differences among societies in terms of the perception of the outdoor experience.

5. Factors affecting carrying capacity

In order to understand the concept of tourist carrying capacity, it is necessary to explore the main factors involved and their relationships. It is proposed that the concept of carrying capacity could be understood in terms of three basic dimensions: natural environment, man-made environment and social environment, to which a fourth one, time, should also be included. The carrying capacity of the natural environment cannot be easily measured in quantitative and qualitative terms although individual natural resource parameters can be quantified and measured. However, ecological indicators are not so easily estimated in quantitative terms and even if this were possible, the dynamics of their interaction are so complex and, to a great extent, so unknown that any attempt to achieve an optimum quality/quantity level seems futile. In spite of these limitations, a certain level of an area's natural environment carrying capacity can be measured. To a certain extent, technology could be employed to extend the limits imposed by the carrying capacity of the natural environment (for example, through a sewage treatment plant, which can reduce pollution levels).

The carrying capacity of the man-made environment is even less well defined. Its distinguishing characteristic is that it can be improved to a great extent (for example, by means of new facilities, investment in new products, incentives, management techniques, etc.). Obviously, technological and institutional solutions can be employed to extend the carrying capacity of the man-made environment (for example, by improving the existing infrastructure or allowing lower densities).

The natural environmental carrying capacity imposes usually much more binding and less flexible limits than the man-made environment carrying capacity. When the latter reaches the natural environment carrying capacity of an area, it usually indicates that the natural resources (which influence tourist demand) have become sufficient to meet the needs of both residents and visitors, while if the man-made environment carrying capacity exceeds the natural environment carrying capacity then the natural resources become insufficient to meet these needs and indeed, in this case, the threat of environmental hazards appears imminent.

The social carrying capacity is even less well defined than the other two categories, since it involves subtle and relatively unexpected relationships between man and the environment, both natural and man-made. Although certain relationships and feedback mechanisms can be expressed qualitatively or quantitatively, defining the social carrying capacity is still difficult because of the tremendous variability and diversity of these relationships. Furthermore, strong behaviour adaptation mechanisms seem to exist, which can alter the perception of crowding or saturation. In spite of the fact that quantitative and/or qualitative expressions of the concept require models which have not been constructed, yet because of the complexity of the interactions and the number of variables involved, the concept has an operational validity in policy-making since it provides a conceptual tool which can serve as a basis and as an indicator of whether the man-environment system is getting better or worse and which can be used as a vehicle to specify policy options. In this respect, it would be useful if the concept of carrying capacity could be integrated into a single operational definition, including ecological, socio-economic, institutional and policy components (De Vries, 1989, Parpairis, 1993).

For the above three distinct categories and their main components (natural, economic, social/psychological, cultural/political, and man-made) of the environmental resources affecting carrying capacity, it is recognised that a single and absolute measure of the carrying capacity of an area is difficult to estimate since the factors involved are not all quantifiable or even measurable, although a variety of methods and techniques have been occasionally employed.

6. Models and techniques for determining carrying capacity

Most of the research on carrying capacity has been approached mainly through one particular discipline, while the complex and dynamic phenomenon of tourism requires a combined view. The emphasis in some studies has been on the description and explanation of processes accounting for variations in the carrying capacity of different areas. There are many studies which attempt to establish the carrying capacity of the natural environment for various types of ecosystems (Laursen, 1981), but the results cannot be easily or universally applied because management objectives and the characteristics of the natural environment vary from site to site. Research on both natural and man-made environment carrying capacity demonstrates the way in which multiple measurement techniques can be used in data collection (combined field studies with direct observation, social surveys, behavioural inquiries and simulation). Over the past twenty years, several methods of estimating carrying capacity have been developed and applied with limited success. These methods have not been commonly used and may not be generalised to other situations. However, some aspects of each method can be applied independently or combined with conventional methods for particular problems (Gold, 1980). There is no common dimension to these methods except for an attempt to be less arbitrary than intuition in estimating carrying capacity. The following approaches and techniques have often been used:

Ecology. Cause and effect relationships (COAP, 1970), associated with different land uses and the specific activities under each land use. The final product is a set of charts, which relate specific actions to possible adverse environmental impacts, which are in turn classified according to initial conditions, consequent conditions and final effects. Surveys determining conflicts between recreation and flora-fauna (Wolf, 1984). Matrix analysis classifying the elements of environmental quality and the range of possible recreational impacts which can be used to assist the identification of problems and indicate which resources in certain areas are approaching their carrying capacity. Behavioural studies analysing attitudes and behavioural patterns of recreationists using interviews in order to assess the various factors that contribute to an awareness of environmental carrying capacity. The questionnaires allow respondents to provide spontaneous responses to both satisfactory and unsatisfactory elements of their experiences and to assess the degree of frustration resulting from congestion and crowding. The Travel Behaviour Model (Fisher and Krutilla, 1972) is similar to modelling techniques used for transportation planning and migration; this Markov model was used to establish visitor distribution programmes (entry rates and quotas were derived from the study used to simulate lake-to-lake movements by campers). Physical resource and human activity studies concentrate upon identifying satisfactory development densities and actual carrying capacity of an area. The resource component leads to mapping and evaluation of the capacity of the land and water to sustain different uses. Overlays of different capability maps allow resource use suitabilities to be compared against existing and potentially desirable patterns.

Economics. Since the impacts of tourism on the economic environment can be (directly or indirectly) quantified and statistically evaluated (Mauren, 1979), the concept of carrying capacity can also be determined. Some of the conventional economic analysis methods used in this respect include: cost-benefit analysis; cost of congestion (Cichetti, 1976); single and multiple-use patterns (Smith and Krutilla, 1976).

Social-Psychology. Quantitative Measurement Studies (Kreimer, 1977), based on the assumptions that the visual and other characteristics (e.g. noise) of the environment can be measured and described adequately in terms of a limited set of specific quantitative parameters, that people's preferences are generally clear-cut and fixed, that there is isomorphism between the real environment and people's perception of that environment and there is also isomorphism between the real and the simulated environment. However, the social/psychological carrying capacity is only very partially quantifiable and thus its evaluation is more subject to subjective factors depending on each evaluator's value system.

Cultural Studies. As a conceptual analysis of intercultural interaction between tourists and locals (Nederlof, 1989), the research will produce some theoretical outlines, presented in a model, concerning the different interactions in the field of tourist experiences. Obviously, the evaluation of the cultural carrying capacity depends entirely on totally subjective criteria, so it is not quantifiable at all.

Built Environment Studies. The carrying capacity of an area can be constrained by the availability of basic resources, like water, power supplies, drainage system, waste disposal, telecommunications capacity, etc. For most of these resources, there are levels determining the carrying capacity in relation to the needs of the tourists and the local population. Attempts have also been made to determine the carrying capacity of an area in relation to accommodation, services and facilities, supporting tourist and population demand. A WTO report (1983b) proposes planning and capacity standards to be used as indicators as well as optimum capacity levels, in different areas, expressed as hotel density (from 13–35 m^2 per person) and overall resort density (from 20–100 beds per hectare up to 200–1000) for urban type resorts, a set of densities which also depend on the image, land availability, cost parameters, building heights and allowable density, etc.

In the work of OECD (1980), reference is made to the absorptive capacity of a tourist area using such indices as: tourist capacity in relation to population, beds, land area, restaurants, commercial licences; and tourist density in relation to number of hotels, normal population of the area, number of nights spent by foreigners in each hotel and in the area. Tourist density indicates the accommodation potential of an area expressed in terms of places available per km^2 or surface area, overall or by hotel category. Such indices can be used for the classification of tourist regions. For example, a tourist intensity index has been estimated at the national and regional level in France producing a six-fold classification of French Communes (Boyer, 1972; Pearce, 1979).

A tourist function index has been taken as a measure of tourist activity or intensity also (Defert, 1966). It is derived by comparing the number of beds available to tourists in an area with the resident population of the same area. Plettner (1979) advocates use of tourist nights rather than bed capacity and proposes a tourist density index, which is the quotient of tourist nights to the local population. The tourist comfort index (Mirloup, 1974) is based on a formula which distinguishes the quality between different types of accommodation using certain criteria. The concentration index (Girard, 1968) is an attempt to determine the degree of concentration of tourist activity (hotels, restaurants, etc.) and the occupied land area. The attractiveness index, derived by comparing the number of bed nights between international and domestic tourists, can be used in order to evaluate a region's profile in attracting specific types of tourism overall or by category.

However, apart from the above attempts to identify models for calculating the carrying capacity for individual aspects of the environment of a given area, it is also useful to review those techniques and methodologies developed in order to determine an overall measure of carrying capacity (including physical and human environment variables). Ricci (1976) provides an operational overall carrying capacity estimate by combining critical factors and resources, which are the principal determinants of social and economic life. These determinants include renewable and non-renewable resources, population dynamics as well as the feedback mechanisms, which relate to these. His approach indicates that the dynamic nature of carrying capacity is affected by its previous and present values, non-linearities, thresholds and the unexpected disturbances imparted on the entire system by the real world.

Cicchetti and Smith (1976) developed the concept of congestion and its implications by modelling variables determining the cost of congestion. The main purpose has been to develop a methodology for measuring the effects of congestion so that the cost associated with it might be taken into account in public allocation decisions. It is well known that the problem of pollution and congestion is now recognised as among the most challenging ones for the efficient allocation of resources. The congestion model was also developed by Smith and Krutilla (1976) in the form of a large-scale traffic simulation model which, in fact, provides the required technical data on use patterns and the associated expected levels of encounters.

Nijkamp (1977) developed the impact structure matrix combining environmental elements and the range of possible impacts on these elements from the development of tourism to a certain level (carrying capacity level). To fill out the structure matrix a set of tools is needed such as: comparative studies; user satisfaction survey; natural resource parameters; ecological indicators; certain criteria and values; environmental impact studies; multiple measurement techniques; field studies; social surveys; behavioural inquiries; norms and standards determining the human environment carrying capacity, etc. This multidimensional model aims at providing an integrated rather than a partial picture of the carrying capacity concept to be used operationally.

Parpairis (1993) proposed a methodology and a system of certain indicators and criteria which can be introduced even in a segmented approach, in the context of the area's carrying capacity, to express tourism/environment relations together with guidelines (UNEP, 1996). The system was tested in the case study of Mykonos (Coccossis and Parpairis, 1995; Coccossis and Parpairis, 1996). Among the main indicators were the ratio of tourists to local population, the ratio of residential buildings to tourist buildings, the ratio of historical village stock to recently built areas, changes in building use and in land use patterns throughout time. Two basic groups of criteria controlling the concept were used; those affecting the indigenous environment and those affecting the tourist image/product. The criteria include ecological aspects (acceptable level of visual impact, point at which ecological damage occurs, the need for conservation of wildlife and marine life), economic benefits (level of employment providing optimum economic benefits, level of employment suited to the local community, etc), social aspects (the volume of tourism that can be absorbed in the social life of the area), cultural aspects (the level of tourism that will help maintain heritage, monuments and cultural traditions without detrimental effects), and resource availability (public utilities, water, transport facilities, etc.).

The second group of criteria, that of tourism image/product, include physical characteristics and issues (climate, pollution, attractiveness, quality of accommodation, etc), economic (cost of holiday), social/cultural (the quality of local attractions, community culture, etc.), and resource availability (standards for services, infrastructure, etc). Accordingly, the overall carrying capacity of the case study appears to be the result of the combination of the above indicators and criteria under the methodological framework where perception studies, investigation instruments and other related tools were employed.

7. Conclusions

Some important lessons emerge from the brief review of the concept of carrying capacity attempted in the preceding sections. It becomes obvious that all methods described have deficiencies although some of them are better for specific purposes than some others. As a general observation:

- more research is needed on the qualitative aspects of the carrying capacity concept before any of these methods is employed in planning
- carrying capacity can be estimated more easily on a case-by-case basis using an approximation method oriented towards a specific resource while general limits should be expected to be difficult and dependent on a variable mix of the three main components
- due to particularities of each area and in view of the uncertainties involving about the interactions of environmental factors, it is necessary that the limits imposed will be considered as adaptive, flexible, and open to reappraisal

Some related concepts, such as sustainable development, can be explored in conjunction with carrying capacity. For example, De Vries (1989) proposed five perspectives on sustainable development without sharp boundaries between them:

1. *technological* perspective (technology regarded as the major driving force)
2. *resource economic* perpective (system dynamics with an emphasis on a succession of limits which will constrain exponential growth of population and material output, natural resources, energy analysis based on physical

constraints on energy supply or land productivity, ensuring that the cost of material inputs and outputs reflect the cost to maintain or restore the natural environment)
3. *anthropological* perspective
4. *ecological* perspective
5. *cultural* perspective

All these relate to a great extent to the concept of carrying capacity.

The concept of carrying capacity has already been employed in several outdoor recreation projects and can serve as a useful conceptual tool in tourist studies, especially when it is seen as a means to encourage tourism planners and others to give greater consideration to environmental matters, to qualitative aspects such as the experiences of both hosts and guests and as a supportive tool for the specification of goals and objectives. At present, the possibility of establishing quantitative limits through carrying capacity expressions as absolute figures seems rather remote. In any case, the concept of tourist carrying capacity is not the panacea it may seem to be.

The development of an operational tool for tourism planning and management seems to be a necessity. This need was expressed in the Manila Declaration (WTO, 1983b):

"Tourism resources available in the various countries consist at the same time of space, facilities and values. These are resources whose use cannot be left uncontrolled without running the risk of their deterioration, or even their destruction. The satisfaction of tourism requirements must not be prejudicial to the social and economic interests of the population in tourist areas, to the environment or above all, to natural resources (which are the fundamental attraction of tourism), historical, and cultural sites."

Accordingly, tourism development should be planned in such a way and at such levels that it meets the natural and human/physical carrying capacity requirements of the area under consideration to avoid any unforeseen cause and unacceptable environmental degradation. This chapter intends to stimulate further thought towards this end.

References

Boyer, M. (1972) *Le Tourisme.* Éditions du Sevil, Paris.
Burton, R. (1979) *Cannock Chase Recreational Carrying Capacity Study.* Keele University Library, occasional publication No. 11. Keele University, Keele.
Chali, M.A. (1977) *Tourism and Regional Growth.* Martinus Nijhoff, 's Gravenhage.
Cichetti, C.J. (1973) *7he Cost of Congestion: An Econometric Analysis of Wilderness Recreation.* University of Wisconsin, Madison.
Conrad, T. (1983) In *Adaptability – The Significance of Variability from Molecule to Ecosystem.* Plenum Press, New York, pp. 37–43.

COAP, (1970) *Comprehensive Ocean Plan.* State of California.

Coccossis, H.N. (1987) Planning for Islands. *Ekistics* **323/324**, 84–87.

Coccossis, H.N. and A. Parpairis (1995) Assessing the interaction between heritage, environment and tourism: Mykonos. In H. Coccossis and P. Nijkamp (eds) *Sustainable Tourism Development.* Avebury, London, pp. 107–125.

Coccossis, H.N. and A. Parpairis (1996) Tourism and carrying capacity in coastal areas. In G.K. Priestley and H. Coccossis (eds) *Sustainable Tourism.* CAB International, Wallingford, pp. 153–175.

Coccossis, H.N. and A. Parpairis (1993) Environment and tourism issues: preservation of local identity and growth management: case study of Mykonos. In D. Konsola (ed.) *Culture, Environment and Regional Development.* Regional Development Institute, Athens, pp. 79–100.

Coccossis, H.N. and A. Parpairis (1991) The relationship between historical/cultural environment and tourism development. In *Proceedings of the International Symposium on Architecture of Tourism in the Mediterranean.* Constantinople Vol. 12, pp. 331–352

Defert, P. (1966) *La Localisation Touristique: Problèmes Théorique et Pratiques.* Éditions Gurten, Bern.

Fisher, A.C. and J. Krutilla (1972) Determination of optional capacity of resource-based recreation facilities. *Natural Resources Journal,* Vol. 12, University of Kentucky, Kentucky, 80–89.

Girard, P.S.T. (1968) Geographical aspects of tourism in Guernsey. *La Societé Guernesiaise Rapports et Transactions* **18**:2, 185–205.

Gold, S.M. (1980) *Recreation Planning and Design.* MacGraw-Hill Book, New York, pp. 163–177.

Kreimer, A. (1977) Environmental preferences: a critical analysis of some research methodologies. *Journal of Leisure Research,* 88–98.

Leiper, M. (1984) Tourism and leisure: the significance of tourism in the leisure spectrum. 12th N.Z. Geography Conference N.Z. *Geog.Soc.,* pp. 249–53, Christchurch.

Lotka, A.J. (1932) Growth of mixed populations. *Journal of Washington Academy of Science* .**22**, 461–469.

Lursen, K. and S. Wolft (1981) *Ecological Effects of Tourism in the Wadden Sea.* Free University of Amsterdam.

Mauren, J.L. (1979) *Tourism and Development in a Sociocultural Perspective.* Free University of Amsterdam.

May, R.H. (1973) *Stability and Complexity in Model Ecosystems, Monograph in Population Biology,* No.6. Princeton University Press, Princeton, New Jersey.

Miossec, J.M. (1977) Un modèle de l'espace touristique, *L'Espace Géographique* **1**, 41–48.

Mirloup, J. (1974) Eléments méthodologiques pour une étude de l' équippement hotelier. *Norois,* **73**, 433–452.

Nederlof, S. (1989) *Anthropological Research Project: Cultural Identity and Social Change* (unpublished). Katholieke Universiteit, Leuven.

Nielsen, J. (1977) Sociological carrying capacity and the last settler syndrome. *Pacific Sociological Review* **20**:4.

Nijkamp, P. (1974) *Environmental Attraction Factors and Regional Tourist Effects.* Paper 14, Free University of Amsterdam.

Nijkamp, P. (1977) *Operational Methods in Studying Tourist and Recreational Behaviour.* Paper 67, Free University of Amsterdam.

OECD (1980) *The Impact of Tourism on the Environment: the Spanish Example.* Tourist Capacity and Tourist Density Case Study, OECD, Paris.

Parpairis, A. (1992) The evolution of the life cycle of a tourist product. In *Proceedings of the 3rd International Conference on Environmental Science and Technology.* Vol. B. University of the Aegean, Lesvos, pp. 673–689.

Parpairis, A. (1993) The Concept of Carrying Capacity, Ph.D. thesis, Department of Environmental Studies. University of the Aegean, Lesvos, Greece.

Pearce, D.C. (1987) *Tourism Today: A Geographical Analysis.* Longman Scientific and Technical Publishers, New York.

Pearce, D.C. (1979) *Tourist Development.* Longman Scientific and Technical Publishers, Halcrow.

Pearce, D.C. (1989) *Tourist Development.* Longman Scientific and Technical Publishers, Halcrow.

Plettner, H.J. (1979) Geographical Aspects of Tourism in the Republic of Ireland. Research paper No.9, Social Sciences Research Centre, University College, Galway.

Ricci, F.P.. (1976) *Carrying Capacity, Implications for Research.* University of Ottawa, Ontario.

Smith, U.K. and J.V. Krutilla (1976) *The Structure and Properties of Wilderness; User's Travel Simulator.* Johns Hopkins University Press, Baltimore.

Stankey, G.H. (1982) Recreational carrying capacity research review. *Ontario Geography,* 57–72.

UNEP (1996) *Guidelines for Carrying Capacity Assessment for Tourism in Mediterranean Coastal Areas.* Priority Action Programme, Regional Activity Centre, Split.

Vernicos, N. (1985) Three basic concepts: man as part of the environment, carrying capacity, conservation, some further considerations. In D.O. Hall, N. Myers and N.S. Margaris (eds) *Economics of Ecosystem Management.* W. Junk, Dordrecht, pp. 41–45.

Vernicos, N. (1987) The study of Mediterranean small islands: emerging theoretical issues. *Ekistiks* **323/324**, 101–108.

Volterra, V. (1931) *Le Con sur la Théoric Mathematique de la Lutte pour la Vie.* Gauthier-Villars, Paris.

Vries, H.J.M. de (1989) *Sustainable Resource Use: an Inquiry into Modelling and Planning.* Rijks Universiteit, Groningen.

Wolf, W.J. (1984) *The Effects of Recreation on the Wadden Sea Ecosystem.* Free University of Amsterdam.

World Commission on Environment and Development (1987) *Our Common Future.* Oxford University Press, Oxford.

WTO (1983a) *Prospects for Restructuring Tourist Flows, Destinations and Markets.* Madrid.

WTO (1983b) *Risks of Saturation of Tourist Carrying Capacity.* Fifth Session, Madrid.

RURAL TOURISM AND RURAL DEVELOPMENT

MICHAEL KEANE
Department of Economics
University College Galway
Galway
Ireland

1. Introduction

There is a renewed degree of interest in rural development. Despite this interest, and some positive action at various levels, there is little by way of a successful blueprint for rural development. As one reviewer of research (in Europe) has stated "durable generalisations (about rural development) are scarce" (Whitby, 1986). This scarcity of results may be due to the relatively recent origin of the strategies adopted and to the fact that much of what is taking place is ad hoc, often co-existing alongside a more traditional institutional and policy framework and in all cases conditioned by the particular economic, institutional and socio-economic context of each rural area.

The term rural development itself is amenable to a range of definitions and interpretations. One widely quoted definition is that of Jasma *et al.* (1981) who define rural development as "an overall improvement in the economic and social well-being of rural residents and the institutional and physical environment in which they live." This definition seems to include everything that is relevant to rural development, yet it has been criticised for not including any reference or concern for the level of well-being of non-rural residents. The criticism is by Hodge (1986) who makes the point that, at least in the European context, many rural areas are quite densely settled and have strong urban links. He cites the recreational, environmental and residential location roles that rural places fulfil. Consequently, his argument is that any analysis of rural problems must take greater account of urban interests. There are other problems with definitions, like the issue of what is rural, what is local, etc. where it is difficult to find agreement. Perhaps the most important point on which all definitions will agree is one which recognises that rural development must be far wider in scope than merely dealing with issues linked to agriculture and food production. The focus on rural development must, for example, be increasingly centred on issues of environmental preservation and enhancement. Similarly, the policy framework must go beyond the narrow sectoral approach and become more global. This means giving support for all rural activities, whether agricultural or non-agricultural. It also means that structures and policy approaches must be flexible enough to accommodate new and innovative local initiatives.

The need for rural diversification at the farm level is a pressing one. The relatively open ended commitment to support quantities of food production not required by the market, which characterised the Common Agricultural Policy (CAP) during the 1970s and early 1980s, has now ended and the CAP has been in the process of reform since 1984. The imposition of the quota restrictions on milk production, with penalties on over-production, was a market-led adjustment. This change in EC policy represents a major and relatively sudden departure from a guaranteed milk price for an unlimited supply. The policy shift has been particularly severe in Ireland given its farm product mix, with its special dependence on dairying, and the lack of alternative enterprises, which are financially attractive. A variety of alternative enterprises are being mooted for farmers, including forestry, agro-tourism, set aside proposals, direct income aids and so on. There is a more explicit concern for the rural environment in all of these different alternatives. However, it is unlikely that farmers will be attracted into these schemes unless the terms are sufficiently generous (Mowle, 1989). There is a need for convincing evidence that can demonstrate the viability, suitability and economic potential of different enterprises.

2. Outline of the chapter

This chapter looks at rural tourism as a potential enterprise in rural areas. It also examines how differences in the way in which the organisation and the supply of the rural tourism product are approached can affect the kind of outcomes that are achieved in a given area. Rural tourism is not to be interpreted as a narrow rural option that simply complements traditional agricultural activities. Its potential is far greater than this; consequently, it should be considered in very broad terms and in relation to general local development (Greffe, 1990). The capacity of rural tourism to contribute to the resolution of the many problems faced by rural areas will depend on how a number of critical issues are resolved, issues like how rural tourism is organised in an area, who gets to participate, and the kind of structures that are put in place to actually develop and promote the rural tourism product. It is difficult to separate these issues from the broader issues concerning the thrust and content of overall local development efforts. A theme running through this chapter is that rural tourism can make a maximum contribution to rural development if it is developed as part of a locally integrated development plan. It is this integrated and co-ordinated approach which distinguishes rural community tourism from other types of rural tourism initiatives. This integrated approach can also serve as an important mechanism for the management of the tourism-environment relationship at the local level.

3. Definitions

There are a variety of terms used to describe tourism activity in rural areas: agri-tourism, farm tourism, rural tourism, soft tourism, alternative tourism, and many others. Despite Nordbo's (1985) suggestion that the term "agri-tourism" was the best market description of tourism in rural areas, none of the above terms have won common acceptance throughout Europe. An added complication is that these tourism terms have different meanings from one country to another. Thus, it is difficult to avoid some of this confusion in relation to labels and definitions. However, a basic distinction can be made between farm-based tourism (such as agri-tourism, farm tourism) and the remainder, which covers all tourist activities in rural areas. The term rural tourism has been adopted by the EC (EC, 1987) to refer to the entire tourism activity in a rural area.

The terms agri-tourism, rural tourism and rural community tourism are used throughout this paper. These different terms reflect some basic differences in the way in which tourism is developed and organised in any given area. Agri-tourism is farm-based tourism, rural tourism refers to the entire tourism activity in an area. The rural community tourism label is used to represent situations where tourism development takes place in an integrated and co-ordinated manner at the local level. The actual products and general supply structures involved in rural community tourism may in one area be dominated by farmers and farm-based activities, while in another area they may be more broadly based, both in terms of the product and in terms of those who participate, In both instances, the distinguishing feature is that there is local co-operation and community involvement in the planning and implementation process.

A community-based approach to rural tourism is specifically supported by Thibal (1988, p.7) who suggests that "when a rural area chooses to bank on tourism as part of a local development programme, every inhabitant is potentially an interested party, not only because he may happen to have a project of his own, but above all as a member of the local community and potential beneficiary of the expected collective development." Thibal argues that spontaneous initiatives are a thing of the past and that rural tourism must be everyone's business if it is to become, in small rural areas, simultaneously an incentive to the establishment of collective leisure facilities available to local people as well as tourists, a stimulus to local income and trade, an opportunity for job creation, a factor for the development of the area's economy and an instrument for re-awakening local culture, and for managing the environmental resources of the countryside.

4. Rural development

At the heart of the rural development problem is the lack of economic diversity, not just in the agricultural sector, but in the local economy generally. If a community successfully manages to diversify the local economic base it will find itself in a better position to retain population and cope with economic change and shocks. Diversification will give some degree of stability as well as an opportunity for the area to grow organically. This notion of diversification is at the heart of the difference between economic development and economic growth. The two concepts do not mean the same thing; they involve different processes (Flammang, 1979). Economic growth can be defined as a process of simple increase, implying more of the same, while economic development is a process of structural change, implying something different, if not something more. In many ways development is something that occurs because of necessity. We change things – institutions, structures, etc. – because we have to. Thus, a rural community seeking to develop is basically looking for an "ecological niche" for itself whereby it can cope with various economic realities through the use of its resources and strengths. The range of strengths include space, physical environment, a sense of community, natural resources and people – all of which can be used to create new economic opportunities. The contribution that rural tourism can make to the diversification and development of rural areas will only materialise if rural tourism can capitalise on these attributes and successfully compete with other forms of tourism. Competitiveness does not mean that rural tourism must seek to match the high-rise hotel buildings in the cities or match what coastal resorts offer. The strength of rural tourism lies in the special product that it has to offer – a product that is essentially the result of the natural and human environment that one can find in the countryside.

A significant diversification of the rural economic base represents development but so also does the putting in place of significant new capacity for positive change – new investment resources or new local structures. By increasing a community's capacity to make decisions and to take organisational action, development can mitigate decline and the negative impacts of projects. It can also expand options for the local community and in the process reduce its economic vulnerability (Douglas, 1989). All elements of the rural base must be involved: private entrepreneurs, community interests, local government, state agencies and the farming community. All these players have different kinds of strengths and, consequently, may occupy different roles in the rural development process. Individual decision-makers in the local area will pursue market driven decisions of their own, whereas collective or community-based initiatives will be motivated by some socio-economic goals. All initiatives will require the help of external agents.

The stage of development and the degree of dynamism in the rural economy determine the scope and the prospects for local development strategies. In many rural areas, the economic fabric is very thin and there are few opportunities that can afford a sufficient rate of return to allow significant private investment. Often, the only response available to development needs is of that of a collective or community-based initiative. Such initiatives are defined as including all of those designs and activities set in place and fostered by rural communities, or groups within communities, to enhance the economic wellbeing of the community as whole (Douglas, 1989). Douglas differentiates between community-based economic development and other sources of development, i.e. those provided by private individuals or by local and national governments or their agencies, on a number of important accounts. In community-based economic development, the community is the target, the main beneficiary and the decision-making body. Motives of profit and return on investment may loom large, but the overriding motive is community betterment. Community-based economic development attempts to be highly integrated. By design, economic initiatives are linked to other businesses and economic activities in the area and are tied directly to social objectives (e.g. relieving youth unemployment, employment for married women). A holistic approach is taken (Clark, 1981). The community welfare and participation ethic introduces a strong distributional element into community-based economic development initiatives. Involvement and maximising the spread of beneficial effects are important. Accountability is important as the process is usually characterised by a large number of participants, shared resources, trust and strong social interaction. Feedback, discussion and accountability are therefore significant dimensions of the process. Getting the balance right between all of these dimensions is the difficult task that those who choose to participate must face.

There is the danger with these different dimensions that economic criteria and outcomes somehow get disavowed along the way. Shaffer's (1989, p.9) viewpoint "that community economic development is not an attempt to exploit resources to yield the maximum economic return", but is more about helping communities to achieve a broader set of goals on their own terms is indicative of the dilemmas faced by the scientist. It is difficult, however, to see how economic issues can be ignored.

The basis of local economic development, irrespective of how we might care to define this development, is the placing of capital and capital has its own, well defined rules, which direct the form, the amount and the places where that capital goes. These rules are described in terms of market signals about profitability and rates of return. The problem in many rural areas is that there are not enough profitable opportunities that can offer a sufficient rate of return to attract private capital. There is also a shortage of entrepreneurs in these areas. Thus, the scope for profitable, or market-led ventures is limited. Most of the ventures undertaken by local groups are motivated by the view that there are profit-making opportunities within their local areas and also by a sense of frustration with the fact that nobody seems prepared to grasp these opportunities. What local people are essentially saying is that the market is not working well in their local area i.e. it is not transmitting the signals that will encourage individuals to act. A more correct analysis, and course of action, might be that suggested by Michaelson (1979) – to focus more on the quality and the content of the market signals, rather than complain about the actual lack of signals. His market-failures analysis of is a useful one that can help to define various roles in local economic development. His categorisation of market failures is presented in Table 1.

Table 1. Categorisation of market failures

Type of market failure	Description
Market Failure 1 (MF1)	occurs when the capital market fails to place capital, where it makes a maximum internal-monetary return through ignorance of the opportunity.
Market Failure 2 (MF2)	occurs when the capital market does maximise internal-monetary return but, by other criteria, the capital allocation is 'wrong'.
Market Failure 3 (MF3)	occurs when the capital market is aware of opportunities to increase its internal-monetary return but is constrained from doing so.
Market Failure 4 (MF4)	occurs when the capital, whether maximising only internal-monetary return or social return, given the resources available to it, fails to produce the return it would, without additional costs, through a change in some legal form such as ownership control.

Source: Michaelson (1979)

Market Failure 2 (MF2) is, perhaps, the one that is the best known and is usually discussed in relation to the environmental externalities of economic activity not accommodated by the market system. The MF4 argument is particularly relevant to rural tourism development. The development of rural tourism is a potent vehicle for local development, economic recovery, social progress and conservation of the rural heritage. Consequently, it should be thought out, organised, marketed and managed in accordance with the characteristics, needs, limitations and potentials of the receiving community, as they are the potential beneficiaries of any development. Similarly, the rural community can most efficiently carry many of the costs, with assistance from the government or its agencies. The costs of marketing the product, for example, is a problem in that an individual operator or a commercially oriented agency cannot expect to recover benefits that will be in any way commensurate with these costs. But, unless these market development costs are incurred, there will be no significant market and the potential economic benefits from rural tourism to local communities and local operators will be minimal. A community type approach can internalise these costs, but to do so it will need suitable support from the government and its agencies. This support must be structured so as to facilitate the integrative approach at the local level. Local participation can best be encouraged through community structures. Such structures can mobilise resources which otherwise would not be available to the single individual. Furthermore, a community structure can internalise many of the risks and uncertainties which, again, individuals on their own would have to face.

5. Rural tourism issues

The rural tourism product has always been available, to some extent, in all rural regions. Different activities, services and amenities have been provided but with little or no integration or co-ordination and with little overall impact.

There is now a widespread view that rural tourism can be a beneficial mechanism for rural development (European Commission, 1987; European Parliament, 1988; European Commission, 1990). Potential economic benefits from tourism in rural areas include:

- the creation of local income and employment
- the effects that tourist spending has on various sectors of the local economy
- the valorisation of the countryside and natural phenomena to which no economic value is attached
- the creation of a demand for craftwork and labour-intensive products and services which can be met in rural areas
- the multiplier effects of tourist spending in the receiving area and the potential developmental effect which this can have

These economic benefits will be greater if it is possible to persuade visitors to spend a maximum amount of money in the local area. Multiplier effects of different types of autonomous spending at the local level tend to be rather small because of the significant leakages that usually occur. There is the problem that incomes created on the basis of this autonomous spending may not be recycled locally, and this can dilute the expected economic development effects in the longer term. However, rural tourism, when it involves a highly varied and local set of inputs, alongside a set of labour-intensive commercial activities, some of which may not involve any significant amount of investment, can have a large local impact. There are opportunities for economies of scope (Greffe, 1990) where local tourism product providers can provide a variety of products and services and thereby benefit from any economies associated with joint production. Varied economic strategies become possible with rural tourism when it is organised to be a joint provision of several services and products in a relatively well-defined location. The benefits from tourism are equal to, if not greater than, the benefits from other economic activities. One Norwegian study (reported in Greffe, 1990) demonstrates the importance of tourism at the local level. The study shows how tourist activity created as much local value-added as industry for a given intermediate consumption, and much more than conventional distribution activity. This example, and other evidence on tourism multipliers in the literature (see Mathieson and Wall, 1982), confirm the view that tourism represents a local development potential that is comparable with that of industry and that is much greater than that of other service activities. O'Cinneide and Keane (1989) have estimated a local income multiplier with a value of 1.27 for Irish language colleges in the Galway Gaeltacht (Irish language-speaking areas). Irish language courses are a major element of the tourism industry in the Galway Gaeltacht. The activities involved are not unlike some of the activities envisaged in the rural tourism product. Irish colleges generated revenue of over Irú 2.7 ml. in the Gaeltacht area in 1987. This "cultural tourism" income multiplier can be compared with an income multiplier of 1.2 estimated by O'Connor *et al.* (1981) for the employment impact of a major power station in rural County Clare. This comparison once again suggests that tourism-type activities in rural areas compare very favourably with other forms of development activity in terms of their ability to create income and employment.

Setting up new rural tourism systems to capture these economic benefits presents a number of organisational difficulties. There are practical problems in modernising the supply structure and the production of tourist services in an area, there is the problem of promoting these structures and of ensuring that the tourism product meets customer demand. There are also fundamental questions like what kind of development is involved, what is the nature of local participation, what types of benefits are created, and how these benefits are distributed. Tourism also requires management to minimise its negative impacts, particularly in relation to the environment and its social carrying capacity. Hodge (1986) suggests a planning approach that includes a high degree of local participation as a solution to these difficulties. Local participation in the management and development of tourism can increase its acceptance in rural areas (Gannon *et al.*, 1987). The supply structures require local community involvement. Furthermore, the production of tourist services should not be confined to farmers. Tourists must be made feel welcome in social contacts such as shops, post offices and pubs, while non-farm based rural residents may have access to resources and be potential entrepreneurs in adding to the rural tourism product. Participative structures facilitate the development of different small-scale development projects and allow for co-ordination and dovetailing of activities in the local area.

France and Austria are two countries that have experience in co-ordinated rural tourism development. In 1976, France introduced a new policy to exploit the tourist potential of rural areas, the. 'Pays d'Accueil' Cointat (1988) describes the policy as a back-up to local initiatives and efforts to develop the economy through tourist activity. Three essential conditions must be satisfied for a geographical area to be recognised as a 'Pays d'Accueil': demonstrate a willingness to implement concerted tourist development programmes, involving all those locally involved; enjoy a number of tourist assets, such as quality of the environment or natural habitats, potential for developing a range of leisure activities, and an active cultural life; meet the cost of employing technical support to organise the provision of accommodation and leisure activities for holidaymakers and to ensure the co-ordinated development of tourism. During the period 1976-1980, ninety 'Pays d'Accueil' schemes were established with basic assistance from the state. From 1980-1983, the idea of a contract between the state and the 'Pays d'Accueil' was explored with the establishment of multi-annual programmes to exploit leisure facilities and to organise tourism accommodation. The emphasis was on the terms on which tourism was marketed, the extent to which resources were taken up and the identification of client groups (the need to tap all sectors of the community). In the period 1983-1988, the French government established planning contracts, which resulted in the formation of more than 200 extra 'Pays d'Accueil'. Rural tourism with a high degree of local involvement has been successfully developed in Austria also, particularly in the Innsbruck and Tyrol regions (for details see Gannon *et al.*, 1987; Needham, 1989; Lyons, 1989).

6. Rural tourism projects in Ireland

In Ireland, a number of rural tourism projects are in place at present (Keane and Quinn, 1990). The remainder of this chapter looks at some of the outcomes from one particular pilot project, Ballyhoura. The tourism co-operative in the Ballyhoura area is a community model of development. It was established in 1986 but for two years it achieved very little. The coop did no marketing. It depended on its members to voluntarily co-ordinate responses to enquiries. There was little capability amongst its tourism interests to carry out medium and long-term planning. In 1988, the local community drew up an integrated development plan for the area, which centred on the development of a rural tourism industry worth £ 1 ml. annually. The goal of rural tourism in the plan is to help solve the problems of low-income farmers. A business development programme was put in place and a full-time co-ordinator was appointed. The present analysis looks briefly at two outcomes; involvement and economic results. The main data are presented in Tables 2, 3, 4, 5 and 6.

A number of agri-tourism studies (Jones and Green, 1986; Winter, 1986; Neale, 1987) agree that it is the medium to large farmers that become involved in tourism. These farmers command more resources and are more likely to have unused buildings. A major reason for involvement in agri-tourism is to supplement farm income, but the different studies have highlighted that returns from agri-tourism are low. A logit regression was performed to identify the different variables which are thought to have some influence on agri-tourism involvement in the Irish pilot area. The results (Table 2) show that the model performed reasonably well, with two variables, farm size and farm type, highly significant and producing the expected effect on agri-tourism entrepreneurship. Smaller farmers are more prepared to avail of opportunities to supplement incomes through rural tourism in Ballyhoura, as are farms where less intensive systems of agriculture are practised.

Table 3 describes the range of tourism products provided by farms. The findings shown in Table 4 are consistent with the findings from other studies. Tourism is not, at least not yet, a major source of income for the majority of farmers. Table 3 also shows the type of tourist product offered on the different sized farms. Self-catering and guesthouse accommodation are likely to yield more income than student accommodation and activities on farms as they are more highly priced than the latter. They also require more investment to enter. Although the respondents indicated a relatively low income from tourism, ten of the fifteen farms of less than 100 acres involved in tourism had developed enterprises which are potentially significant income sources. Tables 5 and 6 describe the investment and maintenance costs of the different tourism products.

Table 2. Logit Analysis of Agri-Tourism Entrepreneurship

Variable	Coefficient	Standard Error	T Ratio	Prob t > x
Full time/Part time Farming	-0.28554	0.92981		
Farm Size	3.90141	1.58827	2.546	0.01403
Farm type*	-2.30614	1.04856	-2.199	0.02786
Family circumstances*	1.14382	1.04417		
Education level	-1.97560	1.38860		
Agricultural Training	-0.36504	0.91183		
Foreign language	1.40235	0.90220	1.554	0.12010
Previous generation Farmed in area*	-0.78535	1.18032		
Number of bedrooms in farm house	0.44947	1.00598		
Community membership	-0.99978	1.36686		
Community leadership	1.48869	1.14960		

* Indicates significant variables and level of significance. All other variables are not significant.

Table 3. Agri-Tourism Products in Ballyhoura

Tourism	Proportion of operators	Product number by farm size (acres)			
		0-30	30-100	100+	Total
Self-Catering	20%	3	0	1	4
Guesthouse Accommodation	40%	0	6	2	8
Student Accommodation	30%	2	1	2	5
Farm Activities	10%	2	1	0	3

Table 4. Tourism income as a percentage of farm income

Income percentage	Farm size			Total
	0-30	30-100	100+	
0- 5%	6	4	5	15
5-10%	0	1	0	1
10-20%	0	2	0	2
20-50%	0	0	0	0
51%+	1	1	0	2
Total	7	8	5	20

Table 5. Start-up costs of main agri-tourism enterprise

Start up Costs £	Enterprise				Total
	Self catering	Guest accomm.	Student accomm.	Farm activity	
0–100			4	1	5
100–1,000	1		1		2
1,000–5,000		2			2
5,000–10,000	1				1
10,000–20,000		3			3
20,000+		2	3	2	7

Table 6. Maintenance costs of main agri-tourism enterprise

Maintenance Costs £ Annual	Enterprise				Total
	Self catering	Guest accomm.	Student accomm.	Farm activity	
0–100	4	1	4	1	10
100–500		4	1		5
500–5,000		3			3
5,000+			2		2

It must be stated that outcomes have been modest. However, the process and the structures that have been put in place are potentially good ones. The level of involvement in rural tourism in the area, although small in absolute terms, has increased significantly since 1986. The community co-ordinating framework has facilitated and encouraged a wide degree of participation. Economic returns have not been that significant so far but, with the proper attention to training, standards and marketing, there is potential for larger economic impacts to occur.

The modest outcomes reflect the kind of difficulties that all local development initiatives have to contend with. Locally based development, by definition, must always start small. It also involves a process that is fairly long-term from the point of view of achieving results. Smaller farms are getting involved in rural tourism in the area as seen in the data in Table 2. It is interesting to contrast this with what is being achieved by the national agri-tourism scheme. The national agri-tourism grants scheme launched under the revised programme for western development (Bord Failte, 1989) confines its support to farmers. The objectives of the scheme are to provide incentives to farmers towards the cost of providing agri-tourism facilities, which will supplement their incomes from farming. The scheme only applies to farmers located in the area designated as less favoured within the meaning of Directive 85/350/EC. By the end of 1989, the total number of applications made for support under the scheme, if approved, would have taken up less than 2% of the budget allocated.

There are many reasons for the low take-up of this national scheme. Applicants are quite constrained in terms of project size or the combination of projects that can be grant-aided. The level of grant support for the accommodation elements of projects is only 20% and is subject to accommodation expenditure not exceeding 25% of the entire project cost. The maximum investment on which grant aid will be paid is £4,000, and the maximum amount of investment by a farmer, which will be eligible for consideration is £15,000. Participation is also limited by the fact that the scheme is confined to the less-favoured areas. There is also the point that tourist organisations are responsible for administering and promoting this scheme to a farmer client group, of which these organisations have limited experience and understanding. It can be argued too that the outcome indicates a reluctance on the part of individual farmers to invest in agri-tourism. The scheme offers no support for community co-ordination and very little support for marketing. The outcome from the scheme contrasts unfavourably with rural tourism investment in Ballyhoura, the pilot area, which has a community co-ordinator. Despite being ineligible for grant support under the agri-tourism grants scheme (Ballyhoura is not a less-favoured area), rural tourism investment planned by individual residents in Ballyhoura far exceeds that involved in the applications under the agri-tourism grants scheme for a large part of Ireland up to the end of 1989.

7. Discussion

Smaller farmers are getting involved in rural tourism in Ballyhoura where there is a full time co-ordinator in place. This is in contrast to participation in the official grants scheme just described, a fact indicating that individual development objectives can best be achieved where there is a community support structure for rural tourism. Rural tourism is essentially a community product, a major component of which is co-ordination – co-ordinating product suppliers in preparing packages to attract customers and servicing the customers when they arrive in the area. The future development of rural tourism depends on supports being available to community-based approaches rather than individual developments.

However, while a community-based approach is essential to the development of rural tourism, there are inherent weaknesses in the approach, which must be guarded against. Communities are at different stages of preparedness to engage in economic initiative. Breathnach (1986) highlights the operational problems of community-based economic initiatives, such as low levels of business expertise, lack of capital, too much dependence on a manager for policy, low commitment and an unsympathetic and unsuitable policy framework. The key elements in providing the necessary local capacity for local tourism initiatives to achieve commercial viability are member training for business and hospitality skills and the employment of a full-time manager/co-ordinator.

Many of the costs associated with the full-time manager are costs, which do not yield a direct income to meet them. The opportunity for rural tourism is to achieve economies of scope rather than economies of scale by maximising opportunities for tourists to spend money in rural areas. While tourists may perceive the value of co-ordinated information about all of the possibilities available in a rural area, they expect this information to be provided free in destination areas. Ultimately, the product providers in local tourism will bear the costs of a full-time co-ordinator. It is unrealistic, however, in the early stages of market development to expect product providers to meet all of these costs. The full-time co-ordinator needs access to a working capital fund to develop basic infrastructure such as information guides, maps, walks and tourist-attraction-related signposting.

Another major requirement in the development of a significant rural tourism sector is access to marketing supports. There is usually a long lead time in developing the tourist product as new operators try to build up credibility in the marketplace. The long distance between the supplier and many of the markets is also a factor in establishing worthwhile contacts. It does not make sense for every community to incur these costs. It is essential that a central marketing structure for rural community tourism be supported until the industry reaches a level of commercial viability. The role of this central marketing structure is to promote a clear image for rural community tourism in the marketplace and to persuade tour operators and agents to distribute the community product to its main markets. Rural communities may be the main instigators of rural tourism but they cannot be sure of being the main beneficiary. A central marketing structure has the added advantage of enabling rural communities to have greater control of the system.

One of the big difficulties faced by rural tourism is that it does not have a distinct image. A proliferation of names for the product exists, none of which would entice the potential customer to rush out and buy a holiday. A major requirement for success in the market is the development of a credible, easily-recognisable name for rural community tourism in the marketplace.

8. Conclusion

Few rural communities are likely to bring about rural development solely through the mechanism of tourism. A more effective strategy for rural development is to make tourism development a part of a community integrated development plan. There are good a priori economic arguments as well as encouraging pieces of empirical evidence to support this view. This chapter makes the point that economic development and revitalisation in rural areas must extend beyond the responsibility of individuals. The community must be involved, and it has to take on what Gannon (1989) calls the structures and qualities necessary to form an infrastructure that is supportive of entrepreneurial development. A hallmark of rural community tourism is that it is a community product and that it is developed from local structures. A key feature in the development of community tourism is local co-ordination linked to wider product and marketing structures. It is this wider framework that offers the best prospects for tourism development in rural areas.

Rural places are potentially very attractive as tourist destinations. Their appeal relates to the peace and quiet, the values and the natural environment found in the countryside. These rural assets can be used by rural people to generate economic benefits through tourism. Demand for, benefits from and threats due to tourism are best managed through an integrated approach that encourages wide participation and co-ordination at the local level. Such participative structures have the potential to encourage the development of different small-scale tourism projects on farms and in rural communities. Local co-ordination can provide the mechanism that will help minimise and manage tourism environment conflicts at the local level.

References

Bord Failte (1989) *Grant Scheme for Investments in Agri-Tourism under the Revised Programme for Western Development.* Bord Failte,Dublin.

Breathnach, P. (1986) The structural and functional problems of community development co-operatives in the Irish Gaeltacht. In D. O'Cearbhaill (ed.) *The Organisation and Development of Local Initiative.* Proceedings of the 8th International Seminar on Marginal Regions. University College Galway.

Cointat, M. (1988) *The Pays d'Accueil in France.* Paper presented at a Conference on Tourism and Leisure in Rural Areas, St.Peter Ording, Schleswig-Holstein.

Commission of the European Communities (1987) *Rural Tourism in the 12 Member States of the European Economic Community.* European Community, Brussels.

Commission of the European Communities (1990) *Objective 5(b)- Rural Development: Adoption of the Community Support Framework.* Communication of the Agriculture and Rural Development Commissioner, European Commission, Brussels.

Douglas, D. (1989) Community economic development in rural Canada: A critical review. *Plan Canada* **29**:2, 28–46.

European Parliament (1988) *Tourism in the Community.* Official Journal of the European Communities, No. c 49/157. European Commission, Brussels.

Flammang, R. A. (1979) Economic growth and economic development: counterpoints or competition? *Economic Development and Cultural Change* **28**:2, 47–61.

Gannon, A. (1989) *A Strategy for the Development of Agri-Tourism.* Paper presented to the 4th Session of the Working Party on Women and the Agricultural Family in Rural Development, Rome.

Gannon, A., B. Hennelly and C. Fox (1987) *Agri-Tourism in Austria Study Tour Report.* ACOT, Dublin.

Greffe, X. (1990) *Rural Tourism, Economic Development and Employment.* Paper presented at the Conference on Enterprise and Employment Creation in Rural Areas, Paris.

Hodge, I. (1986) The scope and context of rural development. *European Review of Agricultural Economics* **13**, 271-282.

Jasma, D.J., H.B. Gamble, J.P. Madden and R.H. Warland (1981) Rural development· A review of conceptual and empirical studies. In L.R. Martin (ed.) *Economics of Welfare, Rural Development and Natural Resources in Agriculture, 1940's to 1960's.* Survey of Agricultural Economics Literature, Vol.3, University of Minnesota Press, Minneapolis.

Jones, W.D., and D.A.G. Green (1986) *Farm Tourism in Hill and Upland Areas of Wales.* Aberystwyth: Department of Agricultural Economics and Marketing, The University College of Wales.

Keane, M.J. and J. Quinn (1990) *Rural Development and Rural Tourism.* Research Report No. 5, Social Sciences Research Centre; University College Galway.

Lyons, M. (1989) *Agri-Tourism in Lermoos.* Paper presented to Agri-Tourism in Austria Seminar. Kilmallock Co. Limerick.

Mathieson, A., and G. Wall (1982) *Tourism: Economic, Physical and Social Impacts.* Longman, New-York.

Michaelson, S., (1979) Community-based development in urban areas. In B. Chinitz (ed.) *Central City Economic Development,* Abt Books, New York.

Mowle, A. (1989) Changing countryside: land use policies and environment. *Geography* **73** (321), 318–326.

Neate, S. (1987) The role of tourism in sustaining farm structure and communities on the isles of Scilly. In Bouquet and Winter (eds.) *Who from their Labours rest? Conflict and Practice in Rural Tourism.* Avebury, Aldershot.

Needham, C. (1989) *The Marketing of Agri-Tourism in Austria.* Paper presented to Agri-Tourism in Austria Seminar, Kilmallock, Co. Limerick.

Nordbo, T. (1985) *Tourism in the Rural Districts. A Threat to Agriculture or a Possibility?* Paper presented at the International Joint Course in Agricultural Education, Dublin.

O'Cinneide, M. S., and M.J. Keane (1989) *Local Socio-economic Impacts Associated with the Galway Gaeltacht.* Research Report No. 3 Social Sciences Research Centre: University College Galway.

O'Connor, R., J.A. Crutchfield and B.J. Whelan (1981) *Socio-Economic Impacts of the Construction of the ESB Power Station at Moneypoint, Co.* Clare.Economic and Social Research Institute, Dublin.

Shaffe, R. (1989) *Community Economics: Economic Structure and Change in Smaller Communities.* Iowa State University Press, Ames.

Thibal, S. (1988) *Rural Tourism in Europe.* Paper presented at a meeting of the International Organising and Steering Committee of the European Campaign for the Countryside. Council of Europe, Strassbourg.

Whitby, M. (1986) An editorial postscript. *European Review of Agricultural Economics* **13**, 433–438.

Winter, M. (1986) Private tourism in the English and Welsh Uplands. In W. Bouquet and M. Winter (eds) *Who from their Labours Rest? Conflict and Practice in Rural Tourism.* Avebury, Aldershot.

THE ECONOMIC VALUE OF NATURE

JAN VAN DER STRAATEN
Department of Leisure Studies
Tilburg University
P.O. Box 90153, 5000 LE Tilburg
The Netherlands

1. Introduction

As has been argued in the first chapter of this book, nature and the environment are used in different ways in tourism and tourism activities. A crucial question in this respect is to what extent it is possible to assign an economic value to nature and the environment. If this is possible to do, the decision-making process regarding the use of nature and the environment in tourism activities can be based on sound economic information. This facilitates the decision making process regarding the use of nature.

An important topic is the distinction between nature and the environment. In many publications and in everyday language, nature and the environment have approximately the same meaning and are used interchangeable. If the economic valuation of nature and the environment is at stake, we cannot overlook the differences between the two concepts. In most cases, the environment is defined as the abiotic part of the ecosystem, which implies that water, air and soil pollution are the main issues involved. An additional issue is the use of natural resources such as iron ore, natural gas, wood and plastics. The latter are economic goods, which normally have a price, as they are exchanged in a market. Water, air and soil pollution does not have a price, however. But with regard to water, air and soil pollution people are often economically affected. These costs for consumers and producers can be measured.

With respect to nature, the context is different. Nature, being the biotic part of the ecosystem, particularly in tourism, is not traded in a market, which means that it is not possible to assign a suitable price to represent the economic value of nature. Beautiful beaches, mountain scenery and attractive Mediterranean landscapes, for example, are very important for tourists. They can be seen as an input in the production process of the tourism industry and have a high economic value. However, it is not clear how to measure these economic values.

This chapter will provide more insight into this difficult question. In the first part, the principle questions regarding economic value are discussed, while in the second part the various methods are discussed which have been developed to tackle this problem.

2. Total economic value

Pearce (1993) states that the total economic value of environmental assets such as a tropical rain forest consists of two broad components: use value and non-use value. Use value has three subcategories such as direct value, indirect value and option value. With respect to the rainforest, timber products, recreation, education and human habitat can be seen as forms of use value. Indirect value includes topics such as watershed protection and air pollution control. The option value can be regarded as future use and non-use value. Finally, non-use value includes existence and heritage value. We assign a value to tropical rain forests, for example, as we are of the opinion that these ecosystems have to be protected, regardless of the use we make of it.

The use of an environmental asset is sometimes linked to price forming in markets, as is the case with timber products. On the other hand, in most cases the use of a rainforest in tourism is not linked to price forming. When dealing with non-use values, price forming is completely absent. This implies that we have only prices in a limited number of cases.

This brief overview points to the conclusion that, given the usual economic instruments, we have a low level of information regarding the economic value of an environmental asset such as a tropical rainforest. This leads to a crucial question. Does it make sense to develop new economic instruments, which enable us to give more and more accurate information regarding the economic value of nature? Many economists (for example, Willis and Corkindale, 1995; Hanley and Spash, 1993; Pethig, 1993; Mitchell and Carson, 1989) are of the opinion that it is necessary to develop new instruments to facilitate the assignment of economic values to environmental assets. We take a somewhat different position. In the first place, it is undeniable that many of these economic instruments can provide more information regarding the economic value of the environmental asset than we have now without their use. On the other hand, these instruments have many shortcomings, which raises the question of what this new information can contribute in cases where environmental protection is jeopardised. This problem will be discussed in more detail after the discussion of the instruments.

In this problematic field, we should not forget that in normal everyday life, we are often confronted with economic goods that do not have a price without fundamental problems regarding the decision-making process. For example, San Marco Square in Venice is a marvellous location for a skyscraper with many prestigious offices. It is certain that many large companies would be willing to pay very high prices for an office at such a location. Therefore, one could argue, from an economic point of view, that the opportunity to use San Marco Square differently in the future can bring great benefits compared to its current use. However, nobody has the slightest intention to monetarise the existing value of San Marco Square as it is currently used vis-à-vis its use as a location for high ranking offices. That is out of the question. Moral, ethical, and cultural ideas form, in fact, a complete barrier to use the square for any other purpose.

Of course, economists can argue that the square as it is currently used yields a higher economic value than could be obtained through any other use. But this is too simple, as many others would argue that these considerations have nothing to do with economic value. They would argue that San Marco Square could not be used differently on the basis of any other economic bid. In other words, in this line of thought the use of the San Marco Square falls outside the realm of economics. The same ideas can be found when dealing with the existence of an army, a monarchy, a Rubens painting, a Shakespeare poem, a piece of music by Mozart, the sight of migrating raptors and storks on Gibraltar, an attractive partner, the Amsterdam canals, a Marquez novel, the Matterhorn, Delphi, the Atlantic coast of Portugal, a nation, human rights, or an educational system. There is no price for these goods and that, in this view, is the best option. Few economists would claim that it is better to calculate prices for these goods in one way or another.

Of course, the counter argument could be presented that people are willing to pay money to see these assets and to read these books, etc. This is true, but that is different from saying that the price people are willing to pay to visit the Amsterdam canals, the Matterhorn, or the Van Gogh Museum is the economic value of these assets. In this chapter, we take up an ambiguous position, as we are aware of the strength of the economic argument in a world where many decisions are based on the analysis of market costs and benefits. On the other hand, we cannot overlook the validity of the claim not to monetarise all assets, which do not have a price, including nature and the environment. In the next section, the various instruments used to monetarise environmental assets that have been developed in the last decades will be discussed. At the end of the chapter, we pay special attention to the extent to which an expansion of the economic argument into a more integrated approach can provide better insight into the decision-making process involved when environmental assets are at stake.

3. Travel Cost Method

The Travel Cost Method is based on the idea that people are only willing to spend time and money to travel to a certain location if they value that site as interesting. In addition, it is assumed that people rank interesting sites and that they calculate the benefits or the utility of the trip compared to the costs they have to make to visit these recreational sites. In fact, it is assumed that a recreational trip is a 'normal' economic good, which implies that consumers only visit a particular site if the benefits of the visit are higher than the costs. In this method, the costs of travelling by car, bus, train or airplane, and the hours involved in travelling are seen as costs to be made to visit the area.

This method has been developed in the USA, where certain groups claimed that government spending for forests and national parks were high while benefits were seen as fairly low or uncertain. Employees of the US National Park Service tried to develop new methods, which could make it clear that a national park would have high benefits for visitors and claimed that these benefits could be measured as economic benefits. As early as 1958, Wood and Trice wrote an article on the measurement of recreational benefits. Later, in 1966, Clawson and Knetsch elaborated on that concept in more detail. In the USA, and later also in the UK, this method was widely used to demonstrate the beneficial effect of the establishment of national parks and other types of recreational areas.

There are various problems associated with the application of this method. In the first place, what is the value of one hour of travelling time? It is often the case that many tourists like to travel to a location of high natural beauty such as a national park. Cesario (1976) suggested using one-third of the value of the hourly wage rate for the calculation of travelling time. Several proposals have appeared in the course of time. Chevas *et al.* (1989) made a distinction between the opportunity cost measure of travel time and the 'commodity' value of travel and visiting time. However, in all these proposals and discussions, the fact that the value of travel or visiting time is only relevant as far as consumers have a job for which they are paid is completely overlooked. In the Netherlands, for example, less than 40 per cent of the total population have a job, which implies that this method can only calculate the benefits of recreational activities for the labour force; for most other members of the population (e.g. elderly, children, housewives) the method is not valid. This problem is related to the method's starting point, which takes consumers' market behaviour as a basis for economic calculations. People, who do not have a paid job, do not display market behaviour regarding the use of their time.

Other problems, which are often mentioned in the literature, concern the fact that people frequently make multi-purpose trips. Hence, the difficulty is how the different components of travel time can be distinguished. Furthermore, there is the question of residents and tourists. Perhaps the choice of visitors from abroad to visit that country was based on the possibility to visit a given national park. In such a case, it is quite difficult to include the costs which have been made to visit the country in the Travel Cost Method.

More serious problems, which are hardly discussed in the literature, are linked to the travelling process itself. During the 1992 United Nations Conference in Rio de Janeiro, global climate change was given full attention. The main topic in the global climate change debate was and still is the use of fossil energy. At the 1998 Kyoto Conference, participating countries agreed on a reduction in the use of fossil energy in order to limit the danger of global climate change. This implies that from that moment on, the reduction in the use of fossil energy became an environmental policy goal. However, national parks that are far away from the main urban centres will have a high economic value if many people visit these remote located areas.

The more people visit Antarctica, the Andes, or the Himalayas, the higher their economic value. This implies that high economic values of national parks can be calculated on the basis of non-adherence to environmental policies. This is a strange conclusion, which makes the method rather questionable.

A related problem is the location of the natural areas. The national parks and nature areas located near urban centres are visited, ceteris paribus, by more people than those located in remote areas. Therefore, we can calculate higher economic values for these areas. Many people visit the tropical rain forests in Costa Rica. Is this the result of the high ecological value of these forests? On the contrary, the majority of the visitors from abroad do not have the slightest idea about the ecological value of these forests in Central and South-America. They do not know whether the forests of Costa Rica, Surinam, Guatemala, Columbia, or the Amazon have a low or a high ecological value. They go to Costa Rica because there is an excellent tourist infrastructure compared to other countries in the region. In addition, Costa Rica is a safe country, something which cannot be said for other countries in that region.

Finally, the Travel Cost Method takes the current income distribution as a starting point and guiding principle. If areas of high ecological value are located in poor countries where people cannot construct modern tourist facilities, tourists will not visit the country. The people of this country also are poor and, therefore, do not have the economic means to visit natural areas in their own country. This implies that the use of the Travel Cost Method can lead to low economic values, even in cases where nature reserves are of high ecological value.

This leads to the conclusion that this method has limited possibilities, particularly in locations outside the first world, and many shortcomings. Its main advantage is that the actual behaviour of people is taken as a starting point; no assumptions have to be made regarding what people could or should do. The method deals with the observed preferences of tourists and recreationists. However, the relationship between ecological and economic value is often quite weak.

4. The Contingent Valuation Method

The Travel Cost Method takes actual behaviour of consumers as a starting point in the model. By doing so, the 'revealed preference' paradigm of neo-classical economic theory is followed. In the Contingent Valuation Method, a hypothetical market situation is created. Consumers are asked what is their willingness to pay to protect or maintain a certain environmental asset; or consumers are asked what *is* their willingness to accept a certain level of environmental pollution. When the same level of environmental protection is considered, the willingness to accept, in principle, should be the same as the willingness to pay. It is assumed that consumers are able to rank environmental assets in such a way that they can assign a monetary value to each in relation to the other. This situation is merely hypothetical, as, of course, consumers do not need to pay. In fact, it is a type of survey, which can be used to assign or calculate a value to environmental goods. Of course, respondents have to be given full information regarding the environmental asset in question. Davis developed the method as early as 1963. It became quite popular in the 1980s, particularly in the USA and the UK.

Many research projects using this method have been realised during the past few decades. The method has been used to assess a great array of environmental problems. Turner *et al.*, (1992) for example published a study of the benefits of a coastal defence project in the Norfolk Broads in England; Bergstrom *et al.* (1990) published about the economic value of wetlands-based recreation; and Ruitenbeek (1992) paid attention to the economic value of rainforests. In this volume, some examples of this assessment technique are given.

Many aspects of the CVM method are discussed in the literature. These highlight many shortcomings, such as the hypothetical character of the survey, the so-called part-whole bias, the actual differences between willingness to pay and willingness to accept, the potential free-rider behaviour of respondents, the difficulty of giving information to respondents, etc. Hanley and Spash (1993), Mitchell and Carson (1989), and Pethig (1993) give good overviews of all these problems. The issues discussed by these authors deal mainly with problems with which researchers are confronted when they apply the method. Here, we will not discuss the 'technical' problems, which are associated with the method itself, but will pay attention to certain theoretical problems, which are not given a high profile in the literature.

The first difficulty is that the method takes the current distribution of income as a starting point. This is not a problem when it does not influence the outcome of the projects. However, this is often not the case. In the case of the EXXON Valdez oil spill on the Alaskan coast some years ago, the Contingent Valuation Method played a significant role. The state of Alaska brought EXXON to the court, arguing that they had to pay for the costs of cleaning the coast, the drop in the fishermen's income along the coast, and also for the damage to the intrinsic value of nature. The Contingent Valuation Method was used to estimate the losses in intrinsic value. The population of the USA was taken as a reference group and a group of people was asked what was their willingness to pay to protect the Alaskan coast. As these were US citizens and their average income is one of the highest in the world, the result was very high. There was a long dispute about the validity of the method, but the Court decided that EXXON had to pay a very high sum to compensate for the damage they caused to the intrinsic value of Alaskan nature.

At first glance, this decision can be regarded as a success demonstrating the possibilities of this method. This was the first time in history that a Court assigned a monetary value to the intrinsic value of nature. However, had the same oil spill taken place on the coast of Gambia, Denmark, or Nicaragua, the outcome would have been completely different from the decision of the Court in the EXXON Valdez case, particularly if the same lines of reasoning had followed. Denmark is a small country with only 5 million inhabitants and even with a relatively high income, the outcome of the method would be a relatively low figure. With Gambia and Nicaragua, two small countries with a low level of income, the outcome would have been considerably lower than in the Danish case, and would have been a very small fraction of the Alaskan result. One should not overlook the point that these differences in outcome are effective in a situation where the same ecological damage has been realised.

The strange result that a similar ecological damage can lead to different monetary outcomes draws from the fact that the current geographical situation and the current income distribution are used as guiding principles in the method. In principle, these paradigms do not pay attention to the ecological damage as such. It is not the ecosystem that is taken as the central issue, but the market, the boundaries of countries, and the current income distribution. Using these concepts as starting points will undoubtedly lead to different outcomes in situations where the same ecological value is at stake. This framework of economic reasoning cannot cope with ecosystem questions that have a completely different framework.

Another related fundamental problem is the question of property rights. When the property rights of an environmental asset are defined, then it is relatively easy to calculate monetary value. Is the coast of Alaska the property of the people of the USA, or is it the world's heritage? What about the Danish coast: is that the property of the Danish people, citizens of the European Union, or the property of the world? The answers to these questions hinge on value judgements and, therefore, they are value-laden. In the Contingent Valuation Method the actual situation is taken as a starting point, but the outcome of that approach as regards the monetary value of an ecosystem is significant. This again is a strange conclusion, when we start from the ecological situation. If we take the ecological value of an ecosystem as the crucial guiding principle, the outcome of monetarising its economic value cannot be fundamentally influenced by the outcome of a discussion on property rights, which is a concept alien to the ecological value of an ecosystem.

The property rights discussion started as early as 1968 with the publication of Hardin's article 'The Tragedy of the Commons'. Hardin argued that the commons (*viz*, the common property resources) are always frustrated by the same type of problems. He used the example of shepherds using a common pasture. Every shepherd is inclined to increase the number of sheep he wants to take to the common. By doing so, he has the benefit of a substantial increase in personal income from the higher output of the animals, whereas the negative effect of potential overgrazing is shifted to all other shepherds using the common. In Hardin's analysis, this process will continue until all the commons are overused, resulting in the system's collapse.

In Hardin's approach, common property resources are always overexploited and overused until the resource is exhausted. However, particularly in the case of common pasture land, historical evidence counterdicts his view. In many countries there are common property pasturelands. Let us take the Alps as an example. For many centuries, the use of common pasturelands (the Alps) was normal practice. However, common property can only exist if the common property regime is defined. Without a definition of common property, it does not exist, as was clearly demonstrated by Bromley (1991). This means that access to a common property is not open to everybody, which is the case in Hardin's example. The most typical characteristic of a common property right is that some persons (in most cases people from a village) are allowed to use the pastureland to the exclusion of others. If everybody is allowed to use the resource, there is no clear definition of property rights, and, therefore, a situation of nobody's property results, which means open access to everybody. In all cases of traditional common property, there are regulations stating what is permitted to the users of the resource and what is not. One may therefore conclude that Hardin did not realise the difference between common property and open access.

Hardin's article is often referred to argue that common property rights lead to overuse of the resource; and that, as a result, the solution to environmental problems is privatising common property. This is not the point of Hardin's article. Furthermore, this idea frustrates public policies aiming to limit the use of common resources and to protect the environment. The examples of the Alps clearly demonstrate that, generally speaking, 'normal' practice has always been regulation of the use of natural resources. The main issue associated with environmental problems is the rapid change of modern society where the use of natural resources changes rapidly, but the old institutions that regulated its use for a long period did not change sufficiently to cope with the new problems. For that reason, it is argued that instruments and methods are needed to evaluate the monetary value of environmental assets, something that was not necessary many decades ago.

5. Conclusions

- In principle, the integration of ecological values with economic theory is only possible to a certain extent. The main problem lies in the different lines of reasoning which are used in the two disciplines. This implies that the ecological approach (in which the value of ecosystems is based on their ecological value as such) is, from the theoretical point of view, superior to the economic approach (where ecological values can only have economic relevance to the extent that a certain type of monetary value can be linked to the ecological issue).
- The possibilities related to the assessment of costs and benefits in the field of nature and the environment is strongly linked to the definition of the value of these issues. The main point is that the value of nature and of the environment cannot be determined precisely because many of their elements are not traded in markets. It is sometimes said that energy or the ecosystem should be taken as a measure of value. However, as has been made clear by Martinez Alier (1987), this argument is not valid. It constitutes a subjective choice, which cannot be made without making a number of political decisions as well. This leads to the conclusion that the value of nature and *of* the environment depends on political ideas and decisions.

- This being the case, we cannot overlook the point that economic decision-making is based on the assumption that we know the economic value ex ante, otherwise we cannot calculate the costs and benefits of environmental damage and environmental regulation. This brings us to the conclusion that this concept is strongly related to political decisions.

- In general, economic theories do not take into account sufficiently the problems of nature and the environment. They focus more on labour and capital issues; choices to be made in economic and environmental policies reflect this situation.

- All developed methods to give economic values to nature and environment issues have their limitations. Some of them, such as the Travel Cost Method, cannot cope with future situations; others, such as the Contingent Valuation Method, can but are based on hypothetical situations. Additionally, this method can only be used to assess well-defined and relatively unimportant problems.

- When a method to assess costs and benefits must be chosen, the best approach is a pragmatic one. Researchers have to realise what the limitations of each method are, given the problem in question. It is the type of problem, which facilitates the choice of the most appropriate instrument. Other instruments such as an environmental impact assessment or a multi-criteria analysis can provide additional and sometimes better information in the decision-making process.

- In discussions dealing with the use of nature and the environment for tourism activities, these difficulties and limitations have to be given full attention. For the tourism industry nature is a very important input in the production process. The sector has often to compete regarding its use of environmental resources and services with industrial development, the construction of harbours, and agricultural activities. Sometimes there is a conflict with nature conservation. In all these cases, it is possible to criticise the basis of information about the economic or the monetary value of these ecosystems.

- The property rights discussion makes it clear that current definitions of the property rights of nature and the environment are often inadequate due to the rapid change in the technological development of production processes. From the theoretical point of view it is important that public policies aiming to protect nature and the environment to redefine property rights, giving full attention to actual damage to ecosystems. By doing so, changes are made in the definitions of monetary values. Other definitions of property rights will result in other monetary values.

- This brings us to our final conclusion. New instruments such as Contingent Valuation Methods can provide important additional information, which is relevant in the decision-making process. However, the more complex the ecological question is, the more likely it is that this information is of minor importance. The definition of economic and monetary costs is strongly related to and influenced by the way in which we have constructed our society. Political decisions, geographical boundaries resulting from previous wars and current income distribution strongly influence the outcome of many economic calculations. The intrinsic value of nature and the environment has a meaning in itself, independent of these considerations.

References

Bergstrom, J., J. Stoll, J. Titre and V. Wright (1990) Economic value of wetlands-based recreation. *Ecological Economics* **2**:2, 129–148.

Bromley, D.W. (1991) *Environment and Economy: Property Rights and Public Policy.* Basil Blackwell, Oxford.

Cesario, F. (1976) Value of time in recreation benefit studies. *Land Economics* **55**, 32–41.

Chevas, J.P., J Stoll and C. Sellar (1989) On the commodity time value of travel time in recreational activities. *Applied Economics* **21**, 711–722.

Clawson, M. and J. Knetsch (1966) *Economics of Outdoor Recreation.* The Johns Hopkins University Press, Baltimore.

Davis, R. (1963) Recreation planning as an economic problem, *Natural Resources Journal* **3**:2, 239–249.

Hanley, N.and C. L.Spash (1993) *Cost-Benefit Analysis and the Environment.* Edward Elgar, Alderschot.

Hardin, G. (1968) The tragedy of the commons. *Science* **162**, 1243–1248.

Martinez Alier, J. (1987) *Ecological Economics.* Basil Blackwell, Oxford.

Mitchel, R.C. and R.T. Carson (1989) *Using Surveys to Value Public Goods; The Contingent Method.* Resources for the Future, Washington DC.

Pearce, D. (1993) *Economic Values and the Natural World.* Earthscan, London.

Pethig, R. (ed.) (1993) *Valuing the Environment: Methodological and Measurement Issues.* Kluwer Academic Publishers, Dordrecht.

Ruitenbeek, J. (1992) The rainforest supply price: a tool for evaluating rainforest conservation expenditures, *Ecological Economics* **6**:1, 57–78.

Turner, R.K., I. Bateman and J.S. Brooke (1992) Valuing the benefits of coastal defence: a case study of the Aldeburgh sea-defence scheme. In A. Coker and C. Richards (eds) *Valuing the Environment.* Belhaven, London, pp. 77–100.

Willis, K.G., and J.T. Corkindale (1995) *Environmental Valuation: New Perspectives.* CAB International, Oxon.

Wood, S. and A. Trice (1958) Measurement of recreation benefits. *Land Economics* **34**, 195–207.

SUSTAINABLE TOURISM IN MOUNTAIN AREAS

JAN VAN DER STRAATEN
Department of Leisure Studies
Tilburg University
P.O. Box 90153, 5000 LE Tilburg
The Netherlands

1. Introduction

Many countries pursue tourism development policies. In most cases, the sole aim of these policies is to promote tourism as it is believed that it has a positive effect upon value added, income and employment. The many negative effects upon nature and the environment resulting from recreation and tourism are, unfortunately, very often ignored by national and regional authorities. These effects may be so serious that the unhindered development of tourism has a negative effect upon tourism itself.

One example is the erosion of mountain slopes in skiing areas in the Western Alps resulting from the cutting down of woods, the levelling of steep slopes, and the use of artificial snow to provide skiing facilities. These effects are not noticed by skiers in winter because the snow covers everything in the landscape. But in summer, these areas are not attractive to summer tourists, since the eroded mountains arc very ugly. This development proceeded without a sound policy to protect the environment. In many cases certain economic interests have the power to translate their own economic interests into a social aim for the area as a whole. In Western and Central Europe, the interests of the environment are organised in weak organisations, compared to the interests of labour and capital. As a result, there does not exist an optimal allocation of the production factors nature and environment.

In most countries of Europe, criteria which should be used by authorities when they seek to implement a sound environmental policy have not been established yet. In the Brundtland Report, the concept of sustainable development was introduced. The idea is that consumption and production should take place without diminishing the possibilities for future generations to use natural resources. The market mechanism, which is a tool to realise an optimal allocation of production factors, will not work in this direction. Intervention by authorities is the only way to achieve sustainable development of the economic process. One of the most important steps is the development of instruments with which an optimal development of tourism can be realised.

In the following, the problem itself is described. Secondly, instruments used in traditional welfare economics are examined. Lastly, a more integrated method, which has been developed in a previous study, is introduced (Dietz and Van der Straaten, 1992).

2. Tourism in the Alps

The development of mountain tourism cannot be seen independently of the total economic development of the mountain valleys. For a long time, the people in the mountain valleys were farmers, who bred cattle in the northern parts of the Alps and sheep in the southern parts. They used the mountain meadows above the treeline as feeding grounds for cattle. The woods on the slopes were used for timber and as fuel. Generally speaking, these woods were used carefully, as they were the only barriers against avalanches from the higher slopes in narrow mountain valleys. The valley itself was used for fruit trees, vegetables, corn and potatoes. Tourism started in the course of the nineteenth century when British mountain climbers came to mountain villages such as Zermatt, Chamonix and Sölden. Mountain huts were built on the mountains and new hotels were built in the villages. Income and employment increased for a long period. This development had a positive effect upon welfare in these areas.

After the Second World War, a great increase in income and free time was realised in many countries of Europe. The number of tourists increased during winter and summer. The infrastructure in the mountains had to be adapted to accommodate all these tourists. Winter tourism, in particular, caused many problems. Lifts were built to bring the skiers higher up on the mountains. An ideal ski run had to have a certain angle of inclination, so when mountains were too steep, they were modelled with the help of bulldozers and dynamite. In certain French ski resorts, it is normal that more than 30 metres of ground and rocks are blown up to get the desired angle of inclination. When vast mountain woods grow on the slopes, parts of it are cut to realise a ski run from the mountain top to the valley, where the skier can be transported high up the mountain again with the help of lifts using a lot of electric energy. Many skiing villages in France and Italy are surrounded by ski lifts and ski runs which are cut in the landscape as if giants had used their knives. The effect on vegetation is disastrous. The ski runs are influenced by heavy erosion, which brings the rest of the fertile humus to the valleys and to the sea.

Traditional farmers are often forced to give up their work in the mountains as costs are high and yields are low. Competing with the agricultural products from the plains is very difficult. So, without tourism there is a decrease in employment, which results in the migration of people from mountain areas to the plains. The abandonment of mountain meadows has a negative effect upon the vegetation. The grass remains on the meadows, so avalanches are more common than before.

In conclusion, mountain areas suffer either from the effects of winter tourism or from the migration of farmers to the cities and the plains (Bätzing, 1988). The exodus of people stops as soon as the development of tourism takes place. In many cases, tourism development is argued for by pointing to the positive effects upon the number of people living in mountain areas. But it is very often not mentioned that, from an ecological point of view, this influence is very negative.

3. Analytical instruments in traditional welfare economics

When an authority wants to get some insight into costs and benefits of a public project, a simple instrument which can be used is Cost Benefit Analysis. All costs and benefits associated with the project are summed up and quantified in monetary terms. Such an approach is only possible when a great deal of costs and benefits can be quantified. For public projects from which many environmental problems arise, this method cannot be used at all.

As there is no market for most environmental goods, there are no prices; thus, costs and benefits cannot be quantified. In this situation, authorities will try to develop methods to give a price to the unpriced environment. It is a peculiar situation that this demand for prices is restricted to environmental problems. The European Community states, for instance, "The Commission will endeavour to develop methods of assessment which will facilitate this task (to make a balance-sheet of the positive and negative economic and employment effects of environmental policies and actions) and which will, as far as possible, ensure the preparation of an adequate cost-benefit analysis of a basis for environmental proposals" (European Commission, 1987).

This demand for a real cost-benefit analysis is seldom made when dealing with investments in education, health care or military projects. Nearly always it is assumed, explicitly or implicitly, that investments in these public fields are more or less productive. But it will be very difficult to prove these assumptions. When investments are more or less necessary in the field of nature and the environment, there is a general belief among authorities that there is an implicit bias towards too much investment in environmental protection. This explains why authorities stress the necessity of making cost-benefit analyses when dealing with environmental problems.

These ideas among authorities have brought about a strong demand for methods of economic evaluation of environmental measures. One of the most important methods is the Contingent Valuation Method. A second one is the Hedonic Pricing. The Contingent Valuation Method is based on the traditional neo-classical approach of the measurement of the willingness to pay. When a product is bought and sold on the market, it has a market price. When the market is in equilibrium, the equilibrium price is a reflection of the consumer's willingness to pay. When no market exists, we can ask the consumer how much he or she wants to pay for an improvement. With the help of a questionnaire, a kind of hypothetical market of environmental goods can be constructed. This method has been used for the analysis of many environmental problems. In the USA, in particular, this method is rather popular. Mitchell and Carson (1989) gave an overview of more than a hundred case studies in the USA in which this method was used. Hoevenagel (1990) gave a list of eleven case studies in the Netherlands.

4. The usefulness of the Contingent Valuation Method

This method cannot be used in every situation. Generally speaking, it is necessary that the researcher has the convinction that the answers given by consumers are related to the problem itself. Cummings *et al.* (1986) give some preliminary remarks about the scenarios which are used in this type of investigation. They state that valid results are likely when:
- familiar environmental goods are used
- respondents have some prior experience about valuation regarding the environmental change in question
- there is little uncertainty
- the questions have something to do with the willingness to pay
- the values which are obtained have a minimum of ideological content

Other conditions are formulated by Mitchell and Carson (1987), such as:
- the scenario must be meaningful to respondents
- the scenario must be realistic for respondents
- the scenario must be plausible for respondents

Other authors have discussed the possibilities of some potential bias by respondents. This problem will be addressed later on. At this point we pay attention to the conditions mentioned by Cummings *et al.* (1986) and by Mitchell and Carson (1987).

In the case of the disruption of the landscape by skiing and the use of a great deal of electricity for the ski lifts, a Contingent Valuation scenario can be constructed. From a strict theoretical point, there are two groups: in the first place, skiers, who enjoy the pleasure of skiing in a beautiful white landscape. They will argue that they derive benefits from this situation. The second group is the summer visitors. It is impossible to ask this group anything about the negative impact upon their welfare position, caused by the destruction of the landscape, as in most cases there are rarely summer visitors to typical ski resorts. This will result in a high value of the current situation of skiing leading to damages of the landscape. If we want to meet the necessary conditions for valid results, the only possibility for applying a Contingent Valuation Method is to give sufficient information to skiers about the destruction they cause to the landscape.

Firstly, it is necessary to inform skiers about the effects on the landscape. However, it is difficult to demonstrate the effect of the use of a lot of electricity. In the Alps, a considerable amount of electricity is produced by white coal, which destroys many mountain valleys. Apart from the small amount produced by nuclear energy, the rest of the electricity is generated by traditional power plants in which coal or oil are burnt and which causes acid rain and the dying off of woods. The effect of the use of electric energy can be demonstrated by visual information about ruined mountain valleys and dying woods. Lastly, the tourist should get information about the ugliness of ski resorts in summer. This can be done by showing pictures of these landscapes in which the erosion, caused by skiing and the cutting of woods, should be clearly visible.

After giving this information to the respondent, he or she knows the effect of skiing on nature and the environment. Then the question should be "How much are you willing to pay to prevent the destruction of nature and the environment caused by skiing?" When we look at the conditions, which should be met, there is some doubt. In the first place, it is not certain that there is a familiar environmental good. One may doubt the ability of most people to recognise the dangerous effects of erosion on mountain slopes, for instance. Besides, many people go to skiing areas and, generally speaking, many of them will argue that the situation is exaggerated by the investigator who asks the question. We will deal with this problem when we shall discuss the notion of certain bias.

The same will be true in many cases when investigating the uncertainty condition. The influence on the landscape will be uncertain for many persons. Skiers are accustomed to the landscapes of mountains in winter. They often believe that this is the 'real' landscape. They cannot imagine that an area, which was so lovely and smooth in winter is so ugly in summer. They will point at the uncertainty that is always a factor when dealing with environmental problems.

The condition related to the ideological content is rather difficult to meet. Skiing and the use of ski lifts that consume a lot of electricity is so generally accepted that this situation itself is full of ideological contents. When the investigator gives information about the effect of erosion on mountain valleys, most skiers will experience this information as loaded with ideology. The responses to questions in such a situation are hard to interpret.

The same is true of the conditions formulated by Mitchell and Carson. Most consumers cannot see the scenario as meaningful, realistic or plausible. The way a normal skier in mountain valleys thinks and feels is so associated with freedom, sports, nature, sun and a lot of snow that any information about erosion, dying woods, destruction of mountain vegetation and the use of energy will be rejected as a form of exaggerated ideas of people who want to protect nature at any price.

A more fundamental problem arises when one looks carefully at the background of the idea of the willingness to pay. These ideas are based upon neoclassical paradigms, which start with the possibility of choice. Every consumer has, in this view, to choose one good or another. But when dealing with the erosion of mountain valleys, one may ask whether this starting point can be used. Acid rain in mountain areas, partly caused by the emission of NO_x by motorcars of skiers, trucks crossing the Alps, and electric power plants, has a very negative effect upon the vitality of mountain forests. It has already resulted in the dying off of mountain forests in many parts of the Alps. If this process continues, disastrous erosion in many valleys could be the result. This implies that the effect of cutting trees for ski runs will have a cumulative effect upon the erosion of mountain valleys. Furthermore, higher temperatures caused by the enhanced greenhouse effect increased the snowline in summer. This results in considerable erosion of mountain slopes, as the ice no longer keeps the stones on the steep slopes together. This process of heavy erosion will result in valleys no longer being fit for habitation. In such a situation there will be no opportunity to choose, which is one of the fundamental hypotheses of the neoclassical paradigm. Hence, this is a typical situation in which the paradigms of neoclassical welfare theories are not fulfilled.

5. The bias of the Contingent Valuation Method

We speak about bias when there is an expected tendency towards systematic errors which means that the results of the method cannot be used. The method is used in a hypothetical situation. Perhaps there is a bias, caused by this hypothetical starting point. Freeman (1986) is of the opinion that precisely this hypothetical situation leads to random measurement and not to bias. Other researchers, such as Randall *et al.* (1983), have argued that it is possible that respondents indicate a lower level of the willingness to pay. Unfortunately, it is hardly possible to demonstrate the existence of such a hypothetical bias.

Many possibilities for the occurrence of bias are discussed in the literature. One of the most important is the part-whole bias. It is questionable whether respondents are able to separate general situations from a certain localised problem. So, it is possible to ask respondents what they are willing to pay to prevent the erosion of mountain areas caused by skiing. It is also possible to ask the question for a certain location.

Tversky and Kahneman (1981) believe that people think in terms of 'mental accounts' when they make decisions about their budget. Kneese (1984) pointed out the problem that people, when asked about some hypothetical situation, are inclined to allocate everything in their environmental account and, thus, they do not pay attention to other possibilities to improve the environmental situation. With regard to erosion in mountain valleys, the misspecification of the amenity will be a reality. Skiing is not the only cause of the erosion problem in mountain areas. The separation of these causes will be difficult for many respondents. We cannot discuss this problem in detail here. More information can be found in Rae (1982) and Slovic (1972).

Based on the previous discussion, we can conclude that a Contingent Valuation Method will not offer real solutions to the problem of the unpriced environment when dealing with tourism in mountain areas.

6. Hedonic Pricing and Travel Cost Method

When Hedonic Pricing is used, the assumption is made that the price of a product is related to all the characteristics of this product. By differentiating the price of that product, the price of a certain characteristic can be isolated. This method is used to evaluate the value of a certain landscape. One investigates the differences in the prices of a certain type of house at different locations, which differ only in one characteristic, namely the landscape. The value of the landscape is isolated and then included in the price of houses.

This method can be used when the deterioration of the landscape is caused by the production process in sectors such as agriculture. In this case, the visual component of the landscape is affected by a sector other than tourism. In the case of skiing, which affects the landscape and the living possibilities in the long run, the valuation of houses is complicated by many factors. So, the effect of skiing can hardly be isolated. Besides, there is a problem in this case about the value itself. The price of housing in skiing areas is, generally speaking, high, due to the availability of skiing facilities. Skiing apartments are seldom used in summer. So, the deterioration of the landscape in summer has little influence on the price of apartments, houses and hotels in skiing areas. This means that this method is not appropriate for dealing with these types of environmental problems.

The Travel Cost Method is based on the assumption that people may spend large amounts for travel expenses when the destination area is attractive. By evaluating the differences in travel costs, one can get some insight into the willingness of tourists to pay to visit certain areas of outstanding beauty.

Here we have the same difficulty as with Hedonic Pricing. Skiers visit resorts with a lot of skiing facilities and this causes erosion of mountain slopes. Furthermore, during winter, erosion is hardly recognised by the majority of winter tourists. The application of these methods would lead to the result that areas with a high number of skiing visitors and therefore, *ceteris paribus*, a high level of damage to the landscape, will have the highest economic value. This is an unacceptable outcome.

7. An ecological approach

From the above arguments it is concluded that a more general approach is necessary, in which more attention is given to ecological conditions.

Figure 1 can provide the necessary insight into the complexity of the problems (Dietz and Van der Straaten, 1992). The model in this figure describes the relations between the social system, of which economic decisions are a part, and the natural system. The aim of this model is to offer some insight into the mechanisms by which economic decisions and activities in industrialised countries disrupt the ecosystem.

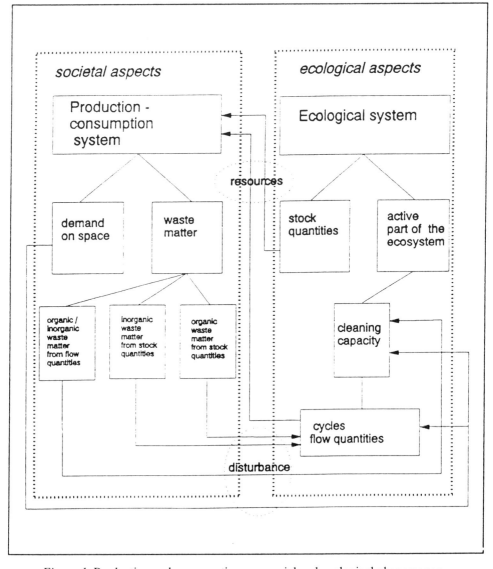

Figure 1. Production and consumption as a social and ecological phenomenon

The working of this conceptual model can give information about the ecological limits to be set upon production and consumption, when a situation of sustainable development (WCED, 1987) has to be guaranteed.

A system of human production is based, among other things, on the need to use natural resources from ecological cycles, the active part of the ecosystem. These natural resources are in principle inexhaustible. For about 200 years, man has also used fossil natural resources on a large scale. However, these resources are exhaustible, as is natural oil. The fossil part of the ecological system is hardly, if at all, affected by the flow of waste products originating in the economic system. Pollution of the environment occurs in that part of the ecological system where existing cycles develop, these cycles being disturbed by the discharge of waste products. There is a great difference between the dumping of organic materials and the dumping of inorganic and synthetic-organic materials into the ecocycles. Organic materials belong to the normal functioning of the cycles, while the others are foreign to them. Among the latter are the waste products from fossil resources. They cause a fundamental disturbance when dumped into the cycles, which have no mechanisms to cope with these waste products by way of processing or decomposition. On the other hand, decomposable organic material does not cause such a disturbance. Such material is already part and parcel of natural cycles and can be decomposed by bacteria as a matter of course.

Pollution by fossil material is worse than pollution by organic, decomposable material. Whereas the latter occurs locally and is likely to be neutralised after some time, pollution by non-decomposable stock material cannot be reversed. It is almost impossible to restore the cycle in this case. Matter foreign to the environment is stored up within the cycle, causing the effects of such dumping to be felt across a large area.

Attention should be given to the use of land, which is of great importance when dealing with skiing in mountain areas. As we have seen, this disrupts the ecosystem and, thus, threatens the cycles. This effect is different from that of the discharge of waste products in that this threatens the functioning of cycles much faster and more directly, without complicated intermediary processes. The analysis of environmental problems with the use of this model can be applied to the case of tourism in mountain areas. The first criterion from this model is the closing of ecocycles and the reduction of the use of fossil fuels. A second criterion is to use the land in a way that guarantees ecological sustainability.

These criteria will be applied to two situations, which occur in the same region. In the Western Italian Alps, the regions of Lombardija and Piemonte have developed a long distance footpath through the mountains – the Grande Traversata delle Alpi (GTA) – to promote tourist development in this area. On the other side of the frontier, there is the French ski resort of Isola 2000, which has also been developed for tourism.

8. The Grande Traversata delle Alpi (GTA)

The GTA is a long distance footpath, which goes from Carnino in the south, some 40 kilometres north of Ventimiglia, to Cannobio at the western part of Lago Maggiore. This long distance footpath was not built recently for tourist purposes. The organisers used existing footpaths, which were built many centuries ago by the farmers living in the mountain valleys. The route chosen is east of the French-Italian border, except in the Valle d'Aosta, where it goes directly to Monte Rosa. This mountain region has many more or less parallel valleys, which run from the mountains to the plain of the Po in the east and the south. Some valleys have an old industrial development, based on mines of minerals in the mountains. Most of these mines are exhausted. Employment in the valleys is mainly in the agricultural sector. In some valleys there are some tourist facilities, mainly used by Italians from the cities in the plain. These facilities are mostly for summer visitors, skiing is important only in a few villages.

Industrial employment is found in the cities of the plain, like Torino, Cuneo and Ivrea. In the lower parts of the valleys, workers commute to these cities. In the upper part of the valleys conditions are difficult. Mechanisation is always difficult, the production costs of farm products is high leading to high product prices and, consequently, mountain farmers cannot compete with farmers of the plains where production is cheaper and easier. Therefore, people are inclined to migrate to the cities in the plains. When this process starts, several village facilities are no longer available. Schools and shops close. The social climate worsens, and, therefore, more people want to migrate. At the end of the process, villages high up in the mountains are abandoned or only old people, who have no possibilities to go elsewhere, remain in the villages (Chiaretta, 1982; Chiaretta et al., 1983).

There are only a few ways to stop this process. One is tourist development on a large scale. But these valleys are not connected with each other; so international tourism is hindered by transport or accessibility problems. Another possibility is to intensify agriculture, which is being done in certain regions of the Upper Po Valley. But the result is that the sources of River Po are already polluted, due to intensive pig farming (Roggeri, 1990). Furthermore, intensification of agriculture will lead to a rise in labour productivity, which means that the same quantity of work in agriculture can be done with fewer workers. A surplus of labour is the result, as the area of agricultural land is limited. Again this will lead to migration.

The Grande Traversata delle Alpi is another example of how the migration of farmers from the upper valleys can be stopped. A system of posto-tappa's was built, which are accommodations, mainly located in the villages, where people can sleep and have meals. In most cases, buildings, which are no longer used, were transformed into posto tappa's. At present, 90 posto tappa's are in use. Food can be bought either in the village shops, or it is possible to eat in many small restaurants, which are rather abundant in these areas. The number of available beds for the whole GTA is 2,550 (GTA, 1989). This system offers a real possibility to bring value added and employment in the mountain valleys. This is why this project has been supported by the authorities of the Region Piemonte and Lombardija.

9. Isola 2000

Not so long ago, Isola 2000 was only a small mountain village high up on the mountains of France some 70 kilometres north of Nice. But it has been developed as a skiing resort on a level of 2000 metres. There is always snow during the winter. It is a typical skiing resort with four hotels and a large number of apartments. There are 22 ski lifts. The construction of these lifts and the ski runs were accompanied by severe destruction of the landscape. For years, workers were busy with dynamite to model the landscape to the 'right' inclination for skiers. When one enters the village from the Italian border one sees everywhere around the village the deep incisions of 10-40 metres, which started at the level of the village at 2,000 metres till 2,600 metres high up on the mountains. At the entrance of the village there is a car park for 550 motorcars (Office du Tourisme, 1988).

The city of Nice is the place where the tourist can get transportation from the airport to Isola 2000. This can be done by helicopter, bus or taxi. This village is used by skiers during the winter. In summer, nearly everything in the village is closed. Then it is a ghost town. The old village no longer exists. There are some old farmhouses, but they are no longer used. Few people living in the mountain valleys work in the hotels and shops. Nearly all the workers come from outside the region.

There are approximately 1,500 beds in Isola 2000. From interviews with people working in the villages, the author was informed that the whole project was financed by banks outside the region. So, a great part of the value added will not be spent in the village. Furthermore, the salaries of the people working in Isola 2000 is only partly spent in the mountain valley region, as these people live in Isola 2000 only in winter. Food, materials and the merchandise in the shops are bought outside the region.

The two developments described above should be related to the model of sustainable growth, which we have sketched in Figure 1. The following conclusions can be drawn:

- The development of the GTA is more related to existing infrastructure. Investment costs are low compared with Isola 2000, where a totally new infrastructure has been built. The return to investment could, therefore, be higher in the GTA.
- The GTA is a project which is stimulated by regional authorities. Isola 2000 is a private project, financed by banks. Therefore, there are great differences between the two projects regarding the aims and the results. The GTA has been realised to improve the economic and social climate of the people living in the mountain valleys. Isola 2000 is a project with the aim of gaining money from a tourist project.

- The influence upon the environment is completely different. Isola 2000 seriously disrupts the ecosystem of soil and water. The use of land makes it impossible for the ecosystem to function properly. The influence of the GTA upon the environment remains within the normal limits of the local ecosystems.

- In Isola 2000, a lot of energy is used for the functioning of the total tourist infrastructure. This is especially the case with ski lifts. The GTA project is a low input project with respect to fossil materials and fossil fuels use. Taking the model of Figure 1 as a starting point for sustainable development, the GTA project rates much better than Isola 2000.

- In the GTA project, value added and employment benefit the inhabitants of the mountain valleys. This is hardly the case in Isola 2000. Employees come from outside the valley while few products are bought from the mountain valley.

- The GTA project leaves the social infrastructure intact and strengthens it. Isola 2000 destroys the existing social and economic infrastructure.

- The general conclusion is that the GTA project gives important impulses to the solution of social and economic problems occurring in the mountain valleys. From an environmental point of view, Isola 2000 has a disastrous effect.

- Some criticism should be made, however, regarding the organisational aspects of the GTA. The capacity of the Posto Tappa's differs strongly. It is possible that the capacity of a Posto Tappa in one valley is 12 beds and the next valley 65 beds. Because of this, the total infrastructure cannot be used efficiently.

Furthermore, there is a great difference between the two projects regarding marketing. Information about Isola 2000 can be obtained easily, but information about the GTA is inadequate. This means that the project is not used by a sufficient number of tourists, due to a lack of knowledge about all the possibilities offered.

References

Bätzinger, W. (1988) *Die Alpen, Naturbearbeitung und Umweltzerstörung.* Beck Verlag, Zurich.

Chiaretta, F. (ed.) (1982) *Grande Traversata delle Alpi 1982.* Centro Documentazione Alpina,Torino.

Chiaretta, F., M. di Maio and R. Genre (1983) *Grande Traversata delle Alpi 1983.* Pricoli and Verlucca, Ivrea.

Cummings, R.G., D.S. Brookshire and W. Schulze (eds) (1986) *Valuing Environmental Goods:a State of the Arts Assessment of the Contingent Valuation Method.* Rowan and Allanheld, Totowa, N.Y.

Dietz, F.J. and J. van der Straaten, (1992) Rethinking environmental economics: the missing link between economic theory and environmental policy. *Journal of Economic Issues,* March 1992, 27–51.

European Commission (1987) *EC Fourth Environmental Action Programme (1987–1992).* Official Journal of the European Communities, December 1987.

Freeman, A.M. (1986) On assessing the state of the arts of the contingent valuation methods of valuing environmental changes. In R.G. Cummings, D.S. Brookshire and W.D. Schulze (eds) *Valuing Environmental Goods.* Rowan Allenheld, Totowa, N.Y.

GTA (1989) *Punti Apoggio e Note Logistiche.* Centro Documentazione Alpina, Torino.

Hoevenagel, R. (1990) *The Validity of the Contingent Valuation Method: Some Aspects on the Bias of Three Dutch Studies.* Paper presented at the European Association of Environmental and Resource Economists Conferenze in Venezia, Italia, 17–20 April 1990.

Kneese, A.V. (1984) *Measuring the Benefits of Clearer Air and Water.* Resources for the Future, Washington, D.C.

Mitchell, R.C. and R.T. Carson (1987) *Evaluating the Validity of Contingent Valuation Studies.* Discussion Paper QE 87-06. Quality of Environment Division, Resources for the Future, Washington, D.C.

Mitchell, R.C. and R.T. Carson (1989) *Using Surveys to Value Public Goods: the Contingent Valuation Method.* Resources for the Future,Washington D.C.

Office du Tourisme (1990) *Guide Isola 2000,* Isola 2000.

Rae, D.A. (1982) *Benefits of Visual Air Quality in Cincinatti.* Report to the Electric Power Research Institute, Charles River Associates, Boston.

Randall, A., J.P. Hoehn and D.S. Brookshire (1983) Contingent valuation surveys for evaluating environmental assets. *Natural Resources Journal* **23**, 635–648.

Roggeri, G. (1990) *Dove il Fiumo Incontra i Monti* (Where the River Meets the Mountains). Airone, April 1990, 47–52.

Slovic, P. (1972) *From Shakespeare to Simon: Speculations - and Some Evidence - About Man's Ability to Process Information.* ORI Research Monograph, Volume 12, Eugene.

Tversky, A., and D. Kahneman, (1981) The framing of decisions and the rationality of choice. *Science* **211**, 453–458.

World Commission on Environment and Development (1987) *Our Common Future.* Oxford University Press, Oxford.

TOURISM AND RECREATION SURVEYS: THE PROBLEMS OF SAMPLE SELECTION AND TRUNCATION BIAS

K.G. WILLIS
Department of Town and Country Planning
University of Newcastle upon Tyne
Newcastle upon Tyne NE1 7RU
United Kingdom

G.D. GARROD
Department of Agricultural Economics and Food Marketing
University of Newcastle upon Tyne
Newcastle upon Tyne NE1 7RU
United Kingdom

1. Introduction

The analysis of tourism, and subsequent policy recommendations, are often heavily dependent upon the reliability of sample surveys. Whilst considerable attention is devoted to obtaining a reasonable sample size – so that statistically significance of sample means can be assessed – and to minimising other errors common to both complete counts and surveys[1], relatively little recognition is given to the problems of sample selection and truncation bias in tourism and recreation surveys.

Surveys of some types of tourists, and of visitors to recreation sites, result in a sample being drawn from a truncated distribution. The distribution of frequent tourists to an area is truncated because a sample is drawn from tourists who visit the site within a time period: typically a year, the period over which a tourism sample is usually drawn. Tourists are typically asked how many times they have visited the area over the past year or 5 years; and first time visitors are recorded as having made one visit over the time period. Hence no observations are recorded for tourists

---

[1] See Deming (1950). Non sampling errors concern such issues as the failure of the questionnaire (failure to recognise difficulties of acquiring certain types of information; lack of clarity in definitions; emotional words and leading questions; etc.); faulty instructions and definitions to interviewers; bias arising from non-response; errors in response (e.g. memory bias; refusal to answer; 'prestige bias', etc.); interviewer bias; and errors in data entry and analysis.

who visit an area less than once per year (or other time period): hence no sample is drawn below one visit per period of time, and less frequent tourists are effectively eliminated from the study.[1]

The sample selection and truncation bias problems vary in importance according to the type of tourism and recreational visits undertaken. Tourism surveys of long-haul tourists may be least prone to this bias, because the trip is a 'once in a life time visit', and the questionnaire survey may elicit this information. But for within country 'weekend tourists' who frequently visit the capital city of a country, or a popular tourist area such as the Lake District in Britain; or who regularly camp or caravan in an area; then sample selection and truncation bias in surveys will be a potential source of error for many types of analysis. The sample selection problem and truncation bias issue is perhaps most pronounced in surveys of visitors to recreation sites, who may be day visitors or tourists on holiday, where the frequency of visits is recorded over the past 12 month period. For this reason this chapter explores the sample selection and truncation bias problem in relation to visitors surveys of recreation sites. It also illustrates the effect of ignoring sample selection and truncation bias in relation to estimating the value recreation sites through a travel-cost method.

The travel-cost method (TCM) values a recreation site by observing how much people are prepared to pay in terms of transport costs and travel time in order to gain access to the site.[2] TCM is a revealed preference method: it estimates recreational demand based upon people's actual consumption choices.[3] The study illustrates the effect of truncation in a survey of informal recreation along canals and inland waterways in Britain. Inland waterways embrace both commercial and 'public good' type recreational activities. Commercial recreation extracts some form

---

[1] The sampling procedure can give rise to apparent perverse results, and the development of new theories to explain these supposed 'anomalies'. For example Bell and Leeworthy (1990) noted that the recreation demand by tourists for beach days in Florida increased with distance travelled and with the total cost of the trip. In contrast economic theories, and travel-cost models, postulate that demand declines with increasing cost. Bell and Leeworthy suggested tourists respond differently to on-site price (for beach access: a negative relationship with price) and travel-cost per trip (a positive relationship with the number of beach days). However, this result is probably an artefact of sample selection: sampling tourists to Florida during one year, and regarding this as the relevant time period. Relating number of beach days in Florida over the life of a tourist in relation to cost would reveal fewer beach days in Florida, compared to those at a local beach at a lower cost.

[2] Travel-cost models have also been used to explain ecotourism to a tropical rain forest (Tobias and Mendelsohn, 1991), and long-haul tourism to Madagascar (Maille and Mendelsohn, 1993).

[3] The Department of Interior in the USA, which oversees the Comprehensive Environmental Response, Compensation and Liability Act (CERCLA), 1980, under which non-priced tourist and recreational activities lost through accidental pollution incidents are valued, regards revealed preference techniques using people's actual behaviour as more reliable than expressed preference techniques. Expressed preference ask people themselves what value they place on recreational activities lost.

of payment, for example cruising, which requires mooring licences and fees, or fishing which needs a permit. However, for other types of recreation, such as walking along a towpath, viewing a canal scene or watching boats pass through locks, there is commonly no charge. In these cases charging a fee might be impractical because of (i) the high transactions costs of collecting the fee in relation to the revenue generated, and (ii) its non-optimality (in the absence of congestion it costs no more to admit another individual to walk along a towpath or river bank).

To illustrate truncation effects in a recreation survey, this study estimates the utility or economic benefits associated with informal, non-priced, or public good forms of recreation along selected inland waterways and canals. Given that no price is charged for access, considerable benefits from non-priced recreational benefits may accrue to individuals by ways of consumer surplus benefits.

The British Waterways Board maintains inland waterways and canals for the carrying of commercial freight and for recreational uses. Over the years the relative importance of freight has declined, so that the principal benefit canals now provide is in terms of their recreational use. This is reflected in the operating accounts of the Board (British Waterways, 1989), for the year the recreation survey was undertaken. In terms of revenue from waterways operations, commercial leisure uses generated £5.416 million in 1988/89 compared to £3.334 million from commercial freight, including water transfer fees and sales. Whilst revenue from freight traffic increased by 3.8 percent between 1987/88 and 1988/89, revenue form pleasure craft, licences, etc., increased by 10.5 percent, and revenue from angling by 28.9 percent.

The cost of maintaining the British Waterways system far exceeds the revenue generated from commercial, leisure, and other sources. Total operating costs amounted to £49.897 million in 1988/89. The deficit between revenue and waterways operation, maintenance, and capital expenditure, is covered by grants from central government funds. These amounted to some £44.566 million in 1988/89 for the canal system as a whole.

However, the benefits derived by individuals from informal non-priced recreation associated with inland waterways and canals are not included in the financial accounts. They are though real economic benefits (in the sense that they affect people's utility) which ought to be set against the resource costs (government grants) of maintaining the canal system. The purposes of this study is (i) to estimate the value of the non-price informal recreational use of waterways, through a random sample of visitors to selected canals and recreational points along them, (ii) to provide some indication of the social value of waterways, (iii) to set alongside operating revenue generated by these waterways and the cost of maintaining them, and (iv) to assess whether such economic benefits offset some, or all, of the subsidy towards maintaining the canal system.

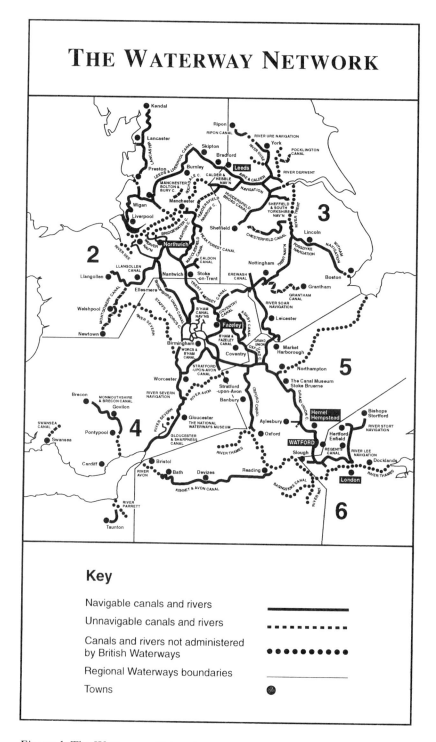

Figure 1. The Waterways Network

2. Visitor Surveys

A random sample of visitors was interviewed at a number of sites on various inland waterways and canals throughout July, August and September 1989, on both weekdays and weekend days. The sample survey was organised by British Waterways Board, who regularly undertake recreation surveys using trained interviewers. A total of 1,502 completed questionnaires were obtained.

The sites were chosen to provide a representative selection of different types of canals and recreational experiences. The sample surveys covered sites on (1) various narrow canals in the West Midlands (Coventry, Grand Union, Trent and Mersey, and Worcester and Birmingham Canals): at Fradley Junction near Lichfield north of Birmingham; Gas Street Basin and Farmers Bridge Lock in Birmingham; Hatton Locks near Warwick south of Coventry; Hawksbury Junction near Nuneaton north of Coventry; and Tardebigge near Bromsgrove north of Worcester; (2) the Gloucester and Sharpness Canal: at Saul Junction; Frampton Village; Patch Bridge near Slimbridge; and Purton Bends near Sharpness; (3) Trent Navigation: Newark; (4) the Weaver Navigation: Winsford; Vale Royal Locks; Hunt's Lock at Northwich; Saltersford Locks; Dutton Locks; and the Anderton Lift between the Weaver Navigation and the Trent and Mersey Canal also at Northwich (Figure 1).

3. Travel-Cost Method

The travel-cost method (TCM) estimates consumer surplus for a public good by the link between the public good and the market for a related private good. Consumer surplus is the utility an individual derives from a good, which is over and above the price he or she pays for the good. Where there is complementarity in consumption between the public good (e.g. open access non-priced recreation) and the market good (e.g. transport necessary to consume the public good), then expenditure on the private good (transport) can be used as a proxy for willingness-to-pay (WTP) in demand estimation. Thus expenditure on transport is the price that an individual has to pay to consume the recreational good. Since he is willing to pay this amount, the utility the marginal individual derives from the good must be at least as great and this expenditure. However, intra-marginal individuals will derive utility which is greater than their expenditure on transport. This additional utility is consumer surplus, and is the *net* benefit individuals derive from the recreational good.

The individual travel-cost model (ITCM) is based upon the relationship between visits to a site by individual i over some time period, and a number of other variables such as price (cost of transport) and income that determine the number of visits. The general ITCM is:

$$V_{ij} = f\ (TC_{ij},\ D_{ij},\ M_{ij},\ AL_{ij},\ P_j,\ SC_{ik},\ AC_{ij},\ IN_i,\ N_i,\ e_i),$$

where:

V_{ij} = the number of visits made by individual i to waterway or canal site j.

TC_{ij} = the travel cost faced by individual i to visit site j (£s).

D_{ij} = a vector of dummy variables used to reflect the activities undertaken by individual i at site j: for instance fishing, walking a dog, boating, shopping, walking, etc.

M_{ij} = dummy variable: whether visit by individual i to site j is the main purpose of the trip.

AL_{ij} = dummy variable: whether individual i specifies a local alternative to site.

P_j = dummy variable: whether there is a particular building or structure of tourist interest at site j.

SC_{ik} = rating given by visitor i to place j for each of five categories k.

AC_{ij} = number of specified canal activities which individual i has participated in.

IN_i = income rating of individual i.

N_i = number of individuals in the party of individual i.

e_i = error term.

Only a handful of ITCMs have been developed in the USA and Europe, and those only since the mid 1980s, so experience in applying them are extremely limited. Major problems in deriving values for ITCMs centre on imperfect knowledge about the individual's utility function, the large variances associated with frequency data on individual visits for any particular distance to the recreation site, and the extent to which frequency of visit data over some time period reflects an individual's demand for that recreation.

Since the ITCM is based upon an individual's visits, its dependent variable may be discrete: individuals may make any number of visits in a year, but most make either 1 or 2, though a few may make weekly or daily visits. In addition, the dependent variable for the ITCM is truncated to 1 (zero visit individuals, and those who visit less than once over the time period, are necessarily omitted from the sample).

The ITCM is well suited to modelling the responses and estimating recreational benefits along a linear feature such as a canal; and, in addition, permits the disaggregation of visitors by type of activity they indulge in and by the sort of visits they undertake. This information allows the respondents to be divided into a series of sub-groups each of which can be analysed separately using a reduced form of the model. In this way consumer surplus estimates can be generated for each category of user, from dog walkers to day trippers, giving an insight into how each group values the waterway with reference to its particular activity. The method can also be used with a smaller respondent sample size than that required for the traditional zonal travel-cost model (Willis and Garrod, 1991a). Clearly, specific models relating to a site or activity may only use a fraction of the variables outlined, but it is important that they are all included in the general model for purposes of identification.

Four classes of variables are used in the model: site variables, respondent variables, interaction variables, and ratings variables. Site variables like P_j provide information about the canal site j where the survey is conducted, while the respondent variables relate specifically to the person being interviewed: for instance the variable N_i records the size of the respondent's party as a possible measure of the non-financial organisation costs involved in the visit. The ratings variable SC_{ij} depends on a subjective appraisal by the visitor of five site variables: staff, facilities, upkeep, information provided, and enjoyment value. Visitor ratings can fall into three broad categories: good, average, and poor, and provide information relevant in modelling the number of visits from an indication of visitor satisfaction. Finally there is the most common group, those variables such as TC_{ij}, D_{ij}, and M_{ij}, which are the consequences of interaction between the visitor and the site, providing some information about both.

In the calculation of TC_{ij} it was decided that it was reasonable to assume that the time spent travelling to and from a canal site had some opportunity cost, either in terms of wages or some alternative leisure activity foregone. This meant that time costs could be included in the travel cost. In an appraisal of the value of non-working time, the Department of Transport (1987) advocated a standard average value of non-working time at 43% of earnings, with appropriate reductions for children and non-working individuals; and these 'official' values were adopted as the costs of travel time. Thus, for visits made by car, the travel cost was estimated as the sum of full car running costs plus time costs. Full car running costs (cost of petrol plus other variable costs related to mileage such as depreciation, maintenance or service costs) were derived from Royal Automobile Club estimates, based on 1300-1600cc engine sizes with petrol at £1.75 per gallon (1989) price. The costs were adopted on the basis that they are the real costs that road users face. This is borne out in a study of forest recreational values (Willis and Garrod, 1991a) which indicated that the perceived estimated travel costs for visitors in cars were about three times their petrol costs and only just below the estimated full running costs of their cars. Exclusion of time and full running costs is likely to reduce any estimates of consumer surplus considerably as demonstrated by Willis and Garrod (1991a). As for other users, walkers and cyclists were adjudged to incur travel costs equal only to their time costs, while visitors travelling by public transport had a travel cost estimated as their cost incurred by using public transport plus a time cost. A vector of interaction variable, D_{ij}, specifically describing a respondent's range of activities at a particular site was included in order to model consumer surplus benefits generated by each particular activity at all sites.

After the variables to be included in the model have been identified, its functional form must be determined. There is no set technique for achieving this, but certain theoretical and practical considerations provide a rule a thumb for the identification of a suitable functional form. In order to estimate the consumer surplus (CS) for the recreational benefits of each site, it is necessary that the model chosen not only fulfils certain technical requirements, but in addition meets some standard as to what is an acceptable CS estimate. Technical consideration may arise from experience in earlier studies, which indicate, for instance, which independent variables should be significant in the model and the direction of their coefficients. Failure to meet these expectations may throw some doubt on the validity of the model. Further, there are considerations with regard to goodness of fit and statistical significance, which are important in determining the optimal functional form. From the basis of comparison it is deemed unacceptable to use a model which gives estimates which differ in order of magnitude from those yielded by other travel-cost studies at recreation sites in the UK (e.g. Willis and Garrod, 1991a, on forestry). Another way of judging the estimates derived from different functional forms is to compare them with those obtained from other valuation methods such as contingent valuation (CV).

The data were first modelled using ordinary least squares (OLS) regression. Of the forms tested, only one, the linear, was in any way satisfactory. Previous research (e.g. Willis and Garrod, 1990) has shown that ITCMs based on the semi-log (independent) form may occasionally suffer from heteroskedastic disturbances[1], making them unsuitable candidates for OLS regression. A Breusch-Pagan test confirmed the presence of heteroskedasticity, so this model was rejected. Several transformations were made on the data in an attempt to relieve this problem but no significant improvements were made. As often happens with the ITCM, the double-log model generated infinite consumer surplus, indicating that demand for canal visits was inelastic. This situation may be alleviated by adding unity or some other positive value to the dependent variable, though as Price (1990) points out, this expedient is entirely arbitrary and gives different answers depending upon the unit in which the visit rate is measured. The semi-log (dependent) model, much favoured in zonal TCM studies (see Willis, 1990), gave a consumer surplus in excess of £128 per visit. Comparing this latter estimate with those from similar ITCM studies (e.g. Smith, 1988), and with CV estimates reported earlier in this chapter, suggests that it is unreasonably high. This is probably due to the combination of a high concentration of single visits with the asymptotic nature of the model. The linear form, on the other hand, whilst not perhaps providing such a flexible modelling framework as others, does not exhibit any asymptotic behaviour and when used to estimate consumer surplus for a canal visit gives the not unreasonable estimate of £3.01.

---

[1] Heteroskedasticity occurs when variance increases as the value of the variable increases. This violates one of the assumptions of the classical linear regression model, that the disturbance terms all have the same variance (see Kennedy, 1979).

It may be argued that then true functional relationship between individual visits and travel cost is non-linear, and perhaps convex. If, in practice, a convex functional form provided a better approximation of the data, then any estimates derived from the linear model could be shown to be lower bounds of actual consumer surplus per visitor. Figure 2 shows that the linear specification implies an under-estimate of consumer surplus for those visitors who made only one trip to a canal. In addition, for those individuals who visited a canal more than once, the estimated consumer surplus of a single visit clearly under-estimates their total consumer surplus. Thus, in this case, the use of single visit estimates to derive total consumer surplus per visitor, regardless of the number of visits made, ensures that the linear demand form always gives lower-bound estimates.[1] Clearly, this may misrepresent the total benefit derived by frequent visitors, which could be far greater than that estimated here.

There is a growing literature on the problem of estimation bias in ITCMs due to truncated distribution arising from the exclusion of non-users within the specified time period over which a visit is recorded (Bockstael *et al*, 1984; Smith and Desvousges, 1985; Balkan and Kahn, 1988; Willis and Garrod, 1991a). These studies have shown that any attempt to use OLS to estimate the parameters of ITCMs will be biased because the demand model will be based solely on the survey of actual visitors to the canal within the time period. These surveys provide no information on individuals who choose not to use the canal; or use the canal less than the time period set to record a visit (typically one year). Thus sample surveys of visitors will be biased, suggesting the measure of quantity demanded will be truncated to one. That is, no account is taken of people who choose not to visit the canal or have a frequency of use of less than one visit per year. This lack of information on non-visitors means that the resulting OLS estimates will be biased towards positive responses, and hence will grossly over-estimate consumer surplus per visitor. In order to avoid this problem, it is possible to fit a truncated maximum-likelihood (TML) model to the data (Maddala, 1983). This can then be used to derive the log-likelihood function and a suitable numerical algorithm used to compute the maximum-likelihood estimates from the regression. From there a more realistic consumer surplus estimate can be derived. As with the OLS approach, only the linear TML model produced a reasonable consumer surplus estimate. For the combined data set as a whole, consumer surplus was estimated at 50.7 pence for a single visit, compared with £84.38 for the semi-log (dependent) model. Similar investigations were carried out on the data for each individual canal, using a dependent variable based on visits over a twelve months period. In all cases the

---

[1] Integrating over the linear equation: $V = a + bC + e$, where V = number of visits, C = cost of visit, and b = coefficient of C, consumer surplus (CS) for an individual making q visits to a site is: $CS = -q^2/2b$. Hence summing visits to a site rather than number of visits over each visitor to a site, will produce a lower-bound estimate of consumer surplus.

linear model was eventually chosen to estimate the regression parameters for both the OLS and TML models. Variables were included on the basis of *a priori* logic of predicted coefficient values, and are significant at the 0.15 level unless otherwise indicated. Consumer surplus estimates for individual visits to each site can then be calculated by substituting values from the resulting linear equations into the appropriate estimator.

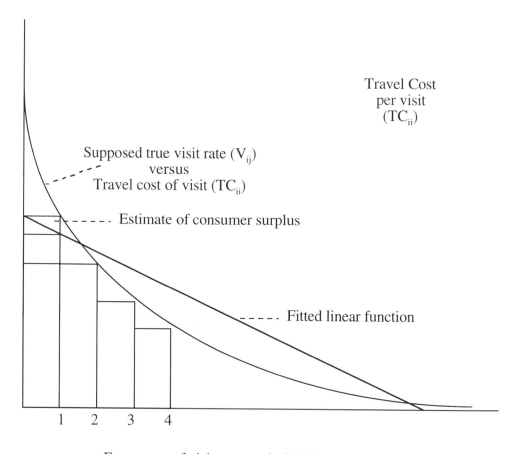

Figure 2. Consumer surplus and frequency of visits

4. ITCM Results

Results for the Weaver Navigation and for Newark Town Lock for both regression techniques are shown in Table 1 (a full set of results for all sites can be found in Willis and Garrod, 1991b). These clearly show the difference between the biased OLS estimates and those derived using TML. As expected, these latter estimates are by far the lower of the two. While OLS estimates range from 298.4p for Newark

Town Lock down to only 18p at the more commercial Weaver Navigation, the comparable figures from TML are considerably lower at 45p and 4.7p respectively. All subsequent analyses, consumers' surplus aggregations, and policy recommendations are based on TML estimates.

Table 1. Individual travel-cost method coefficients and results for Weaver Navigation

	OLS travel-cost coefficient (t-ratio)	Regression consumer surplus	Truncated max. travel-cost coefficient (t-ratio)	Likelihood consumer surplus
All users	-2.78864 (-2.97)	17.9p	-10.6042 (-3.75)	4.7p
Non-casual users	-2.42908 (-2.76)	20.6p	-9.60920 (-3.83)	5.2p
Walkers	-1.36517 (-1.50)	36.6p	-6.67460 (-3.83)	7.5p
Canal scene viewers	-1.94820 (-2.76)	25.7p	-6.90400 (-5.10)	7.2p
Dog walkers	-4.97278 (-1.85)	10.1p	-18.72920 (-4.39)	2.7p
Nature lovers	-1.16368 (-1.03)	43.0p	-9.21345 (-4.39)	5.4p
Drinkers	-4.39790 (-3.00)	11.4p	-10.29250 (-6.84)	4.9p
Local users (<16 km)	-4.47052 (-1.86)	11.2p	-17.76800 (-2.28)	2.8p
Non-local users (>16 km)	-0.78910 (-1.59)	63.4p	-2.29082 (-3.71)	21.8p

As far as the different visitor activities are concerned, the estimates contained in Tables 1 and 2 provide a measure of how much each group values the benefits of each site or group of sites. Respondents indulging in more everyday activities, like walking the dog, value the canal less, in general, than those whose activities are dependent upon it. Variations in results between canals may be attributable to differences in quality of amenity and environment and may provide useful pointers to the provision of improved facilities.

When casual users, such as dog walkers or those using the canal as a short cut, are removed from the analysis, the consumer surplus estimates rise by as much as 53p. Most sites, however, experience a much smaller increase, less than 0.7p for both the Weaver Navigation and Newark. A much more pronounced difference can be seen in the comparison of local and non-local users. Local users, defined as those living within a 16 km radius, have significantly lower travel costs than anyone travelling from further afield. As a consequence the ITCM yields far lower consumer surplus estimates for these locals than for those with greater travel costs. This low valuation is more a reflection of the bias of ITCM against local users, than of an inherently low

Table 2. Comparison of individual travel-cost method consumer surplus estimates per visit and contingent valuation estimates of willingness-to-pay, by distance travelled to canal site (1989 prices)

	ITCM estimates	Contingent all distances	Valuation "locals" (< 16 km)	Estimates "non-locals" (>16 km)
Fradley	0.39	0.43	0.32	0.64
Gas St.	0.60	0.19	0.18	0.25
Hatton	0.29	0.27	0.28	0.26
Hawkesbury	0.07	0.32	0.30	0.75*
Newark	0.45	0.30	0.26	0.32
Gloucester	1.05	0.41	0.40	0.43
Patch	0.65	0.41	0.43	0.40
Purton	0.20	0.38	0.38	0.38
Saul	2.71	0.43	0.38	0.55
Tardebigge	0.07	0.19	0.15	0.33
Weaver	0.05	0.37	0.34	0.50
Anderton	0.24	0.73	0.81	0.63

* only 2 observations

valuation of the recreational facility.[1] Even so, it is probably reasonable to assume that those respondents who make a long journey specially to visit a site do, in fact, value it more highly at the margin than the majority of users who visit the canal more frequently.

The 'ball-park' magnitudes of the ITCM estimates were confirmed by results from a contingent valuation (CV) question in the same survey, which asked the maximum amount individuals would be willing to pay to gain access to a canal to pursue their recreational activity if a price was charged. Table 2 documents the results, recording average willingness-to-pay across all types of visitors. ITCM estimates are included for comparative purposes. However, the correlation between

---

[1] TCMs under-estimate the benefits derived by those who decide to live near a recreation site to avoid travel cost incurred in reaching it.

ITCM and CV estimates by site is weak. CV estimates themselves varied by distance of trip, with CV estimates for local trips being only half those for trips over 16 km for most sites (Table 2). CV estimates also varied by the principal recreational activity in which the respondent is engaged. More local activities, such as dog walking and short cuts, attracted lower CV bids than fishing, wildlife, and other canal attractions. Despite potential biases with CV techniques (Garrod and Willis, 1990), these results appear reasonable and consistent. It is useful to employ more than one technique in any environmental valuation, so that one can be used as a validation check on the other.

5. Aggregation of Consumer Surplus Estimates

The total value of open-access recreation on British Waterways requires consumer surplus to be aggregated in some way. Aggregation from individual experiences and values to a macro-economic value is always problematic in economics, and the current example is no exception. There are a number of ways by which such an aggregation might be attempted, *viz.* by:
1. the total number of visits to the waterways system multiplied by the average consumer surplus per visit
2. the total number of visits of each type to the waterways system, multiplied by the consumer surplus for the type of recreation undertaken during the visit

In the context of the separability of information within individual data records on visits and visitors to canals, and problems with model formulation, the latter aggregation method was not feasible. The questionnaire recorded activities undertaken during a visit to a canal. It did not specifically elicit the major purpose of the visit. Thus, some visits recorded more than one activity: consumer surplus measures for different activities are constructed from all individuals who mentioned that activity. Hence, observed numbers for individual activities exceeded the total number of visitors to any canal; and consumer surplus measures are not mutually exclusive. Separable estimates would have been desirable, for example, of canal viewers; dog walkers; canal viewers who were dog walking; etc., but this was not possible because of the small number of observations recorded. Aggregation by type of visit is therefore, somewhat problematical.

Aggregating the total number of visits to the waterways system, estimated by MAS Ltd in 1989 to be 105.9 million adult visits and 26.6 million child visits, provides through the ITCM, a weighted average aggregate consumer surplus of £58.245 million for adults and £3.724 million for children, which equals £61.969 million in total (Table 3). This is likely to be an under-estimate, with consumer surplus corresponding to that from one (marginal) visit. For those individuals who make more than one visit, total consumer surplus would, as explained earlier, be greater. The frequency of visits to waterways is high (about 18 visits per visitor on average), but mainly concentrated in those recreational activities (e.g. dog walking) for which consumer surplus is low. But even accepting the lower bound estimate of £61.969 million, this net benefit from open access recreation, far exceeds the government grants which make up the difference between the revenue (£11.235 million) and the operating and maintenance costs (£49.897 million).

Table 3: Annual financial cash flows and consumer surplus from open access recreation (£'000s 1988/89 prices)

	All canals	Gloucester and Sharpness Canal
Revenue	11,235	1,095
Expenditure	49,897	1,642
Subsidy	44,566*	547
ITCM estimate	61,969	3,736
CV estimate	40,773	1,458

* grants cover operating costs and capital assets including depreciation

Refinement of these figures and consumer surplus estimates would require further visitor sample surveys, with larger sample sizes to reduce the standard error on the estimates. This is especially important when the data are split up for different recreation visit types. Generally the existing samples were simply not large enough to permit more accurate estimates of consumer surpluses for recreation sub-categories (as the t-statistics on some of the coefficients indicate).

An aggregate consumer surplus value for informal recreation can also be produced from the CV WTP estimate. Average WTP over all canals from the CV question was £0.362 per visit. Accepting this as representative of all canals and types of visit, and aggregating by total number of annual visits, suggests a consumer surplus of £38.366 million for adults and £2.407 million for children, or £40.773 million in total. This confirms the 'ball-park' estimate produced by the ITCM. However, CV may be subject to biases so that stated WTP does not reflect true WTP. In the survey the CV question elicited maximum WTP as an entry fee or car parking charge to be collected by British Waterways. Respondents regarding maintenance of canals as a government duty, or to be supplied through general taxation, are likely to under-bid; a feature noted in a similar CV of open access to forest recreation (Willis and Benson, 1989). In the latter study, 12 per cent of respondents were revealed as registering a protest bid or bidding strategically through a follow up question eliciting why their declared bid was zero.

The only canal for which non-priced open access recreation benefits could be compared with financial revenue and cost data, because a comprehensive visitor survey was undertaken at a range of places along the canal, was the Gloucester and Sharpness Canal. Table 3 documents the revenues and costs from operating the Gloucester and Sharpness Canal, and the operating deficit of £0.547 million. The annual number of informal recreation visits to the canal was estimated to be 4.045 million of which 2.706 million were at Gloucester Docks, the home of the British Waterways museum and other recreational attractions. Clearly, with a consumer surplus of £1.05 per visit, the average for the Gloucester and Sharpness Canal, aggregate consumer surplus exceeds the financial deficit. Even accepting the lowest consumer surplus of £0.20 results in an aggregate consumer surplus considerably more than the financial deficit. Similarly, accepting the contingent valuation estimate (£0.41) of net benefit on the Gloucester and Sharpness Canal, results in an aggregate consumer surplus of £1.458 million; again substantially more than the financial deficit.

Undoubtedly, many other canals occupy a similar position to the Gloucester and Sharpness Canal, with non-priced social benefits far exceeding financial operating deficits. However, for some canals, recreational activity (and revenue) may be so low and costs so high, that consumer surplus from open access recreation will be less than operating deficits (Willis and Garrod, 1990).

6. Conclusion

Recreation surveys carried out at recreation sites suffer from a truncated sampling distribution. This affects consumer surplus estimates derived from individual travel-cost models (ITCMs) based upon this type of recreation survey data. OLS regression is an inappropriate method to use for ITCMs where data is derived from a truncated distribution. However, reliable and accurate estimates of consumer surplus can be derived by using a maximum likelihood model, which corrects for truncation in the sample. In this study, large differences in the consumer surplus values were observed depending upon whether OLS regression or a truncated maximum likelihood (TML) estimator was used. Correcting for truncation bias shows that reasonable consumer surplus estimates of the benefit of non-priced recreation can be provided. However, the functional form of the ITCM model is probably more important than selection bias in affecting consumer surplus results.

Recreation surveys carried out 'on-site' can also suffer from other types of bias in disproportionately sampling more frequent visitors and 'avid' recreational participants. 'On-site' surveys sample visits, rather than visitors, and this creates the problem. To illustrate this, assume a site has two types of visitor: frequent visitors and non-frequent visitors. Assume that there are 120 frequent visitors who visit 5 times per year, and 120 non-frequent visitors who visit only once per year. Assume visits are evenly distributed over the year: January has 50 visits from frequent visitors and 10 visits from non-frequent visitors, as has February to December. If a 10% sample of visits were randomly selected in any month, the sample would consist of 5 frequent visitors and 1 non-frequent visitor. However, the true proportion of frequent to non-frequent visitors is 1:1. This type of site selection bias can be corrected by weighting the number in the visit sample by its reciprocal to derive the true proportion of visitors: in this case [(50 (1/5)) + (10 (1/1))].

Many recreational surveys over-sample 'avid' participants. On-site surveys that intercept recreationalists are more likely to intercept individuals who participate more frequently. Mail surveys undertaken off-site, even those of the public in general, may also over-sample 'avid' participants because such individuals are more likely to respond to the survey, leading to estimates with sample selectivity bias.

Morey *et al* (1991), in response to limited trip data and sample selectivity, developed a repeated discrete choice random utility model (RUM) of participation and site choice that corrects for the over-sampling of 'avid' participants, and that can be estimated with a data set that reports each individual's total number of trips but the individual's actual destination for only one of those trips. Over sampling of 'avid' participants was corrected by assuming the probability individual i being in the sample is a linear function of the total number of trips undertaken. Sample selectivity bias associated with over-sampling avid participants can then be corrected by replacing the population distribution in the likelihood function with its sampling distribution.

A survey of visits by recreationalists to a recreation site creates particular problems in data analysis. The survey based on visits may not be representative of visitors. The survey may result in an over-sample of 'avid' participants creating sampling bias. The fact that the survey is conducted 'on-site' gives rise to a truncated sample. However, once these problems are recognised they can be controlled. Sample weights can be used to correct for sampling by visits instead of visitors. Rigorous random sampling and minimisation of non-response will avoid over-sampling of 'avid' participants, whilst a maximum likelihood function can be derived for a truncated distribution that will permit the estimation of the correct demand function for the site and its associated consumer surplus value.

Acknowledgements

We are grateful to Mr. G. Millar at British Waterways for supplying financial records and permitting access to survey data.

References

Balkan, E. and J.R. Kahn, (1988) The value of changes in deer hunting quality: a travel-cost approach. *Applied Economics* **20**, 533–539.

Bell, F.W. and V.R. Leeworthy (1990) Recreation demand by tourists for saltwater beach days. *Journal of Environmental Economics and Management* **18**, 189–205.

Bockstael, N.E., W.M. Hanemann and I.E. Strand (1984) *Measuring the Benefits of WaterQuality Improvement Using Recreation Demand Models.* Washington DC: U.S. Environmental Protection Agency, Report CR-811043-01-1.

British Waterways (1989) *Report and Accounts of the British Waterways Board 1988/89.* Watford: British Waterways.

Deming, W.E. (1950) *Some Theory of Sampling*. Wiley, New York.

Department of Transport (1987) *Values of Journey Time Savings and Accident Prevention*. Department of Transport, London.

Garrod, G.D. and K.G. Willis (1990) Contingent Valuation Techniques: A Review of Their Unbiasedness, Efficiency and Consistency. *Countryside Change Working Paper Series. WP10*. Countryside Change Unit, Department of Agricultural Economics and Food Marketing, University of Newcastle upon Tyne.

Kennedy, P. (1979) *A Guide to Econometrics*. Martin Robertson, Oxford.

Maddala, G.S. (1983) *Limited-Dependent and Qualitative Variables in Econometrics*. Cambridge University Press, Cambridge.

Maille, P. and R. Mendelsohn (1993) Valuing ecotourism in Madagascar. *Journal of Environmental Management* **38**, 213–218.

Morey, E.R., W.D. Shaw and R.D. Rowe (1991) A discrete choice model of recreational participation, site choice, and activity valuation when complete trip data are not available. *Journal of Environmental Economics and Management* **20**, 181–201.

Smith, V.K. and Desvousges, W.H. (1985) The generalised travel-cost model and water quality benefits. *Southern Economic Journal* **52**, 371-381.

Smith, V.K. (1988) Selection and recreation demand. *American Journal of Agricultural Economics* **70**, 29–36.

Tobias, D. and R. Mendelsohn (1991) Valuing ecotourism a tropical rain-forest reserve. *Ambio* **20**, 91–93.

Willis, K.G. and J.F. Benson (1989) *Values of User Benefits of Forest Recreation: some further site surveys*. Report to the Forestry Commission, Edinburgh.

Willis, K.G. and G.D. Garrod (1990) The individual travel-cost method and the value of recreation: the case of the Montgomery and Lancaster Canals. *Environment and Planning C, Government and Policy* **8**, 315–326.

Willis, K.G. and G.D. Garrod (1991a) An individual travel-cost method of evaluating forest recreation. *Journal of Agricultural Economics* **42**, 33–42.

Willis, K.G. (1991). The recreational value of the Forestry Commission Estate in Great Britain: a Clawson-Knetsch travel cost analysis. *Scottish Journal of Political Economy* **38**, 58–75.

Willis, K.G. and G.D. Garrod (1991b) Valuing open access recreation on inland waterways: on-site recreation surveys and selection effects. *Regional Studies* **25**:6, 511–524.

TOURISM AND CULTURE

GREG RICHARDS
Tilburg University
Department of Leisure Studies
P.O. Box 90153, 5000 LE Tilburg
The Netherlands

1. Introduction

In the past, studies of tourism and the environment have tended to concentrate on the 'natural' environment; these studies did not pay attention to the role of culture in creating environments for tourism, and mediating the way in which tourists consume environments. Recent critical studies of the tourism phenomenon have begun to redress this balance, by pointing to the way in which the production and reproduction of 'nature' is highly culturally determined (Urry, 1996).

As Van der Duim and Philipsen (1995) point out, the modern production of nature is closely bound up with the growth of a culture of tourism. National parks and nature reserves, specifically demarcated natural areas, only came into being as increasing numbers of people began to appreciate the value of such environments through tourism. The growing pacification and packaging of nature has placed an even greater premium on those places which can still be considered to be wild or inaccessible. Growing demand from tourists threatens the very wildness that tourists come to consume, causing a range of management solutions to be adopted for the conflicts between the needs of wildlife, residents and visitors. National parks are enclosed as access is controlled and managed nature is created. Even the last great wilderness, Antarctica, is increasingly subject to demands for visitor management to combat the negative impact of the growing numbers of visitors on the continent's fragile ecosystem (Ezenbacher, 1993).

Most of the 'natural' areas that tourists visit today are in fact not true wilderness, but rural areas; most tourists are actually consuming agricultural landscapes, the productive spaces of the rural. This is the fact that often leads to conflict between urban based tourists and their rural hosts. The consumption of the former begins to impinge on the productive activities of the latter. At the same time, the drop in agricultural activity is leading to increased provision of services for tourists, such as farm holidays and activity holidays.

Modernisation threatens not only natural or rural landscapes, but urban landscapes as well. The rhetoric once employed against the 'rape of the countryside' is increasingly finding echoes in struggles to preserve 'unique' urban landscapes and elements of cultural heritage located in major urban centres. In rural areas, the expansion and intensification of agricultural production has been seen as a major threat; furthermore, the process of redevelopment in urban areas is seen as a modern scourge currently. Old buildings, monuments and styles of architecture are placed on danger lists in the same way as rare species of birds. Historic buildings are herded into special reserves called 'conservation areas', or, in some cases, these buildings are transported to open air museums. As in the case of natural environments, the creation and maintenance of these environments is increasingly dependent on tourism and leisure consumption.

Interesting parallels therefore emerge between the processes at work in both 'natural' and 'cultural' environments. In particular, the increasing scarcity of certain types of landscape or places imbues them with a certain symbolic value. The symbolic value, which attaches to these locations generates, in turn, real economic value through the commodification process, creating still greater pressure to transform these places for economic purposes. Tourism is often seen as one of the least harmful ways of maximising the economic potential of these symbolic places. Both natural and cultural environments become inextricably linked to a symbolic production process which consumes their natural and cultural components to generate cultural and economic value.

This chapter reviews the relationship between tourism and culture both in urban and rural environments; it analyses the way in which the environment has been transformed into a cultural product for tourism consumption. In doing so, consideration is given to the development of perspectives on nature and culture; furthermore, it is discussed how these have tended to converge as elements of the tourism product. The way in which culture is produced and reproduced for tourism consumption is analysed in both urban and rural contexts, with specific attention being aid to recent trends in cultural tourism consumption in rural areas.

2. Views of nature

The role of culture in mediating our perception of the natural environment can be gauged from the varied conceptions of 'nature' in different cultures. As Yi-Fu Tuan (1974, p.132) pointed out, in western cultures the meaning of nature has narrowed over time, from an all-embracing term to a specific description of countryside and wilderness. In traditional cultures, nature and culture are not separated from each other. As agricultural societies developed, so the culture represented by the tamed or civilised world was contrasted with the wilderness of nature. Nature was a thing to be feared or tamed, it was not until the rise of the 'romantic gaze' in the 19th century that nature began to be appreciated aesthetically (Urry, 1996). Under the influence of romanticism, nature became conceived as 'landscape', fit for visual consumption. The concept of 'landscape' arguably 'shed its earthbound roots' and acquired 'the precious meaning of art' (Tuan, 1974:133). This shift towards the visual consumption of landscapes through tourism was also evidenced by the changing emphasis of routes taken by the Grand Tourists in the 18th and 19th centuries. Whereas in the 18th century, the Grand Tour had concentrated almost entirely on the urban centres of continental Europe, in the 19th century there was a significant increase in journeys through the Alps (Towner, 1985), heralding the rise of Alpinism and the modern development of winter sports tourism. However, the aesthetic appreciation of nature has not developed in the same way everywhere, and even in the West there are differing conceptions of this distinction between 'nature' and 'culture'. In France, for example, 'natural' sites associated with great literature are regarded as national cultural assets (Bauer, 1996).

In Europe and North America, demand for the preservation of nature began to grow in the 19th century as a result of urbanisation. The countryside became valued not just for its aesthetic beauty, but also for being an area of 'freedom' from the exploitation of industrial labour. Such views were central to the campaigns to free access to the countryside, which were highlighted by the mass trespasses organised in the upland areas of Britain in the 1930s. Struggles over the concept of nature have become even more complex with the rise of what Urry (1995, p.222) refers to as the 'new sociations'. The development of organisations devoted to the conservation or preservation of the countryside increased after the 1960s, and membership of such organisations rose particularly rapidly during the 'green tide' of the 1980s. Urry relates this rising concern for nature to the shift from an industrial to a 'risk' society (Beck, 1992). Individuals have become more reflexive about the relationship between nature and culture, to the extent that the notion of 'rights' has become attached to animals, plants and even the earth itself (Urry, 1995, p.225). Tourism has had an important role to play in the development of such reflexivity, bringing people into direct contact with threatened nature, and allowing them to reflect on their own role in its disappearance. Nature has acquired a particular value which is used as a means of distinction by many members of the 'service class' in their consumption of specific types of rural and 'wild' environments (Munt, 1994).

3. Views of culture

One of the thorniest problems in any analysis of culture lies in the definition of culture itself. An examination of the way in which culture has been used reveals two basic approaches (Richards, 1996). 'Culture as process' is an approach derived from anthropology and sociology, which regards culture mainly as codes of conduct embedded in a specific social group. The boundaries of social groups, and therefore cultures, are variable, and can cover a nation, tribe, corporation or those pursuing specific activities. We may therefore talk about the culture of a specific country, or a culture of mass tourism (e.g. Urry, 1990). The 'culture as product' approach derives particularly from literary criticism. Culture is regarded as the product of individual or group activities to which certain meanings are attached. Thus we might try and identify different types of 'urban cultures' reflecting the way in which different groups live in, use and reproduce urban spaces.

The product and process approaches to culture tend not to overlap. However, in the field of tourism, there has been a certain degree of integration between the two approaches. 'Culture as process' is the goal of tourists seeking authenticity and meaning through their tourist experiences (MacCannell, 1976; Cohen, 1979). However, the very presence of tourists leads to the creation of cultural manifestations specifically for tourist consumption (Cohen, 1988). MacCannell (1976, p.25) refers to 'cultural productions' as a term which indicates not only the process of culture, but also the products, which result from that process. In other words, 'culture as process' is transformed through tourism (as well as through other social mechanisms) into 'culture as product'. Thus, the rural 'way of life' as a cultural process is transformed into a product for tourism consumption, so that elements of the living culture become cultural attractions in the same way as physical objects such as buildings, monuments and landscapes. As process and product become integrated through tourism, the scope of cultural tourism broadens, to take in aspects of the way of life of a particular people or region; the cultural tourist becomes more than simply a visitor of attractions as suggested by MacCannell (1976), he or she becomes more a consumer of 'atmosphere'. The atmosphere that tourists are seeking can be attached to the built environment in cities, but can also be found in rural areas, which are often appreciated for their slower pace of life and relative tranquillity. As the 'atmosphere' of a cosmopolitan city or the 'peace and quiet' of the countryside become subject to the tourist gaze, both urban and rural environments become involved in the aestheticisation of everyday life described by Featherstone (1991).

The culture seekers who are consuming not just the aesthetic products of culture but also the way of life associated with specific cultures are increasingly being identified with the emerging 'new middle class' or 'service class'. Walsh (1991, p.127) argues that the service class is a phenomenon which emerged in Britain in the 1980s, marked by participation in 'modes of consumption which enhanced their movement away from dull inconspicuous forms of consumption, towards a consumption of signs which many saw as being signs of difference and distinction'.

In contrast to the old cultural elites, the new cultural elite of the service class is based on a greater diversity of consumption, usually organised in globalised niche markets in which the major consumption spaces are metropolitan city centres. The service class is therefore also often seen as the vanguard of gentrification of inner city areas (Zukin, 1991). Munt (1994) also argues that the new middle class is responsible for the colonisation of rural landscapes as they search for a means of distinguishing themselves from the masses. Attempting to escape the rising tide of global tourism, they seek out the undisturbed beaches and 'tourist-free zones' (Munt 1994, p.115) which ensure the exclusion of the 'lager lout' and other emblems of mass tourism. What distinguishes the new middle class tourist above all is the value placed on their respect for the environment. 'A recurring theme in seeking spatial legitimisation is environmental and, to a lesser extent, cultural sensitivity'. The implication is that only those with sufficient cultural capital to properly appreciate such places should be allowed to colonise them.

The following sections of this chapter examine the ways in which culture interacts with the physical environment to produce commodities for tourism consumption in both urban and rural settings. The division between 'urban' and 'rural' made here is in essence artificial, because as Urry (1996) points out, there is no strict division between these spaces, but rather a continuum between two different modes of production. In the context of this chapter, however, a distinction is made between these two end members of the continuum in order to illustrate how the same processes are at work in both 'natural' and 'built' environments. This approach may also help to balance the over-emphasis on nature in studies of tourism and environment.

4. Urban cultures and tourism

As cultural tourism has arguably become a force for social and economic change, the powerhouses of this development have mainly emerged in large cities (Richards, 1996). As Urry (1996) points out, one of the distinguishing features of the urban environment is a high population density, which provides the economic basis for the means of 'collective consumption'. The aim of such collective consumption is frequently culture. As a consequence of economic restructuring from the 1970s onwards, 'culture is more and more the business of cities – the basis of their tourist attractions and the unique competitive edge' (Zukin, 1996, p. 2). In the economic and social chaos of modern urban life, culture and the cultural industries have stepped into the gap left by a retreating public sector, providing the 'vision' so sorely needed to create order out of chaos and stimulate new sources of economic growth. Zukin identifies the rise of a 'symbolic economy' which is a reaction to the economic decline of many cities, rising levels of financial speculation, the growth of cultural consumption and the advent of 'identity politics'. Not only do 'cultural institutions establish a competitive advantage over other cities for attracting new businesses and corporate elites' (1996, p.12), but they also stimulate the growth of 'art museums, boutiques, restaurants and other specialised sites of consumption which create a social space for the exchange of ideas on which businesses thrive' (1996, p.13).

The growth of the symbolic economy requires two parallel production systems: a production of space for cultural production and consumption, and additionally the production of symbols which give meaning and, therefore, add value to the spaces occupied. The raw materials for the symbolic economy are not just high cultural forms and symbols, but, increasingly, symbols of popular culture and nostalgia. The city is reformed in a 'connoisseurs view of the past' which 'reshapes the city's collective memory' to produce a simulacrum of a better past which never actually existed. This form of 'pacification by cappuccino' (Zukin, 1996, p.28) is found in all major cities in the developed world, and is fed by the competitive search for the consumption power of the mobile middle classes that constitute the bulk of the tourism market (Munt, 1994).

Cultural tourism, therefore, has a significant effect on the urban landscape and the way in which it is consumed. As has been illustrated in an analysis of cultural attractions and tourism flows in European cities (Richards, 1996), it is the established centres of 'high' culture, such as Paris, London, Rome, Venice and Madrid which continue to draw large numbers of tourists, in spite of the overcrowding and high prices which their fame also brings. Attempts to diversify the flow of tourists to 'new' attractions developed in old industrial centres such as Bradford, Glasgow, Antwerp or Bilbao have made few significant inroads into the dominance of the pre-industrial cities (Townsend, 1992; Van der Borg, 1996). The 'real cultural capital' which older cities have accrued over centuries of economic development is now paying economic dividends in the form of cultural tourism.

Alongside the historic city centres, which form the powerhouse of the European cultural tourism industry, a more differentiated postmodern landscape of tourist attractions is being created. Numerous authors have described the emergence of new features in the urban landscape, including 'cultural capital development complexes' (Britton, 1991), 'heritage centres' (Rojek, 1993), 'culture/entertainment complexes' (Ashworth and Tunbridge, 1990) and 'festival marketplaces' (Harvey, 1989). Such 'leisurescapes' vary in form, but are functionally convergent, being based on visual consumption and spectacle. Due to the 'waning of effect' experienced in postmodernity, such developments must become even more spectacular in order to draw the large audiences required for their survival.

The competitive drive to renew and innovate creates a reliance on events and festivals, which can provide a rapidly varying menu of sights and sensations. Particularly those cities lacking the competitive advantage of historically sedimented cultural capital resort to festival and event-based tourism strategies. In the Netherlands, for example, Rotterdam has developed an event-led tourism and arts policy designed to attract wealthy consumers from the 'new urban middle class' to the city. Brouwer (1993) describes how Rotterdam has tried to shake off its image as an industrial city by repositioning itself as a 'cultural festival city', with a particular emphasis on the applied arts, such as architecture, design and photography. A consequence of such developments is that urban spaces are utilised to attract and cater to the needs of an 'up-market' audience, which is targeted by policy-makers, event organisers and developers for its spending power. In this way public space is often converted to a commodified arena for cultural consumption.

These new environments not only shape the physical landscape of cities, but also influence the residents through the social and mechanical reproduction of culture which accompanies them. Local residents are asked to play 'roles', sometimes acting out their old jobs for tourist consumption (Hewison, 1987). Maintaining 'quaintness' may be essential for the tourist, but it can have far reaching consequences for those that have to live in 'quaint' surroundings. In Amsterdam, for example, the decision of the city council to declare the entire historical city centre as 'protected city landscape' means increased maintenance costs for local residents who are required to maintain the outward appearance of historic buildings. For private sector tenants, this can have far reaching implications, since landlords are exempted from rent controls in order to help them meet the costs of preservation. The cost of quaintness, therefore, falls on the local residents (and often the poorest residents), not just in economic terms, but also in terms of lack of double glazing, insulation and other modern comforts which conflict with the nostalgic view of the city.

The hegemony of traditional cultural centres is not complete, however. There is a growing recognition that the 'landscapes of power' developed through the symbolic production system can be challenged by the emergence of popular cultures. In Manchester, for example, O'Connor and Wynne (1992) show how a previously marginalised group of 'new cultural intermediaries' has started to use various popular cultural forms, such as house music, to penetrate the symbolic economy. Discos, clubs, music venues and other popular venues provide a means for young inner-city residents to translate their cultural capital regarding a wide range of popular art forms to economic advantage. The ability of such developments to attract tourists has been recognised in the development of Beatles Tours in Liverpool, and pop music tourism in Manchester. A new landscape of popular cultural venues is emerging in old manufacturing centres, such as Manchester (Northern Quarter), Liverpool (Performing Arts Institute) and Tilburg (the Pop Cluster). The nature of culture in urban centres has, therefore, shifted from being an amenity, symbolic of collective identity, towards being a tool used to shape images to provide commodities for consumer markets. The following section examines the extent to which these processes can also be identified in rural areas.

5. Rural landscapes and tourism

In the same way that urban landscapes are consciously shaped by human activity, in the case of 'natural' environments, the meanings of landscapes draw on the cultural codes of the society for which they were made (Aitcheson, 1996). Specific landscapes are the product of particular cultural practices. For example, the distinctive crofting landscapes in Scotland are based on small-scale landholdings and subsistence agriculture, creating a heroic image of the smallholders' struggle against the elements of nature. Such images are now the subject of tourism consumption, which the crofters gratefully exploit to supplement their incomes. The coincidence of many crofting tasks with the tourist season, however, makes the combination difficult. Tourism may, therefore, represent a threat to the very cultural practices on which the maintenance of this distinctive landscape is based (Macritchie, 1995).

Rural landscapes reflect not only the productive activities associated with agriculture, but also the cultural interpretations of the 'rural' which are associated with different cultures. This cultural formation of landscape has become even more important now that rural areas are an important site of tourism and leisure consumption. As Munt (1994) notes, the growth of rural tourism is a reflection of a middle class taste for 'authenticity' in consumption, related to the search for a lost rural past. This demand has, in turn, been met by a growth in 'real' country holidays (Swarbrooke, 1996), which are specifically designed to meet the needs of tourists in search of the authentic, 'off the beaten track' rural landscape.

Bailey (1996) charts the development of the rural environment in the UK through the interaction of economic restructuring and government intervention. In response to EU quotas and government calls to diversify, farmers have developed a wide range of new activities, including new forms of agricultural production, crafts production, leisure activities and tourism. In spite of the divergence of production processes involved, however, farmers were advised to preserve the 'traditional' form of farm buildings, preserving the aura of rural authenticity in spite of the change in function. A more extreme example of this reconstruction of the rural for tourism is provided by the Irish scheme to develop farm tourism. Guidelines from Bord Failte, the Irish Tourist Board, give explicit instructions to farmers on how to create a 'rural' atmosphere and look to their farms, in which the realities of modern production methods intrude on the desires of the tourists as little as possible (Carroll, 1995). Such 'simulacra' abound in the 'rural' environments consumed by tourists. This cultural construction of the countryside determines the consumption even of those who wish to escape from such 'inauthentic' environments.

As Rojek (1993) demonstrates, the countryside is also appropriated for tourism consumption through the creation of literary landscapes, or places depicted in fiction which are now being created in fact to whet the appetite of voracious readers and television viewers. These landscapes are classified by Rojek as 'escape areas', or places in which people can avoid the rising tide of meaningless which characterises late modernity, in the same way as theme parks or 'black spots'. Prentice (1996) shows how similar fictional landscapes have colonised real locations in France.

Such transformations of the rural environment are part of a general process of commodification and the associated production of spectacle for tourism consumption. Cloke (1993) links such developments in the UK to processes of privatisation and deregulation, which have provided more scope for the commercial sector to exploit rural areas. In developing the concept of a 'rural idyll' for tourist consumption, such developments constitute an 'identity-giving spectacle', in which nature appears only as a theme.

In the same way that Zukin identifies a symbolic economy in cities, the dialectic of space and symbols can, therefore, be found in the countryside. The growing demand for tourism and leisure, particularly in terms of 'real' experiences, leads to the creation of more facilities for rural tourism, including the creation of heritage centres, interpretation centres, gites, holiday centres, visitor farms, theme parks and golf courses. In the same way that central locations in cities vested with large amounts of 'real' cultural capital are desired locations for capital investment, so particular 'rural' locations also become sites of capital investment on the basis of their distinctive value. The Dutch holiday centre company Center Parcs, for example, stipulates the characteristics of the 'natural' environment in which it wants to locate its parks. The surroundings of the parks should be wooded, and afford opportunities for recreation. The tourism and leisure functions of companies such as Center Parcs begin to compete with agricultural production in terms of the return on investment, which can be obtained. As more areas of land are set aside from agricultural production, so the demand from farmers to develop alternative sources of income through tourism and leisure development will become still greater. There are, however, in-built limits to this process, given the distance of many rural areas from the urban centres which provide the major markets for rural tourism.

The addition of 'cultural' elements to 'natural' landscapes to increase their attractiveness to tourists is also becoming evident in areas which might be conceived of as 'wilderness', such as the polar regions of Scandinavia. In addition to the creation of Viking literary landscapes as a 'substitute of nature' (sic) in Norway (Viken, 1996), there is a 'postmodern Santa Claus industry' developing in Finnish Lapland, Sweden, Norway and Greenland (Pretes, 1995). Finnish Lapland has even decided to dub itself 'Santa Claus Land', laying claim to being the one and only original home of Santa Claus. The reason for this post-industrial boom is the fact that 'the cultural and natural advantages of Lapland were insufficient in attracting tourists in their desired numbers' (Pretes, 1995, p.8). If 'wild' nature is not sufficient, then simulacra must be provided to increase the 'edutainment' value of the original resource. Sternberg (1997) demonstrates that this cultural transformation of nature is evident even in cases where the natural resource might be considered a significant attraction in its own right, such as Niagara Falls.

The interaction between economic restructuring and increasing tourist demand for rural environments is therefore producing a series of new environments or 'leisurescapes' inhabited by tourists and their hosts. The shape of these environments is not determined so much by nature as it is, or was, but by ideas of how nature should be, or should have been. What Walt Disney achieved in his theme parks is now being replicated on a larger scale in rural areas through the staging of authenticity for tourism.

The concept of interaction between nature and human agency to produce specific landscape forms is now being more widely recognised by geographers in the concept of cultural landscapes. UNESCO, the Council of Europe and other bodies are recognising the value of agricultural areas as 'cultural landscapes' which reflect centuries of interaction between human agency and nature. In the Netherlands, Van Dockum et al. (1997) have analysed the value of different types of cultural landscapes as part of the international cultural heritage. The cultural construction of the landscape is particularly obvious in the Netherlands, where centuries of struggle to win land from the sea have created internationally important landscapes, e.g. the polders, country house landscapes *(buitenplaatsenlandschap)* and coastal dunes. Landscape is one of the most frequently cited reasons for tourists visiting the Netherlands, and is identified by 25 per cent of incoming tourists as their most important motive for visiting the country.

Similar processes of restructuring and commodification can, therefore, in many respects be identified in both rural and urban environments. Culture plays a key role in these processes, even in areas, which are regarded, largely, as 'natural'. In spite of this convergence, however, some essential differences remain, as the following analysis of cultural tourism consumption illustrates.

6. Urban and rural cultural tourism

Research by the European Association for Tourism and Leisure Education (ATLAS) has sought to define the relationships between tourism and cultural consumption in a range of different contexts in recent years. Surveys of visitors to cultural attractions have been conducted in both urban and rural settings and at a wide range of different cultural attractions and events (Richards, 1996). A comparison of surveys conducted at urban and rural sites provides some insight into the role of culture in tourism consumption in these environments.

Of the 6000 interviews conducted at 44 cultural sites in 8 European countries surveyed by ATLAS in 1997, about half could be classified as being located in 'rural' environments. Even though cultural tourists can be found in both urban and rural settings, there are clear differences in the type of tourists and tourism consumption between the two settings. The proportion of interviewees indicating that they 'normally' took cultural holidays was only slightly higher at urban sites (27%) than at rural sites (23%). When asked to classify the type of holiday being engaged in at the time of the interview, however, the proportion of cultural holidays was almost twice as high in urban settings (27%) as in rural areas (14%). This tends to suggest that urban areas are more closely associated with cultural tourism activities than rural areas. The acquisition of 'new experiences' was rated as the most important motive to visit cultural sites in urban settings, whereas relaxation was rated far more important at rural sites. Urban cultural tourists were far more likely to indicate that their visit was related to their work in the cultural sector, as the level of cultural employment was almost twice as high for urban sites. The urban respondents were also more likely to be under 50 years of age, to be professionally employed and to have a higher education qualification than their rural counterparts.

Tourists surveyed at rural sites exhibited a significantly lower frequency of visits to cultural attractions. Rural tourists visited less than 3 attractions per day, compared with almost 3.5 for urban tourists. Rural locations were significantly more likely to be characterised by visits to monuments and historic houses, and less likely to be focused on museums, galleries, performing arts events and festivals than in urban areas. Rural respondents also indicated that their travel had been stimulated by one specific cultural attraction, whereas urban respondents were more likely to have a range of different reasons to travel, reflecting the concentration of different types of attractions and 'real cultural capital' in urban areas.

The picture which seems to emerge from this analysis is that cultural tourism consumption in urban areas is more characteristic of the 'new cultural intermediaries', or 'taste-makers' who exhibit high levels of cultural capital, high level of cultural consumption and high levels of cultural employment. Rural cultural tourism, on the other hand, seems more characteristic of the 'new bourgeoisie', a class faction high on both economic and cultural capital. The ATLAS research, therefore, seems to support the conclusion of Munt (1994) that these class factions become spatially differentiated in their search for distinctive forms of tourism consumption. The 'new bourgeoisie', in particular, are identified with the consumption of 'eco-tourism', and arguably use nature conservation as a means of spatial legitimation. Rural cultural tourism, therefore, seems to display an integration of 'cultural' and 'natural' elements of tourism consumption, which makes it increasingly difficult to distinguish between categories such as 'cultural' and 'rural' tourism. Rural areas continue to have a strong attraction for cultural tourists, in spite of the apparent lack of 'real cultural capital' in the terms of Zukin (1992). Perhaps the greatest attraction for the rural cultural tourist is the 'living culture', which is now beginning to be packaged for tourist consumption in rural areas.

7. Conclusions

The main point that this chapter has attempted to make is that nature cannot be viewed in isolation from culture in the analysis of tourism production and consumption. The 'demotion' of nature from a designation of 'All' to a concept synonymous with countryside, landscape or scenery (Tuan, 1974) has arguably been prolonged through the transformation of nature into a commodity for tourist consumption. Nature is increasingly constructed or enhanced by a wide range of producers and intermediaries, attempting to meet consumer demand for 'authentic' experiences of nature. Paradoxically, this demand for idealised rural and natural experiences threatens to undermine the character of these areas, as 'traditional' productive practices are replaced by new forms of economic activity geared to tourist consumption. The cultural processes of rural life are replaced by cultural products destined for tourist consumption.

Concepts of sustainable tourism, therefore, have to include cultural sustainability (Bramwell et al, 1996), since the moment a given culture becomes unsustainable, so does the 'natural' landscape on which it is based. The landscape, therefore, changes from a 'living' environment into an artefact, which has to be preserved in a vast museum. Just as the cultural artefacts of the rural environment began to be preserved in open air museums at the turn of the 19th century, so the turn of the 20th century is seeing the growth of national parks and other forms of preserved landscapes, the finance for which is increasingly coming from the visitors – tourists.

As soon as the historic urban landscape, or the quaint rural landscape, has been demarcated as desirable, a process of transformation is set in motion, which generates an idealised product for tourist consumption. 'Authentic' natural or cultural landscapes, therefore, effectively disappear at the moment of their identification, either through the process of designation, which freezes the landscape in time, and dislocates local culture and landscape, or through the process of emergent authenticity. The 'authentic' landscape is no longer 'real', but a simulacrum - an improved copy of the original landscape.

The relationship between culture and nature, in the context of tourism, is, therefore, far from simple. 'Natural' landscapes are not only altered and reproduced by culture, but natural and cultural landscapes are, also, increasingly influenced by the culture of tourism (Urry, 1990). The distinction between 'urban' and 'rural' tourism cultures may be preserved through spatial struggles between different class factions, but there is increasing structural convergence between the two. 'Natural' environments and cultural monuments increasingly owe their existence to the tourists who provide the money required preserving them. Both are likely to remain the subject of cultural struggles – between cultural producers and intermediaries, between tourists and other consumers, and between the tourists themselves.

References

Aitcheson, C. (1996) Gendered tourist spaces and places: the masuclinisation and militarisation of Scotland's heritage. *LSA Newsletter* **45**, 16–23.

Ashworth, G.J. and J.E. Tunbridge (1990) *The Tourist-Historic City.* Belhaven Press, London.

Bailey, C. (1996) England, whose England? The fashioning of a countryside fit for tourists. In M. Robinson, N. Evans and P. Callaghan (eds) *Tourism and Culture: Image, Identity and Marketing.* Business Education Publishers, Sunderland, pp. 1–14.

Bauer, M. (1996) Cultural tourism in France. In G. Richards (ed.), *Cultural Tourism in Europe.* CAB International, Wallingford, pp. 147–164.

Borg, J., van der (1994) Demand for city tourism in Europe. *Annals of Tourism Research* **21**, 832–833.

Britton, S. (1991) Tourism, capital and place: towards a critical geography of tourism *Environment and Planning D: Society and Space* **9**, 451–478.

Brouwer, R. (1993) Het nieuwe Rotterdam: de kunst, het beleid, de zorg en de markt (The new Rotterdam: Arts, policy, care and market). *Vrijetijd en Samenleving* **11**, 31–47.

Carroll, C. (1995) Tourism: Cultural Construction of the Countryside. MA Thesis, Programme in European Leisure Studies, Tilburg University.

Cloke, P. (1993) The countryside as commodity: new rural spaces for leisure. In S. Glyptis (ed.), *Leisure and the Environment.* Belhaven, London, pp. 53–67.

Cohen, E. (1979) A phenomenology of tourist experiences. *Sociology* **13**, 179–202.

Cohen, E. (1988) Authenticity and commoditization in tourism. *Annals of Tourism Research* **15**, 467–486.

Duim, R. van der, and J. Philipsen (1996) Recreatie, toerisme en natuurbescherming tussen romantiek, ecologie en commercie (Recreation, tourism and nature protection between romantic, ecology and commercial interests). *Vrijetijd en samenleving* **13**:(2, 21–40.

Dockum, S. van, S. van Lochem, D. van Marrewijk, H. Renes, R. Smouter, and K. van der Wielen (1997) Nederlandse landschappen van wereldformaat (Dutch world fame landscapes). *Geografie* **6**:1, 24–29.

Droste, B. von, H. Plachter, and M. Rössler (eds) (1995) *Cultural Landscapes of Universal Value.* Fischer, Jena/ Stuttgart.

Featherstone, M. (1991) *Consumer Culture and Postmodernism.* Sage, London.

Harvey, D. (1989) *The Condition of Postmodernity.* Basil Blackwell, Oxford.

Herbert, D.,T. (1996) Artistic and literary places in France as tourist attractions. *Tourism Management* **17**, 77–85.

MacCannell, D. (1976) *The Tourist: a New Theory of the Leisure Class.* Macmillan, London.

Macritiche, C. (1995) The conflicting demands of the preservation of the cultural heritage and the development of tourism and leisure in the Western Isles. In D. Leslie (ed.), *Tourism and Leisure - Culture, Heritage and Participation.* LSA Publication no 51, pp.121–128.

Milestone, K. (1992) Popular music, place and travel. Paper presented at the conference 'Internationalisation and Leisure Research'. Tilburg University.

Munt, I. (1994) The 'other' postmodern tourism: culture, travel and the new middle classes. *Theory, Culture and Society* **11**, 101–123.

O'Connor, J. and D. Wynne (1993) *From the Margins to the Centre.* Manchester Institute for Popular Culture, Manchester.

Pretes, M. (1995) Postmodern tourism: the Santa Claus industry. *Annals of Tourism Research* **22**, 1–15.

Richards, G. (ed.) (1996) *Cultural Tourism in Europe.* CAB International, Wallingford.

Rojek, C. (1993) *Ways of Escape: Modern Transformations in Leisure and Travel.* Macmillan, Basingstoke.

Sternberg, E. (1997) The iconography of the tourism experience. *Annals of Tourism Research* **24**, 951–969.

Swarbrooke, J. (1996) Towards the development of sustainable rural tourism in Eastern Europe. In G. Richards (ed.), *Tourism in Central and Eastern Europe: Educating for Quality.* Tilburg University Press, Tilburg, pp. 137–163.

Towner, J. (1985) The Grand Tour: a key phase in the history of tourism. *Annals of Tourism Research* **12**, 297–333.

Townsend, A.R. (1992) The attractions of urban areas. *Tourism Recreation Research* **17**, 24–32.

Tuan, Yi-Fu (1974) *Topophilia: a Study of Environmental Perception, Attitudes and Values.* Prentice Hall, Eaglewood Cliffs, N.J.

Urry, J. (1990) *The Tourist Gaze: Leisure and Travel in Contemporary Societies.* Sage, London.

Urry, J. (1992) The Tourist Gaze and the 'Environment'. *Theory, Culture & Society* **9**:3, 1–26.

Urry, J. (1996) *Consuming Places.* Routledge, London.

Viken, A. (1996) Tourist attractions as an industrialized experience and substitute of nature. In M. Robinson, N. Evans and P. Callaghan (eds) *Tourism and Cultural Change.* Business Education Publishers, Sunderland, pp. 295–310.

Walsh, K. (1991) *The Representation of the Past: Museums and Heritage in the Post-modern World.* Routledge, London.

Zukin, S. (1991) *Landscapes of Power: from Detroit to Disney World.* University of California Press, Berkeley.

Zukin, S. (1995) *The Cultures of Cities.* Blackwell, Oxford.

LANDSCAPES OF TOURISM:
A CULTURAL GEOGRAPHIC PERSPECTIVE

THEANO TERKENLI
Department of Geography
University of the Aegean
Karantoni 17, Mytilini, Lesvos 81100
Greece

1. Introduction

In recent years, there has been a growing recognition in social scientific circles and in the humanities of the contextualised, positional character of all forms of knowledge and experience (Norton, 1996). Such recognition, however, has not as yet been followed up by adequate empirical research in order to operationalise and substantiates the theoretical propositions put forth. In practice, this has meant that there is to-date no substantial understanding of cultural mechanisms of change. In the ever-growing interdisciplinary dialogue on space and the environment, meanwhile, contemporary geography seems to have an increasingly significant contribution in the specification, in its fullest sense, of the context of human life and activity. Within this context, the relevance of the cultural in spatial analysis is becoming increasingly compelling, since at the basis of any human-environment interrelationship lie ideologically and symbolically charged conceptions of space can be found. Such conceptions grow out of humanity's quest for meaning and identity and obviously point to the centrality of the social and the cultural in the articulation of space through time.

By use of cultural geographic theory and analytical approach, this chapter examines change in one such cultural medium, namely landscapes created for and by tourism. The attempt to investigate impacts and images of tourism in the human landscape originates, in part, in one of contemporary geography's central objectives, the need to re-assess the role of culture in various sorts of time-space transformation, presently occurring in the human habitat. It also addresses the urgency in contemporary social science to investigate new cultural structures and functions, products of a rapidly changing world geography: products, in great part, of the manifold expansion of the social phenomenon of tourism.

Particularity and change in the forms and structures of tourism are produced by different social, economic and cultural influences filtered through the context of the theoretical approach to their study. Undoubtedly, with the advent of postmodernism and the accompanying onslaught of larger cultural changes in western societies, tourism is adopting new patterns and tendencies. For most social theoretical analysts, tourism has become a commodity to be consumed. The "objects of tourism" in modern times become representations and are commodified and packaged for ready consumption, as reality gives way to representation and "history, time and space, as aspects of culture, become commodities" (Pretes, 1995). Culture may, indeed, be viewed as an economic factor, like any other product (Dietvorst, 1994), as it constitutes a critical factor of location for economic and other human activities. More important, though, it represents collective ways of life, shaped and passed down from generation to generation through the centuries. A recent turn in tourism planning is towards the acknowledgement of the organic place of culture in all manifestations of life, and towards a transformation of public space into representations of the local cultural identity, for purposes of tourist consumption. Among such representations, geographical uniqueness manifested in local ways of life and a distinctive 'spirit of place' have become especially prominent with the advent of so-called 'post-tourism'. We place this chapter, then, in current efforts to assess the extent to which tourism contexts have been informed by the essential 'nature' of the tourist phenomenon, as opposed to the social and cultural histories and geographies within which they have arisen.

The first section of the article will negotiate this larger question by probing into theoretical aspects of the 'nature' of tourism and by addressing ways to analyse cultural sides of the phenomenon from a geographical perspective. The article suggests an analytical framework for tourism research, namely the study of landscapes of tourism. This section of the article specifically focuses on distinctive characteristics of tourist landscapes as the cultural battling grounds on which much of today's socio-cultural difference is increasingly created and development negotiated. The landscape, cultural by definition (Meinig, 1979), is selected as an interface of contact between various strands of tourism study, including the tourist and the local perspectives. To this end, this article cannot be seen as a comprehensive attempt to cover the entire multidisciplinary fields engaged in these aspects of tourism research. It simply purports to highlight a geographical way of studying cultural dimensions of tourism, by bringing together various significant aspects of this social phenomenon in their proper historical and environmental setting.

The second part consists of a synthetic reading of selected modern and post-modern tourist landscapes in the history of the western world, in order to trace the evolution of the inter-relationships suggested above, as well as to illustrate the relevance of cultural factors in historical tourist space construction and perception –and vice versa. The third part of the article engages in an architectural/ townscape and behavioural study of tourist settings/ landscapes on the island of Serifos, Greece in light of the above theoretical propositions. Concrete manifestations of tourism-imparted changes are analysed with the aid of empirical survey data, borrowing from cultural and behavioural ecological perspectives. In this way, the article explores dimensions and meanings of tourist landscape change, as tangible manifestations of larger cultural transformations already at play in the western world. Through this brief landscape evaluation, this last section will begin to investigate the impact of tourism as a homogenising force on place authenticity in the Aegean.

2. The cultural landscape revisited: a framework for tourism analysis

Forms and structures of tourism are intricately woven with larger social and cultural schemata in space and time, in the broader context of human experience. In order for landscapes of tourism to be explored in a concerted and coherent manner, the phenomenon of tourism itself must be placed in its larger social context; in other words, it must be defined and delineated in relationship to the other realms of social life. Urry (1994) specifically suggests that identity is formed through consumption and play, and not through work, whether in the factory or at home.

Our position is rather that consumption patterns, and thus identity formation, cannot be shaped in isolation from the larger context and other realms of life, but only in a dynamic interrelation with larger ongoing transformation in all personal and collective spheres of activity. The fact that tourism is widely characterised as a leisure activity (Stebbins, 1997; Moore et al. 1995) presupposes its opposites, namely work and home life. Indeed, the most substantive dichotomy on which the development of tourism occurred was the dichotomy between the familiar and the extraordinary or the exotic. For the traveller, the essence of tourism became being away from home (Urry, 1990; Jakle, 1987), in order to find relief from the pressures of modern life, but also in order to collect new, mostly visual, experiences which will elevate his or her social image in terms of 'cultural capital'. The cultural realm, of course, has its own logic, currency and rate of convertibility into economic and social capital. Specifically, cultural capital is not just a matter of abstract theoretical knowledge, but is defined as the symbolic competence necessary to appreciate not only 'art' or 'anti-art', but any sight as well and thus also points of tourist interest. Differential access to the means of tourism consumption is then crucial to the reproduction and cultural articulation of class – and social relations more generally.

To place our discussion in the larger context of the exploration of the essential «nature» of tourism, one of tourism's distinctive properties is its definition in contradistinction to home and work, delineations highly geographical in nature. A second characteristic refers to its functioning mechanism, which is "travelling from a 'generating region' to a 'destination area' via 'transit routes'" (Moore et al. 1995). Obviously, concepts of space and place, movement patterns in space and spatial systems of mobility occupy a central place in the definition of tourist behaviour, rendering a spatial and geographical understanding of tourism imperative. Third, as tourism exploded in the twentieth century through the alleged globalisation of relations of the 'New World Order', it assumed certain distinctive attributes, such as superficiality, uniformity, intrusiveness and voyeurism into the supposed «real» lives of the Other, and the tendency to stage culture for the purpose of tourist consumption, resulting into a sacralisation of all sites (Urry, 1990).

Among these tendencies, visual character predominates and forms a connecting thread. Tourism has historically always been concerned with the visual (Urry, 1990; Pretes, 1994), though in modern times, as subsequent analysis illustrates, the visual objects of tourism seem to be assuming additionally properties of the spectacular, and thus the artificial. Fourth, the study of tourism, besides its fundamentally geographical nature, must necessarily rest on the common ground of various areas of inquiry. To a large extent, this research interface is cultural, since culture is a most significant underlying factor of spatial differentiation, one that best addresses its broad multidisciplinary nature. A brief comprehensive examination of basic current geographical and other social scientific lines of inquiry in tourism research is thus apt at this point.

While a wide range of social scientific perspectives has been employed in the study of tourism, most of these treatments have tended to approach tourism from two-directional viewpoints. Without necessarily subscribing to such dualistic views of tourism issues, we may nonetheless borrow insights on the complexity and conflictual nature of tourism from such analyses. On the one hand, are what Dann and Cohen (1991) refer to as 'conflict and critical perspectives', emphasising the negative consequences that sometimes result from tourist development (Moore *et al.*, 1995). On the other hand, there are the 'neo-Durkheimian' perspectives that depict tourism as a modern pilgrimage, a means of overcoming the fragments of modern life by combining them into a unified tourist experience (Moore *et al.*, 1995; MacCannell, 1992). These two directions of approach roughly correspond with the more exploitative and the more organic/holistic aspects of tourism, respectively. They need not necessarily be in conflict with each other. They may embrace each other where the tourist preferences and the local needs and ways of life come together in the creation of a distinctive spirit of place, to serve both as a point of tourist interest and appeal and also a context of everyday life, a home for the local population. If equal importance is given to the personal/emotional as to the visual/commodified dimensions of human activity contexts, "in which ideally there should be no difference in quality between the tourist's and the local's experiences of delight at the local street scene" (Soane, 1994), a sense of place identity emerges, one that caters to both sides' needs. As a case in point, concepts relating to 'place identity' are becoming increasingly relevant with the alleged advent of 'post-tourism', partly, as we shall see below, through the return of the 'new vernacular'.

The preceding overview of distinctive features and analytical approaches pertaining to the study of tourism point, we propose, to the critical place in tourism analysis of the concept of 'place identity' as illustrated in the form and function of the cultural landscape. The model of Figure 1 diagrammatically formulates this argument. All recent effort in tourism analysis and planning, we suggest, essentially revolves around the make-up, assessment, planning or preservation of cultural landscapes for purposes of sustainable, profitable, or unobtrusive tourism. Cultural landscapes are routinely gaged in terms of authenticity (Teo and Yeoh, 1996; MacCannell, 1992), even if not explicitly stated as such. On the one hand (left loop of the diagram), tourism marketing managers reproduce discourses about places through the representations of cultural signs, with which the tourist, through a process of experiential re-interpretation of the sign, may assess the authenticity of the sight and validate the meaning of the sight within the discourse. On the other hand, (right loop of the diagram), local ways of life unfold and imbue places with distinctive geographical 'identities', which become the media and the embodiment of cultural change. As easily appropriated through the landscape's visual character, cultural meanings are defined and expressed in place identity formation.

At the concept of the landscape, as the model of Figure 1 suggests, come together home and work realms for the locals on the one hand and sites of leisure for the tourists on the other. Such a tourism research framework not only serves as an analytical tool of the global-local nexus properties of tourism. It also brings together demand and supply on the one hand versus a context of home for the local population on the other, as well as ideas of use value and exchange value of place. First and foremost, it stems from the definition of a landscape as a reflection of place identity, apprehended mostly visually both by the local and by the visitor. As an analytical tool, it complements not only Johnson's 'tourism and the circuit of culture model' (Norton, 1996), but also informs and substantiates the interface surface of Lanfant's 'tourism and culture cycle model' (Shaw and Williams, 1994). The remainder of the article will attempt to substantiate and justify the choice of the cultural landscape as a tool in tourism analysis, by elaborating on the comprehensiveness of the experiential and phenomenological nature of the tourist landscape in a specific geographical case study, focusing on two quintessential properties of modern tourism: a) the predominance of the visual/ artificial and b) the home-leisure dichotomy inherent in its definition.

Landscapes of tourism are apt research mediums in the polysomic and inter-textual decoding that necessarily constitutes tourism analysis. The discipline of geography, dealing by definition with the multi-dimensionality and complexity of various types of spatial concepts and constructs, has proven since the 1920's to constitute especially fertile grounds for the study of the landscape. In the current integration of human geographic and social theory, an emerging north American social geography and a rejuvenated British social geography are joining more traditional cultural geographic pursuits in forming close links to social theory and identifying cultural geographic concerns and cultural change. Much of the recent social geography is explicitly derived from the landscape tradition, with culture being conceived as social practice and landscape being conceived as place (Norton, 1989). But landscape is one step closer to human experience than the more abstract concept of place: it is the concrete backdrop of life; space or place as conceived and appropriated through the senses and power of cognition.

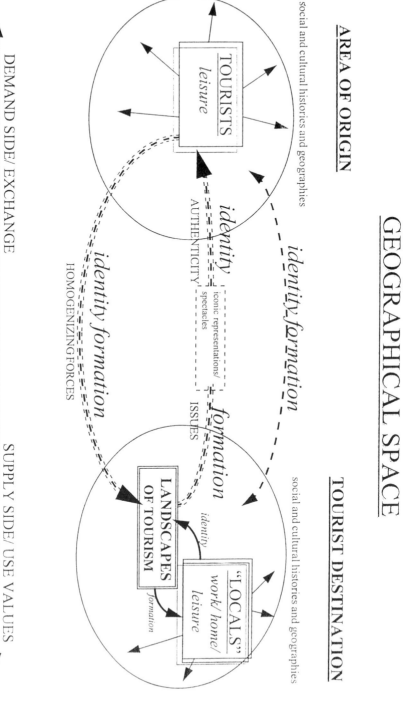

Figure 1. Landscapes of tourism model

Recently, besides a resurgence of interest in the general landscape, the urgency for change in attitudes towards it, as well as a preoccupation with its essentially cultural nature, have been evident in various contexts (Hagerstrand in Germundsson and Riddersporre, 1996; Jones and Daugstad, 1996). Among these have been environmental consultants and researchers, academics, administrators and various sorts of managers, the mass media, and others: "One effect of the aspiration to integrate natural and cultural values in preservation contexts has been that modern preservation work is decidedly oriented to areas instead of to objects.... An alternative approach ...could be to start from a landscape totality and then study the various parts on the basis of their relevance for the whole" (Germundsson and Riddersporre, 1996). From the perspective of professional practice in planning and development or other top-down intervention in space, as well, "the content and character of the landscape have hitherto not generally been seen in relation to regional development or phasing out. Qualities in the landscape and their relationship to social planning are matters that in actual fact we do not know so very much about" (Sporrong, 1996). Nonetheless, they represent an area of increasing convergence in interdepartmental interest. As a matter of practice, however, planners and developers used the term landscape indiscriminately or interchangeably with the term environment. It has been routinely used with a clear leaning exclusively towards its material and tangible aspects and with a lack of theoretical substantiation, merely as a tool of environmental policy formation and implementation – both as affects tourism and local livelihood patterns.

From a geographical perspective, the landscape is such a spatial whole, which exists in historical time and thus exhibits a unique place identity, but it also exists as a system of energy, material and information flows interwoven in real, perceived and symbolic ways. Landscapes represent material constructions which are reflective of the basic organisation of society and economy, and thus they may be read rather like texts (Meinig, 1979). In fact, they constitute inter-textual sites which support unquestioned assumptions about the organisation of society and culture through the naturalisation of particular readings from particular positions (Norton, 1996). Contemporary cultural geography attempts such inter-textual readings of landscapes as three-dimensional realities that stage of our lives. The humanistic tradition in geography also views landscapes as synthetic realities, created through and understood by inter-subjective social and not biological processes and ever-transforming in space-time, reflecting all change in society. In this way, they constitute essentially cultural entities.

Cultural landscapes are selected for the purposes of this study, as the most tangible manifestations of culture in the geographical environment. It is in its immediacy, intertextuality and experiential character that the construct and concept of the cultural landscape is especially apt and useful in post-tourism analysis and consequently in further building on the theory of tourism by advancing our knowledge on its relationship to everyday life. Due especially to its experiential character, the landscape becomes a social interface where local and global perspectives and other dimensions of tourism studies come together in the ready construction and consumption of place identity. In the present-day landscape's composition, image and interpretations, the spatial, the cultural, the visual and the element of consumption figure prominently. These qualities of cultural landscapes have historically been central to the social phenomenon of tourism, as the following section illustrates.

3. Tourist landscapes through time: western cultural formulations

About four hundred years ago, the motive for travelling ceased to be for the sake of religious pilgrimage or establishment of new markets and trading connections. In Montaigne's words, the motive became a desire to "report on the temperament of nations and their ways of life" (Montaigne in Jackson, 1980), a truly geographical motive. Montaigne, the first perhaps to speculate on why people feel the urge to travel (Jackson, 1980), spoke of the world "as a mirror where we must see ourselves in order to know ourselves". These, not surprisingly, happen to be exactly the terms in which landscape analysts, geographers and other social scientists of our days refer to the cultural landscape. Early tourism, of course, was confined to a small though influential class. With increasing self-awareness, sixteenth-century upper classes were motivated to travel with the purpose of eyewitness observation, a scientific and essentially geographical motive, a quest to see the world in order to learn about themselves. And the manner in which they depicted their adventures, either in art or in writing, was so vivid, so compelling and so revealing, according to Jackson (1980), that subsequent generations accepted their view of the landscape as the only authentic one. Until then, geographical contact with the less known world of the time existed in various other ways, such as geographical exploration and expeditions, trade, war, even 'scientific' pursuits.

This classical form of tourism flourished, according to Jackson (1980), until after World War I, and only within the last decades has the Renaissance canon of landscape beauty been seriously challenged. More significantly, this way of seeing the world has left an enormous imprint on the human landscape: not only on urban, suburban, park and highway design and planning; the development of the discipline of landscape architecture itself was based on its principles. The Industrial Revolution opened up this manner of viewing the world to the wide public. With the spread of industrialisation, urbanisation and the emergence of new forms of capitalism, the world came to be known to the Europeans through travelling as a means for the discursive recreation of opposites and others, thereby leading to "an understanding of places, nature, ourselves and mundane social life" (Urry, 1990). The differences between places were very important to the fabrication of identity, which occurs through the "universal practice of designating in one's mind a familiar place which is «ours» and an unfamiliar space beyond 'ours' which is 'theirs'"(Said, 1979). In other words, it was also important in forming a sense of home not only for the travellers, but also for the local populations of tourist destinations.

Inherent in the shift towards the forms of tourism as we know them today was the development of its visual character. 'The tourist gaze', assisted, of course, by the growth of the guidebook industry, promoted new ways of seeing (Urry, 1990). Its growth was stimulated by a nascent European press, thirsty for new horizons of thrills, such as was the discovery of new worlds. Nineteenth-century writer-explorers of Africa constructed representations of Africa for Europeans, opening it up visually and conceptually for their readers as they opened it up economically for their governments. Such descriptions not only partook of the European aesthetic codes favouring panoramic views of Africa, but also portrayed a vast country, rich in resources and empty of people, at the ready disposal of the European visitor (Duncan, 1993). The discipline of ethnography, which arose during the colonial period, played an important role in providing the taxonomic framework of colonial representation, a form of representation serving the double appropriation of the non-European world, both of their material resources and of their identities – which were then substituted with mythical identities (Duncan, 1993). One such form of mythical identity construction that catered to the needs and fantasies of the European public was in the form of exoticism, a major incentive for tourism all along. The cultural landscape, in its visual and experiential character, lends itself especially well in the inscription and communication of the exotic element in tourism analysis.

In the next century and a half, a democratisation of tourism occurred, where all social classes began to travel for reasons unrelated to work or business, aided by the enormously popular new device of seeing, photography. Photography emerged as a socially constructed way of seeing and recording, but also of rearranging space, characterised by some of the following tendencies. Photography is a means of transcribing reality, but more accurately, in some ways, seems to appropriate the object being photographed – in other words, it is a power/knowledge relationship. In fact, photographs are the outcome of an intentional signifying practice in which those taking the photo select, structure and shape the photographed, normally idealising or aestheticising it in the process (Jussim and Linquist-Cock, 1985). The tourist is interested in everything as a sign of itself, as its image. When the tourists see two people kissing in the Cyclades or in Paris, what they capture in the eyes of their mind or on their camera is 'timeless' romance. In this way, everyone becomes an amateur semiotician who collects signs of the cultures visited, and every landscape or other sight thus becomes consumed and trivialised. Travel becomes the strategy for the accumulation of photographs, so that the travellers may convincingly demonstrate upon return that they really have consumed this one more place, by showing their version of its images.

We have here the beginnings of the 'modern gaze' (Urry, 1990), which is one way of seeing and being seen and which represents a planning philosophy rooted in Renaissance travel and landscape depiction, with a profound impact on the construction of modern space. This philosophy, for example, engendered the massive rebuilding of Paris by Haussmann with boulevards that not only facilitated rapid troop movements but also created sweeping vistas where people, indeed every citizen, could see well into the distance and be seen by everybody else. Berman describes the way in which the Parisian boulevards and cafes created a new kind of space, especially one where lovers could be private in public, intimately together without being physically alone. In Vienna, a more functional urban plan prevailed in the form of Otto Wagner's Ringstraße over Camillo Sitte's humanistic community-oriented urban planning. This development, too, was simply another vision of the new, modern use of space (axial landscape), which facilitated the visual appropriation of the modern city to the car or train passenger (Schorske, 1967). These new ways of spatial planning and seeing had already been established in the New World with great emphasis on principles of horizontality and verticality, as in the rectilinear grid of American town and city planning (panoramic landscape) (Jakle, 1987). Later, postmodernity, in its turn, re-introduced or placed higher emphasis on ephemeral landscapes, detail landscapes and landscapes of characteristic forms.

Indeed, it was the renegotiation of boundaries in the landscape, visual boundaries in this case, as well as the possibilities of communication and transportation for the totality of the population that defined a social construction of space based on a broadening of the visual potential. In turn, this newly emergent modern urban space transformed social experience and gave rise to new cultural ways, not just through traffic and contact of people and animals of burden. It permitted privacy amidst the public chaos, thus creating the quintessential romantic setting of modernity; it fostered 'la vie Parisienne', a world-renowned lifestyle expressed in the outlook of Impressionist painters; it created the 'flanneur', that is the stroller (and forerunner of the 20th century tourist), able to move about unnoticed, observing and being observed, but never really interacting with those encountered. It also set the basis for one sort of tourism that developed later, in tandem with the return of the neo-vernacular, often named 'countryside tourism', and which revolves around an experiential relationship with space and landscape, characterised by principles of romance (Urry, 1990).

As tourism increasingly came to consume en masse, the modern way of seeing was complemented by the attributes presented above: superficiality, homogeneity, visual intrusion and the staging of culture as a tourist spectacle. What results is a loss of a sense of history and geography and the emergence of the three-minute culture, whereby the sight-seer, bombarded by the kitsch or intense and dramatic stimuli tends to stereotype the landscape and loses, in time, the ability to appreciate anything more ordinary and common. He or she becomes an insatiable consumer of visual stimuli, strolling among signs of different cultures. Reality becomes a spectacle, in order to fulfil the demands of the tourist who views the surroundings through mental frames, akin to the postcard, the aperture of the camera, the tv-screen, or the coach window. Accordingly, framed landscapes both real and metaphorical (as in strategic placing of arcades and passageways in many European cities and in the name of San Francisco – 'The Golden Gate') became common practice in architecture, landscape architecture and town planning.

Increasingly, meanwhile, the drive for tourist development implies a transformation of sights into spectacles. Whereas sights are experienced mainly visually, and are apprehended as solidity and temporal transcendence, the experience of spectacles is both differentiated and temporally bounded (MacCannell, 1992), in other words, made more palatable for purposes of modern mass tourism consumption. "The difference", MacCannell (1992) writes, "between sights and spectacles is determined by differences in the manner of staging them. Spectacles are bounded in time as humans lives are, and their structure and pace conforms to and reproduces the contours of emotions. The link between spectacular action and emotional response is direct". Spectacles for contemporary tourist consumption additionally take on an attribute mentioned before; they become iconic representations of reality, based on similitude or likeness – another distinctive tendency of the changing cultural makeup of our societies. The tourist, by definition, unable to penetrate into any underlying realities, aims at a superficial experience with place: the modern tourist consumes images or representations of a society, and especially its culture and history. In Lowenthal's (1985) words, "if the past is a foreign country, nostalgia has made it the foreign country with the healthiest tourist trade of all."

Such symbolic accumulation or representation of tourist landscapes either through their visual recording and reproduction or through their self-conscious preservation and planning, combined with learned intersubjective ways of seeing promoted in each culture, contribute to new reformulations of cultural and tourist landscapes and a re-informing of culture itself. As far as tourist landscapes are concerned, travel destinations are chosen from a newly re-enforced sense of 'us' and 'home' on the basis of anticipation, especially through knowledge and fantasy, of intense pleasures either on a different scale or involving different senses from those customarily encountered at home. Such anticipation is constructed and sustained, but also planned through a variety of non-tourist and tourist-oriented practices, such as film, tv, travel and other literature, magazines, audio-visual media, historic preservation measures and tourism-catering professions (Johnson in Norton, 1996). As far as the supply side of tourism is concerned, selective screening, visual appropriation and commodification of tourist landscapes has consequently come under attack from many sides, accompanied by a call for greater weight on local identity and difference, what has been coined 'neo-vernaculism' (Soane, 1994).

A backlash to the homogenising forces of modern tourism's mass-consuming eye and the ensuing placelessness, a new strand of tourism, 'post-tourism' (Feifer: in Urry, 1990), seeks geographical uniqueness. Meanwhile, the weakened collective powers of the working class and the heightened powers of the service and other middle classes have generated, according to Urry (1990), a widespread audience for postmodern cultural forms, particularly as these refer to post-tourism. Feifer describes the profile of today's post-tourist. First, the post-tourist experiences named scenes through a frame, such as the hotel window, the car windshield or the window of the coach, but these can now be experienced at home on tv or video at the flick of a switch, and in this way the distinctiveness and the authenticity of the 'tourist gaze' is lost. In other words, the visual element in tourism continues to constitute the crux of the experience, but it is increasingly staged, made into a spectacle. Second, the post tourist pursues change in the tourist experience and delights in the multitude of choice, moving with ease from one 'delight' to another and gaining pleasure from the contrasts between them: for example, (s)he seeks something now sacred, now informative, now beautiful, now just different. In other words, (s)he seeks varied and full experiences that simulate life, only compressed in a geographical scale that is more accessible to the purposes of tourism. The dialectic home-non-home/ tourist destination that powers the tourism mechanism implodes in space and time and is multiplied in practice ad infinitum, with the creation of rapid successions of alternating familiarity and foreignness in the landscapes of tourism. And third, the post-tourist is aware of the fact that (s)he is a tourist and that tourism is a series of staged pleasure games with multiple texts and no single, 'authentic' tourist experience. In other words, there is realisation of the artificiality of the experience, or a clear understanding of the 'authentic' in the landscape as a context of life. The icon is broken-up, demystified and transcended. The post-tourist is above all well-informed, sophisticated, demanding new out-of-the-ordinary experiences and seeking immediate pleasure rather than interested in holidays that reinforce collective memories.

Similar forces, however, as those that power contemporary tourism, make the satisfaction of post-tourist demands harder in our days and relate to another front of cultural change sweeping the home domains of everyday life. Under the contemporary economic, social and cultural order, with the increasing commodification of resources and human associations, individual identity takes precedence over group identity. Individual ways of life centred around the self and personal habits structure everyday notions of home and of the non-home realm, work, leisure and so on. Having first created a well-established dichotomy between home and workplace, further processes of modernisation and global capitalism entrenched a segregation between leisure and work, as well as leisure and home life. The dichotomy of home-non-home that the Industrial Revolution introduced in western societies is no longer congruent with the dichotomy of private and public: an association that historically served to delineate the ideas of home and the non-home (including work and leisure). If the frontiers between home and the non-home domain of everyday life have become sharper, they do not exclude reciprocal influences. Public life invades our homes, for instance, with the proliferation of mass media and other channels of communication. More and more, people spend most of their waking hours in the public domain, which comes to feel like home, imbued by homelike qualities, and which replaces some of the functions previously carried out at home, in private. In this way, home, spatially defined so far, becomes a state of being, adhering to the latest lifestyle fashions and increasingly centring around the utmost personal realm, the human body itself. Daily life revolves more around the ready fulfilment of bodily needs and pleasures and becomes a state of constant vacationing (Terkenli, 1993).

On the other hand, the falsification of place, time and culture in terms of 'staged authenticity' (MacCannell, 1992) is spreading in scale, nature and diversity through tourism, making it increasingly difficult to talk about tourist environments or tourist resorts, since tourism touches almost every type of place (Shaw and Williams, 1994). This trend is coupled with the trend of greater 'home-centredness' in leisure, as a result of its long-term privatisation: this is linked to the decline of community, to greater individual mobility and to the expansion of the leisure market in sound and vision equipment, computers and technology (Shaw and Williams, 1994). This makes it, of course, much harder to enjoy 'simple' pleasures such as those once found, for example, strolling the landscape of the family seaside resort. It also obfuscates the relevance of place and locality, if every tourist attraction can be found almost everywhere and if leisure and tourist activities are ever present in the course of our home everyday lives. This is the case, for instance, of new shopping centres or malls resembling more and more holiday resorts with amusement parks, restaurants, movie theatres, etc. Until recently, there always used to be some distinctiveness, real or staged, about the geographical properties or the historical or even literary associations of a place or landscape, properties that used to distinguish it from others. One outcome of current economic restructuring, social change, policy intervention and cultural re-evaluation is the transformation of all space into an object of the tourist gaze. This is especially poignant in the case of vernacular settlements and small 'traditional' communities turned into tourist resorts, such as in the South Pacific or in the Caribbean. It also, often, involves a change in the nature of public space, which is now more privately owned, more controlled and policed than before. Another outcome of this larger transformation is that landscapes or cultural sights turned into spectacle seem more original than the original ones they imitate. This blurring between the «authentic» and the staged is, again, more pronounced at the tourist destinations, where touring and daily life domains become interchangeable – i.e. aestheticised domestic architecture of Mykonos.

In these ways, the segregation of leisure from home life that modernisation instilled becomes tentative and irrelevant in the so-called post-modern western society, parallelling the similar trend of a merging of the home with the work domain – as in the current explosion of paid work at home. Landscape experiences, similarly, in which the insider-outsider dichotomy is both conspicuous and essential take up less and less of our daily life and experience, and tend towards irrelevancy. "We are all moving closer to becoming continual tourists and collectors of internal landscapes" (Riley, 1992). What is now tourism and what is more generally everyday life is relatively unclear, as specific pleasures are not place-bound and as the proliferation of objects of visual delight, including through the mass media, are everywhere. What ensues is a society of spectacle (Debord, 1994) with a new collective sense of place based on transcending the geographical barrier of distance and of place. It expresses a de-differentiation in space of leisure, tourism, shopping, culture, education, eating, and so on; it is now possible to experience the world's geography vicariously, as a simulacrum (Harvey, 1989).

4. Place identity in the Greek island tourist landscape: a case study

In 'vernacular postmodernism', 'post-tourism', or otherwise called 'neo-vernacularism', locality is considered central to the tourist experience, and the cultural landscape becomes a most eloquent representation and medium of 'local' culture. At the same time, the commodified landscape that hosts tourism and its imprints is juxtaposed to place as a collective home. The main tenet of this new trend is that the environment "works better [as a pole of tourist attraction] if the people who live, work and play in it are involved in its creation and maintenance. [From a planning perspective] this involves much emphasis on the process of design rather than on the end product, on reducing the power of the architecture vis-à-vis clients, on channelling resources to local residents and communities, and on restoration, or where new building is involved, in ensuing it is appropriate to local historical context" (Urry, 1990). Global processes, however, must engage with resident factors not only because local particularity cannot be completely transcended in the construction of a tourist site.

Cultural particularity has come to occupy a more central position in the organisation of present-day societies, as an important co-determining factor of location, and the trend to transform public space into representations of the cultural identity of its inhabitants is noteworthy in this respect (Dietvorst, 1994). Cultural identity or geographical particularity is often expressed in "locally-rooted traditions, lifestyles, and the arts [...] often compressed into a space and presented in a legible fashion for leisure consumption" (Teo and Yeoh, 1996). At this point issues of 'authenticity' become crucial in the negotiation of place for consumption and, secondarily, for daily life. Obviously, environmental factors play a prominent role too in shaping geographical uniqueness and particularity, by informing cultural possibilities and constraints in the historical evolution of a society (doctrine of probabilism in geography). Although such processes of cultural adaptation to the environment pertain more to pre-modern and pre-industrial societies, local solutions to problems of survival and growth have always been imprinted on the landscape and continue to be part of it up-to-date – imprinted on it in continuous historical process, but rather in leaps and waves (cultural succession in the landscape). The revisiting of the vernacular and the quest for authenticity take us in this last section, to the Greek island vernacular tourist landscape. To a significant degree, authenticity in tourist landscape planning, as we shall see, may be achieved and safeguarded through community participation, an emerging critical social factor in decision-making concerning the shaping of space, place and landscape as a context of high-quality life (Briassoulis, 1997).

The first and most immediate cultural image of place identity to a visitor is in the landscape's visual composition and articulation. Preservation efforts have, at least nominally, sought to protect the unique visual character of the Greek island landscape. Based on a comprehensive typology of 'vernacular' elements of Cycladic domestic and townscape architecture (Terkenli, 1986), an architectural survey of such elements was undertaken on the Cycladic island of Serifos, Greece. Architectural and behavioral participant observation and survey in Serifos initially took place in the summer of 1985. The results of this survey were verified by follow-up research in the summer of 1995. This comparative survey between the more 'traditional' hilltop community of Hora and the modern, touristic port community of Livadi reveals a marked emphasis in the domestic architecture and the townscape of Livadi on visual landscape elements that seemingly preserve the character of the local landscape (Table 1). Specifically, architectural and townscape elements of Hora for the most part conform to the representative Cycladic types, established in the typological scheme mentioned above, as regards all aspects of the house and the townscape, form, function, construction materials, etc.

In contrast, mostly architectural and townscape variables that preserve the visual character of vernacular or folkways of building and situating in space in Livadi conform to the representative Cycladic types. A wide variety of features associated with ancient construction techniques and materials and subtle local particularities, as well as the functional role of specific spaces and objects, have disappeared as no longer 'essential' in serving daily needs or catering to the tourist image of the village. Examples of the latter category are: floor, wall and staircase construction materials, use of the space underneath exterior staircases, trapdoors, the existence of 'samari', and the horizontal division on entrance doors (Sancar and Koop, 1995). On the other hand, a certain 'Greek blue' has become the commonest paint colour of window and door frames in tourist destinations all over Greece, signifying essential 'Greekness' in its connotations of 'traditional' island ways of building. Now, of course, reality imitates myth in the widespread adoption of this particular paint colour, irrespective of the presence of tourism. Soundscapes are becoming a thing of the past, as street traffic conceals traditional street vendors' voices, village bellscapes and horse hooves, wherever these continue to exist.

In the category of visually predominant landscape elements, townscape features predominate, as they cater best to the visual character of the modern way of seeing and sightseeing that larger urban transformations have instilled with the advent of modernity. In the survey, examples of such features were siting attributes of houses, location of churches and public buildings, use of streets and street surface materials, and the existence of semi-public/private spaces at the interface of private with public domain. Especially telling is the obsolescence in recent years of semi-public/private spaces from the landscape of Livadi. Initially, the legacies of industrialisation and urbanisation that underwrote the processes of modernisation created intermediary zones, such as zones of communal space between the private and the public, one manifestation of which are the spaces mentioned above.

Further processes of modernisation and global capitalism, however, evident in the landscape of Serifos as well, wiped out such spaces which served as meeting grounds for a number of large-scale or small scale communal activities: shelling peas, passing on neighbourhood gossip, playing a game of backgammon. In this way, these processes alleviated obstacles to the tourist gaze and resulted in a facilitation of the place-consuming activities of tourism and other modern preoccupations. There is no place, any longer, in the streets of Livadi, for daily informal socialising at the interface of the house and the street, because not only these interfaces are shaped in ways not conducive to lingering or sitting, but also because the public character of the street-house entrance arrangement discourages all such activity. More generally, until recently, landscape architectural styles typical of the Aegean islands could be described as 'romantic' in their spontaneity, organicism, contrapositions and randomness (Jakle, 1987), with some incorporation of principles of 'classical' landscape architectural styles (Norberg-Schultz, 1980) in their austerity, simplicity, harmony and a certain formal repetition. Now these landscapes seem to be in the process of a wholesale transformation into a totally different type of 'classical' landscape: the modern, or even 'postmodern', exhibiting a larger amount of order and uniformity, eliminating elements of spontaneity and surprise and promoting a novel kind of visual openness and accessibility.

The second constitutive element of modern tourist landscapes, as mentioned at the beginning of the article, has been its distinctive non-home nature: the fact, in other words, that the tourist destination normally serves purposes of distraction and change from the familiar obligations of life at home. The fact that tourism, as a leisure activity, presupposes its opposite, namely the drudgery of everyday home and work life, is readily apparent in the context of the comparative study on Serifos. A behavioural analysis of regular everyday activity in open urban spaces (Sancar and Koop, 1995) points to a differentiation of spaces touched by modern tourism in Livadi in contrast to the ones in Hora, which continued to exhibit more 'traditional' behaviour patterns (Table 2).

In specific, there is much more variety in regular activity patterns of outdoor life in Hora than in Livadi; activities that pertain to the whole spectre of life functions (education, recreation, nutrition, religion, health care, personal grooming, politics, and so on). Moreover, the entire range of possible manner of performing these activities (involving various degrees of muscular and mental expenditure) (Barker, 1968) is present here, whereas the inhabitants' levels of personal involvement and leadership are much higher in Hora (the more 'traditional' settlement) than in Livadi (the more 'touristy' settlement). Individual settings in Livadi have become much more specialised, indeed specialised in tourist services. The generic jewelry store has become, for example, a ubiquitous feature of all contemporary Aegean towns. Store-street interfaces have lost their function and significance as nodes of social intercourse, a tradition that dates back to the antiquities, and serve instead as mere physical treads delineating an abrupt transition between public and private. Semi-public/private spaces have lost their multi-purpose identity. Meanwhile, street settings in the Aegean still host-diversified inputs from their inhabitants, and therefore encourage a continuing dependence on 'traditional' forms of architecture and townscape. They have, nonetheless, also become the exclusive domain of the automobile, facilitating fast and uninhibited access by the visitor to all points of interest.

In general, Hora exhibits a far greater richness of individual and collective activity than Livadi, where the function of outdoor public spaces has become much more specialised with the onset of tourism (Sancar and Koop, 1995). Even though Hora has in more recent decades changed somewhat due to outmigration, behaviour mechanisms here are still more diverse and show a more colourful lifestyle than respective ones in Livadi (Table 2).

Table 1. Frequencies of vernacular Cycladic architecture and townscape elements in Hora and Livadi, Serifos

Elements	Hora (%)	Livadi (%)
Domestic architecture		
1. House form based on 'monospito' type	99.16	80.23
2. Flat roof type	99.19	100.00
3. S, SE, SW house orientation	82.11	76.20
4. Fireplace	94.17	82.22
5. Fireplace located in the kitchen	100.00	100.00
6. Toilet	100.00	99.37
7. Toilet located outside the house	82.86	79.40
8. Soil, stone or wooden floor material	77.71	59.09
9. Stone walls	99.15	75.00
10. 'Samari' on roof	100.00	68.00
11. Whitewashed or bare exterior wall surface	100.00	94.15
12. 'Traditional' window and door frame colour	88.49	82.86
13. 'Traditional' flat roof construction technique	82.71	60.26
14. Stone exterior staircase	85.72	57.70
15. Utilisation of space underneath ext. Staircase	88.77	64.28
16. Wooden interior staircase	91.67	100.00
17. Trap door at top of interior staircase	88.15	0.00
18. Soil, stone or mosaic courtyard material	74.32	69.35
19. Elongated 'sala' form	99.00	87.50
20. Wooden window and door frames	97.83	98.25
21. Horizontal division on entrance door	81.04	64.29
Townscape		
22. 'Traditional' siting of houses	95.83	100.00
23. Non-vehicular use of streets	100.00	75.00
24. Stone-paved streets	95.84	72.22
25. Semi-public/ private spaces	88.18	62.50
26. Facade uniformity	82.69	79.55
27. Prominent church location	94.44	50.00

Comparatively larger numbers of population subgroups in Hora's daily activities are due to a relative abundance of locals versus tourists in Hora, as compared to Livadi. Fewer participants in the street, general store entrance and semi-public/ private space settings assume leadership roles in the community of Livadi. It is indeed, the amount of collective and personal control and influence of the local people over their community, and the repetitive, cyclical nature of daily life that inform the notion of home in these Aegean communities for the local inhabitants.

The Aegean landscape has been much romanticised in recent decades through tourism promotional interests as an idyllic insular paradise, isolated and free from the demands of modern life, blessed with perfect climate and constituted by small-scale, intimate settings ideal for romantic adventures, in the land of the 'Greek gods'. The 'four Ses' (Sun, Sea, Sex and Sand) constituted a powerful pole of tourism attraction for the Aegean from the very beginning of its onset, in the 1960's. Landscape elements, both natural (the sea, the beach, sunshine) and human-made (such as the whitewashed cubic houses in real or imitation stone-paved streets), that exemplified and reinforced such images of the Aegean, as well as illustrated its cultural uniqueness of place identity, were preserved and highlighted (motion pictures 'Shirley Valentine' and 'Summer Lovers'). These islands, though, have also been their inhabitants' homes, and thus have to have street access by car, modern sanitary systems, easy upkeep, and all sorts of other contemporary conveniences. In an analogy to the dichotomy presented in Figure 1, Livadi best conforms to the category of modern tourist landscape and Hora of the landscape of home, requiring for its existence a certain continuity of dependence on more traditional landscape forms. For the dual purpose of tourist attraction and local life facilitation, the Aegean facade of the islands' urban landscape was thus retained, but what was considered superfluous in modern life and tourism was dispensed of.

Modern use of space necessitates easier and faster accessibility, which requires efficient visual space appropriation and a favouring of the profane over the sacred, both literally and metaphorically. Thus we encounter motor bikes parked in centuries-old sacred sites, American Express offices in narrow shone-paved island alleys, and advertisements of 'traditional local pottery' – made in Maroussi near Athens and sold all over Greece. Garish advertisements and announcements, trademark signs of tourism, preferably in English, the lingua franca of tourism, abound. In some «preferred» tourist resorts, Greek is often no longer even understood in shops and tourism-catering services (Molyvos in Lesvos, the city of Rhodes). A certain placelessness comes about as a result, defined as "an environment ...[where] the underlying attitude ...does not acknowledge significance in places ...[because of an erosion of] symbols" (Relph, 1976). Where tourism has made sufficient inroads, as in the town of Patmos, for example, in a little over five years, the town landscape is merely a place where passers-by hurry on, whereas in the older, often interior villages of the Cycladic islands, for instance, villagers overwhelmingly continue to associate the whole village community and physical compound with home (Terkenli, 1993).

Table 2. Behavioural comparison by 'general richness index' of activity (Barker, 1968) between Hora and Livadi, Serifos

	Weekdays	Weekends
Street setting		
Number of activity types	H	H
Variety of activity mechanisms	H	H
Levels of local participation	H	H
Total activity duration	-	-
General Richness Index	-	H
House-street interface		
Number of activity types	-	-
Variety of activity mechanisms	-	-
Levels of local participation	H	-
Total activity duration		
General Richness Index		
Small store-street interface		
Number of activity types	H	H
Variety of activity mechanisms	H	H
Levels of local participation	H	H
Total activity duration		H
General Richness Index	H	H
Semi-public/ private space		
Number of activity types	H	H
Variety of activity mechanisms	H	H
Levels of local participation	H	H
Total activity duration	H	
General Richness Index	H	H

Note: 'H' indicates that Hora has a higher richness index of activity. Blank indicates no statistically significant difference between the two place behavioral richness ratings.

In global comparative terms, tourism has, on the whole, a positive impact on Greek life and especially on the Greek economy. Outmigration from the Aegean, of disquieting proportions in the past, has been somewhat curbed in recent decades, due, in part, to an increase in employment opportunities offered by tourism demand, in lieu of disappearing primary and secondary sector activities. In some ways, however, this disappearance of alternative economic venues for the Aegean islanders is a direct outcome of the expansion of the tourist industry. In this way, 'globalising' and homogenising forces of mass tourism are not only affecting the home realm of local societies, but the work domain as well, and by doing so preserve the conditions for their continuation and establishment in the communities of the Aegean. In other words, tourism entrenches schemata of economic dependence in struggling island societies, both in economic/ environmental and in social/ demographic terms. These schemata promote tourism continuity, at the same time as they promote often staged home qualities in terms of self-serving new 'cultural models', while obliterating organic geographical uniqueness – and thus again ensuring their perpetuity. Moreover, tourism's stronghold in the work realm of local life indirectly eliminates relationships of mutuality and reciprocity in social life in the form of communal bonds and of a sense of place and home forged through the centuries out of the performance of common tasks and rites. Tuan (1982) additionally observes that "as important to the forging of such ties are those numerous unstructured occasions when people are simply in each other's presence".

This form of social group cohesiveness, serving mutual livelihood interests and formed out of necessity, even if circumstantial, characterised the social life of Aegean communities in the past. Presently, it seems to be in the process of disappearing both in its manifestations and signs in the local/ tourist landscape and as everyday reality and symbolic imagery. Contexts of strong social allegiance that do persist in the Aegean, however, are religious community, nuclear family and clientilism ties. The traditional family business and store, in their flexibility and personalising touches, have adopted and cater to the demands of the tourism industry especially well. Traditionally, the centrality of the nuclear family in-group in Greek rural and insular life rested largely on the fact that all ordinary operations of farming and fishing were normally performed by the members of a single family.

In conjunction with a strong marketplace principle and skill as a means to success and social acclaim (McNeill, 1978), the nuclear family continues to be upheld by modern life as the primary social unit of consumption and mutuality. Tourist industry supply and demand patterns fit well in the social economy of the Aegean and tend to reinforce it. Solidarity and mutuality still exist in another sociocultural context: almost universal membership in the Greek Orthodox Church. Religious allegiance is illustrated in not only the plethora of religious festivals, processions, churches and roadside shrines in the Aegean landscape, but also in toponyms, as well as in a pronounced inscription of the sacred and the profane in space definition and delineation (i.e. village versus wilderness in folk tales and superstitions). Clientelism, one of the most enduring bastions of older Greece, still polarises whole island communities and manifests itself in heated political debate in the coffee-house and in the striking presence of the political party office in the Aegean village or town square. To conclude, in many of the Aegean islands, the sense of place identity that develops from everyday home life, the diversified landscape where people meet, associate, play, work, assemble politically, worship, and live all matters of life, is in danger of being lost to tourism. The cultural landscape of the Aegean is in danger of being transformed into a theatre of life staged for tourist purposes, made into a spectacle, embellished with foreign qualities that cater to place images that the tourism industry promotes for the Aegean, furnished with elements of convenience for the visitor rather than functional priorities for the local, made more aesthetically pleasing, more accessible and palatable to mass taste. The sense of intimacy that grows in a place that people repeatedly imbue with personal and collective value and go about their everyday lives using as such is becoming obsolete. In Relph's (1976) words, "there is an authenticity in the direct and genuine experience of the place". This trend towards landscape homogenisation, however, is controlled to a large extent by the re-creation of an emerging 'cultural model' for the Aegean, a model that promotes and best caters to tourism and which borrows especially from the appeal of old ways of life – products of indigenous social, cultural, historical and economic life through the centuries.

5. Concluding remarks

This last section has developed from an attempt to make more transparent the centrality of local ways of life on the tourist landscape as these inform place identity with distinctive landscape 'images'. The whole concept of image-making is at the heart of much of the tourism industry, both nationally and locally. "In this respect, regions are labelled and sold to tourists on the basis of single themes acquired through history, novels or television". (Pocock: in Shaw and Williams, 1994). In this process, local forces have always been and are clearly becoming more and more important in tempering market forces from obliterating deliberately or accidentally the heritage and culture of a place (Teo and Yeoh, 1997). Especially critical is the sociocultural pertinence (Fortuna, 1997) of tourism/leisure activities and spaces in the process of identity formation for different social groups and their cultural landscapes, an example of which was presented in the Aegean landscape. One consequence of the fact that post-tourist de-differentiation is becoming more and more apparent in all spheres of social and especially cultural activity, is the emergence of a deeper awareness of the distinction between representations versus reality on the one hand and homogenisation versus place identity on the other. In the tourist experience, all spheres of life implode into each other, while most involve visual spectacle and play. In the process, "the post-tourist knows that they are a tourist and that tourism is a game, or rather a whole series of games with multiple texts and no single, authentic tourist experience" (Urry, 1990). The post-tourist, well-informed and sophisticated, increasingly unravels the cultural icons that constitute the object of his/her interest, enters inside their signs critically and decodifies their symbolic nature. In this way, (s)he transcends them by appropriating them in the search for place identity, gaged in terms of 'authenticity'.

    Similarly, the local, as participant or spectator of the games of tourism, breaks into the code of the tourist icons and spectacles themselves by acquiring a deeper sense of their cultural distinctiveness and by re-developing a sense of place belonging. A new sense of rootedness paradoxically grows out of the cosmopolitanism that tourism instills. Where previous generations readily deserted the island homes of the Aegean in search for better life opportunities, today they reconsider staying and making a go of it in the tourism industry. Consequently, there develops among the host populations a certain pride and increasing sense of stability that tourism affords and which contributes to new, albeit more individualistic, personal growth, by means of a renewed appreciation of local cultural ways and of the significance of their continuity through indigenous group membership. Granted that the societal changes brought about by tourism are not always easy to isolate form other more general 'modernising' influences, its role in this transformation is nonetheless undebatable (Shaw and Williams, 1994).

What we increasingly witness in landscapes touched by tourism is a convergence of the public with the private spheres of life, the infusion of leisure and spectacle in a growing number of local forms and functions, an obliteration or forging of geographical uniqueness and an irreversible homogenisation of such tourism settings, with repercussions critical to both everyday individual and collective geographies and broader human-environment interrelationships. These changes, largely visual, are imprinted on and expose cultural differentiation and change in tourist landscapes, that are meanwhile variously affected by changing relationships between work, home and leisure. Not only, simplistically delineated 'global' and 'local' forces play a role in shaping the cultural landscape of the Greek islands, but all realms of life interweave in the wholesale cultural transformation occurring in our lifetimes and manifesting itself most eloquently in the human landscape.

Given that culture is by no means a constant and the local is in continuous flux, "as an intellectual category, the 'local' seems more protean than primordial. In identifying the typical components of localness," writes Hannerz (1996), "we may also come to realise more clearly that they are not all intrinsically local, linked to territoriality in general or only some one place in particular. That connection is really made rather by recurrent practicalities of life, and by habits of thought. And if somehow these characteristic features of local life... approach the qualities of 'home', then the one and only local would appear to be a rather less privileged site of cultural process". The contact that tourism affords between the less privileged and the more privileged unpacks this contradistinction by encouraging self-awareness and opening up new cultural possibilities for the 'local'. In the homogenising process that ensues between the 'local' and the 'global', what is lost is home, as we have known it thus far and have always been able to return to.

References

Barker, R.G. (1968) *Ecological Psychology. Concepts and Methods for Studying the Environment of Human Behaviour.* Stanford University Press, Stanford, Ca.

Briassoulis, H. (1997) Sustainability indices: A critical bibliographical review. *Topos.* **12** (in Greek), 55–76.

Dann, G. and E. Cohen (1991) Sociology and tourism. *Annals of Tourism Research,* **18**, 155–169.

Debord, G. (1994) *The Society of the Spectacle.* Zone Books, New York.

Dietvorst, A. G. J. (1994) Cultural tourism and time-space behavior. In G.J. Ashworth and P.J. Larkham (eds) *Building a New Heritage : Tourism, Culture and Identity in the New Europe.* Routledge, London, pp. 69–89.

Duncan, J. (1993) Sites of representation: place, time and the discourse of the other. In J. Duncan and D. Ley (eds) *Place, Culture, Representation,* Routledge, London, pp. 39–56.

Feifer, M. (1985) *Going Places.* Macmillan, London.

Fortuna, C. (1997) Cultural tourism in Portugal. *Annals of Tourism Research* **24**:2, 455–457.

Germundsson, T. and M. Riddersporre (1996) Landscape, process and preservation. In I. Margaretha (ed.), *Landscape Analysis in Nordic Countries: Integrated Research in a Holistic Perspective, Proceedings from the Second Seminar of Nordic Landscape Research.* Lund 13–14 May 1994, Swedish Council for Planning and Co-ordination of Research, Stockholm, pp. 98–108.

Hannerz, U. (1996) *Transnational Connections: Culture, People, Places.* Routledge, London.

Harvey, D. (1989) *The Condition of Postmodernity,* Blackwell, Oxford.

Jackson, J. B. (1980) *The Necessity for Ruins and Other Topics.* The University of Massachusetts Press, Amherst.

Jakle, J. A. (1987) *The Visual Elements of Landscape.* The University of Massachusetts Press, Amherst.

Jones, M. and K. Daugstad (1996) Cultural landscape under administration: a conceptual analysis. In I. Margaretha (ed.) *Landscape Analysis in Nordic Countries: Integrated Research in a Holistic Perspective, Proceedings from the Second Seminar of Nordic Landscape Research,* Lund 13–14 May 1994, Swedish Council for Planning and Co-ordination of Research, Stockholm, pp. 162–188.

Jussim, E. and E. Linquist-Cock (1985) *Landscape as Photograph.* Yale University Press, New Haven.

Lowenthal, D. (1985) *The Past is a Foreign Country.* University Press, Cambridge.

MacCannell, D. (1992) *Empty Meeting Grounds: The Tourist Papers.* Routledge, London.

McNeill, W. H. (1978) *The Metamorphosis of Greece Since World War II.* The University of Chicago Press, Chicago.

Meinig, D., (ed.), (1979) *The Interpretation of Ordinary Landscapes: Geographical Essays.* Oxford University Press, New York.

Montaigne, M. (1770) *Journal de Voyage.*

Moore, K., G. Cushmnan and D. Simmons (1995) Behavioral conceptualization of tourism and leisure. *Annals of Tourism Research* **22**:1, 67–85.

Norberg-Schultz, C. (1980) *Genius Loci: Towards a Phenomenology of Architecture.* Rizzoli, New York.

Norton, A. (1996) Experiencing nature: the reproduction of environmental discourse through safari tourism in East Africa. *Geoforum* **27**:3, 355–373.

Norton, W. (1989) *Explorations in the Understanding of Landscape: A Cultural Geography.* Greenwood Press, New York.

Pretes, M. (1995) Postmodern tourism: the Santa Claus industry. *Annals of Tourism Research* **22**:1, 1–5.

Relph, E. (1976) *Place and Placelessness.* Pion, London.

Riley, R. B. (1992) Attachment to the Ordinary Landscape. In I. Altman and S. Low (eds) *Place Attachment, Human Environment and Behavior: Advances in Theory and Research Series,* Plenum Press, New York., pp. 13–35.

Said, E. (1979) *Orientalism.* Routledge, Andover.

Sancar, F. H. and T. Terkenli-Koop (1995) Proposing a behavioral definition of the 'vernacular' based on a comparative analysis of the behavior settings in three settlements in Turkey and Greece. *Journal of Architectural and Planning Research* **12**:2, 141–165.

Schorske, C. E. (1967) *Fin-De-Siecle Vienna: Politics and Culture.* Vintage Books, New York.

Shaw, G. and A. M. Williams (1994) *Critical Issues in Tourism: A Geographical Perspective.* Blackwell, Oxford.

Sporrong, U. (1996) The landscape, a field in which research can be integrated: using landscape research in a wider context. In I. Margaretha (ed.) *Landscape Analysis in Nordic Countries: Integrated Research in a Holistic Perspective, Proceedings from the Second Seminar of Nordic Landscape Research,* Lund 13–14 May 1994, Swedish Council for Planning and Co-ordination of Research, Stockholm, pp. 12–18.

Soane, J. (1994) The renaissance of cultural verancularism in Germany. In G. J. Ashworth and P.J. Larkham, (eds) *Building a New Heritage: Tourism, Culture and Identity in the New Europe,* Routledge, London, pp.159–177.

Stebbins, Rt. (1997) Identity and cultural tourism. *Annals of Tourism Research* **24**:2, 450–452.

Teo, P. and B. S. A. Yeoh (1997) Remaking local heritage for tourism. *Annals of Tourism Research* **24**:1, 192–213.

Terkenli, T. S. (1993) *The Idea of Home: a Cross-Cultural Comparison.* Doctoral Dissertation, Department of Geography, University of Minnesota–Twin Cities.

Terkenli, T. S. (1986) *A Comparison of Vernacular Elements in the Architecture/ Townscape and Behavioural Patterns of Two Settlements on Serifos, Greece.* Master's Thesis, Department of Landscape Architecture, University of Wisconsin –Madison.

Tuan, Yi-Fu (1982) *Segmented Worlds and Self: Group Life and Individual Consciousness.* University of Minnesota Press, Minneapolis.

Urry, J. (1994) Cultural change and contemporary tourism. *Leisure Studies* **13**, 233–238.

Urry, J. (1990) *The Tourist Gaze: Leisure and Travel in Contemporary Societies.* Sage, London.

ENVIRONMENTAL AND CULTURAL TOURISM RESOURCES: PROBLEMS AND IMPLICATIONS FOR THEIR MANAGEMENT

PARIS TSARTAS
Greek Centre for Social Research
Athens
Greece

1. Introduction

The present paper aims at presenting and analysing the problems and implications arising from the use and management of environmental and cultural resources in tourist areas – two groups of resources which directly affect the particular demand and supply characteristics in most tourist areas world-wide. Examples of such resources include: monuments, traditional settlements, cultural events, traditionally produced local products, areas of archaeological, cultural or historic interest, areas of special natural beauty, national parks, ecological parks, wetlands, coasts, mountains, areas with a rich or rare flora and fauna.

The first part of the paper examines the development trends in modern tourism which affect the management practices of environmental and cultural resources in tourist areas. These trends, which portend new facts in tourist demand and lead to changes in the way tourist resources are organised and managed, are: (i) development of 'alternative' tourism, (ii) changes in organised mass tourism, (iii) organisational and economic changes in world tourism, (iv) changes and reorientation of tourist motives, (v) emergence of the notion of 'socially responsible' tourism and (vi) transformation of tourism into a consumer good in modern society. The basic argument of our analysis is that the complex interrelations among these trends lead to:

1. A greater importance attached and a more prominent role assigned to cultural and environmental resources in tourist areas.
2. The organisation and production of new 'types' of cultural and environmental resources, which are promoted and offered in the tourist market.

The second part of the paper examines the factors which affect the development and management of tourist cultural and environmental resources. There are nine factors, which mainly affect the characteristic features of infrastructure and activities offered, namely: (i) the degree of growth of organised mass tourism, (ii) the special characteristics of resources offered by an area, (iii) the 'mass' dimension of alternative tourism development, (iv) the role of tourist marketing and advertising, (v) tourism policy in sustainable development programs, (vi) the definition of 'tourist resource', (vii) the 'tourist dimension' in legislation for the protection and management of culture and the environment, (viii) the organisational development of the country's tourist sector and its position in the international division of leisure and (ix) local actors who engage in tourist development procedures. The interrelations among these factors and their parallel, two-way relation with the trends in tourist demand create new facts for the management and development of cultural and environmental resources in tourist areas around the world. The basic arguments in the second part of the paper are the following:

- The development patterns associated with organised mass tourism affect negatively the ability to organise, protect and manage cultural and environmental resources in tourist countries. The opposite applies in cases where development has more 'sustainable' and 'balanced' characteristics.
- The mass dimension of demand for alternative tourism infrastructure and services has caused many problems in organising and managing resources which are related to culture and the environment.
- The continuous expansion of tourist development leads to the production of many new 'tourist products', most of which are composed of cultural and environmental elements. This situation generates development and management problems, but also demonstrates that all environmental and cultural elements are considered as tourist resources in tourist development.
- The institutional framework for tourist development at the national or local level directly affects the formulation of policies and practices for the management and development of cultural and environmental resources in tourist areas.

In the third part of the paper, the conclusions and proposals point to the following issues:
1. The notion of 'tourist resource' is differentiated and tends to include a wide range of elements, infrastructure and activities related to the culture and the environment of tourist areas.
2. Planning and policy making for the management of cultural and environmental resources in tourist areas becomes all the more complex due to the special characteristics of these resources and their significance for modern tourists-consumers.
3. Culture and the environment will keep on being basic 'fields' for the production of tourist resources which will constitute the poles of attraction of more and more specialised tourist trips.

2. Modern trends in tourist demand and their relation to culture and the environment

The rapid post-war tourist development (WTO, 1993b; Cazes, 1989) was mainly linked to two dominant tendencies in international demand for tourist travel: the first concerns vacation travelling and the second business trips. As regards domestic tourism, during the post-war period up to the present, demand for summerhouses kept increasing steadily, especially in developed countries. Additionally, tourist demand, following the continuous increase in leisure time in modern societies, extends over an increasingly larger number of days on an annual basis. Tourism is becoming a stable, inelastic expenditure of the budget of most families in developed countries, while it constitutes also an important social model for modern way of living. Furthermore, while tourism represented a marginal parameter in world economic development processes in the 1950s and 1960s, it is becoming – especially after 1970 – a decisive sector of the economy (IUOTO, 1975; World Travel and Tourism Council, 1992) on both the regional and the local level all over the world. At the same time, tourism acquires the characteristic features of consumer products, of which demand and supply become specialised systematically in recent years. In this sense, the motives (Dann, 1981; Jafari, 1989; Tsartas, 1996) which direct the 'tourist-consumer' to the selection of specialised tourist travel and activities become a key-element in analysing modern tourist demand and development.

Table 1 summarises the modern trends in tourist demand, which directly affect the management of two important resources: culture and the environment. Those two resources demand a common management approach by competent authorities and specialists.

Table 1. Modern trends in tourist demand which affect the management of environmental and cultural resources

1. Development of 'alternative' tourism	
2. Changes in organised mass holiday tourism	*Management of*
3. Organisational and economic changes in world tourism	*environmental*
4. Changes and restructuring of tourists' motives	*and*
5. Emergence of 'socially responsible' tourism	*cultural*
6. Transformation of tourism into a consumer good in modern society	*resources*

2.1. DEVELOPMENT OF 'ALTERNATIVE' TOURISM

The first trend is linked to the development of 'alternative' tourism (Lanfant and Graburn, 1991; Smith and Eadington, 1992), which (especially after 1980) started to constitute an important special market worldwide, with consumers coming mostly from Europe. This form of tourism resulted from the combined action of many factors, the most important of which were: new motives of dynamic groups of tourist-consumers; critiques by tourist development experts and tourist scholars of the dominant organised mass tourism model after 1950; the search by host countries of development models that would be better integrated in their social and environmental structure. In the context of alternative tourism, the search for 'clean' environment and 'original' culture is taken for granted, which is strengthened by the emergence of tourist destinations which satisfy these needs. At the same time, it is noted that alternative tourism is connected – apart from the special motives that characterise it – with particular organisation features in travel itself: autonomy; touring; better acquaintance with people and places; more activities during the trip, etc. The countryside and areas far form traditional tourist destinations become the main targets for the development of Alternative Tourism.

2.2. CHANGES IN ORGANISED MASS TOURISM

The second trend concerns changes which are observed in organised mass tourism, especially in the policy followed by tour-operators (Bywater, 1992) who control tourist demand. Since 1970, these companies have formed two or three types of 'vacation packages': one of them is 'closed', i.e. tourists follow a fully organised program; other types offer different degrees of autonomy in the organisation of travel. In the second category, higher importance is given to culture and the environment. In this context, the companies organise and promote programs which offer higher quality and more substantial participation in cultural or environmental activities in host areas. In this way, they respond to the general tendencies of demand – especially of tourists coming from Europe – and at the same time they divide the product offered in sectors, which allows them to find 'target groups' of higher quality in the international market. As an indication, it is worth mentioning that many tour-operators promote programs of environmental education and management, as well as programs encouraging local cultural activities. This interesting development – coming from the mass tourism – strengthens even more the need to organise and manage tourist resources which are related to culture and the environment, since many of them are in places offered not merely for a quick guided tour, but in host places where tourists usually stay.

2.3. ORGANISATIONAL AND ECONOMIC CHANGES IN WORLD TOURISM

The third trend is directly linked to organisational and economic developments in the world tourist market. To begin with, there is a dramatic increase in the total number of tourist trips; however, this increase is not always accompanied by an increase in the number of overnight stays in the place of destination. Most travels during the year allow different choices in terms of destination and of more specialised motives. Particularly in Europe (Faits et Opinions, 1987; KONSO, 1987), the share of the population that travels more than twice a year is increasing, resulting in a parallel increase in trips of small duration, which have a dominant special motive, e.g. business, education, culture, the environment, etc. It should be noted that domestic trips play an increasingly important role either because they have smaller cost or because the time available for travel does not allow travelling abroad. At the organisational level, travelling becomes more autonomous, a development, which is linked with the social models of modern tourists as well as with developments in terms of transportation (high percentage of families owing a passenger car) and changes in technology (e.g. ability of tourists to organise their travels through their PCs). In this way, a recovery of touring occurs – mostly by families or couples – which, apart from vacation, aim mostly at culture, education, the environment, etc. The majority of modern touring travel is directed towards the countryside or urban areas, rich in environmental or cultural resources.

2.4. CHANGES AND REORIENTATION OF TOURIST MOTIVES

The fourth trend is related to the considerable restructuring and changes that have taken place as regards the motives behind travelling. To begin with, there is an intense search for quality enriching of the experience of travel, which leads to trips which include many more activities (Krippendorf, 1989) than those offered in the past. Tourists become multi-motivated, as opposed to those moved by a single, dominant motive. Another change in tourist motives is linked to the search for all the more 'energetic' vacations, in contrast to the '4s' vacation travelling (Turner and Ash, 1975), which is considered to be associated with 'passive' vacation. Under this motive of vacation travelling, tourists search mainly for organised infrastructure and services in activities related to the environment, culture, and education. Up to a certain degree, these tourists 'transfer' part of the way their life is organised in the city to tourist travel, in their search for ways of spending their free time. A third – especially important – dimension in changes in motives concerns an increasing general tendency for travelling in the countryside. These trips have complex motives and are related to the 'get-away' mood observed mainly in people living in urban areas. They are related, although not exclusively, with alternative tourism and search for particular settings to visit and stay. Travelling is connected to: love for nature and staying in the countryside; ecological activities and touring; sports activities related to the topography of the area; educational activities; stay in traditional rural settlements; stay in agrotourist units; participation in health and natural life activities; adventure trips; trips or journeys of cultural tourism. The thematic variety of these trips is large, but there exist common parameters in all of them: enjoyment of nature; acquaintance with local culture; activities in the natural environment. These changes in the motives of tourists have led to a 'mass' tourism of a new kind, the development of which affects both the current management practices of tourist resources and the meaning of a 'tourist resource'.

2.5. EMERGENCE OF 'SOCIALLY RESPONSIBLE' TOURISM

The fifth trend refers to the social parameters, which enter the analysis of the tourist phenomenon. These are initially related to the view of socially responsible tourism (D' Amore, 19993; BITS, 1992) which is the product of the intense critique against organised mass tourism and its impacts in host countries. Based on this view, the relation between tourists and locals has to be shaped within the framework of a commonly acceptable chart of rights and obligations, in which the need to respect the local social, cultural and environmental particularities of the host place is dominant. Similar views come from the 'school' of Ethics (Lea, 1993; Hultsman, 1995) which places special attention on social responsibility of human behaviour in all fields of social and economic life. The social dimension of tourist travel was initially restricted to its consumers, the tourists. However, this dimension becomes more and more important in tourist development. This is obvious from the considerable increase, world-wide, in projects and programs for local tourist management which consider the factors of sustainability, protection of the environment and carrying capacity. It is a tendency, which is expected to be more intense in the short term, directly affecting the course of international tourist development.

2.6. TRANSFORMATION OF TOURISM INTO A CONSUMER GOOD IN MODERN SOCIETY

The sixth, especially decisive, trend is the one connecting tourist travel with the modern consumer and the social models of the medium-class in developed countries. Travelling, which was a marginal consumer product concerning a small number of people in the past, has become a product bearing social importance for its user (Graburn, 1977; Cohen, 1984). Through this product and its consumption, tourists form social behaviour patterns, send messages of social upward mobility and shape the general consumer tendencies of the market. As a consumer product, tourist travel follows a series of rules in the operation of the market: it is promoted through advertisement, it has a specific price, it is specialised, it is thematically classified. To respond to this rapidly increasing and thematically specialised demand, the tourist sector (and, more specifically, tourist enterprises) continuously 'produces' new products or destinations. In this context, culture and the environment play a crucial role and appear in various ways: travel in exotic or virgin destinations, which possess both environmental and cultural resources; planning of cultural or ecological trips in the countryside or in natural parks; increase in the number of thematic parks which are related to both of these resources; creation of infrastructure or accommodation complexes which offer a particular cultural experience to the tourist (Alcatraz prisons, war prisoners' camps in Singapore, tours in studios, thematic parks, water-lands, technological parks, etc.), or the artificial creation of a natural environment which has the characteristics of an exotic country (Fache, 1992). It is obvious that this dynamic tendency will bring about more problems in the future, not only in managing environmental and cultural resources, but also in the 'production' of new, innovative or technologically advanced products which, in their turn, will constitute the 'new generation' of tourist resources, as demanded by modern tourist-consumers.

The previous discussion makes clear that the trends of demand in the past twenty years strengthen the already great importance given to environmental and cultural resources in the 1955-1975 period, which was dominated by the organised mass tourism vacation model in the modern world. However, this development leads to considerable problems in terms of management and organisation of these resources and, at the same time, contributes to the 'production' and 'supply' of new products and services by the tourist sector. Finally, as will be pointed out in the third part of this paper, a new semantic approach to the term 'tourist resource' is deemed necessary.

3. Environmental and cultural resources: factors affecting their development and management in host areas

Tourist areas, responding to the trends in tourist demand described above, aim at taking advantage of their environmental and cultural resources, in order to promote their development, especially within a context of intense international competition. However, they face many problems in the implementation of their plans. Table 2 summarises the most important factors affecting the development and management of environmental and cultural resources in tourist areas, which are detailed in the following.

3.1. DEGREE OF GROWTH OF ORGANISED MASS TOURISM

The particular features of the development model of an area heavily affect the way its cultural and environmental resources are developed. Where the model of organised mass tourism is dominant, environmental and cultural resources are often used and consumed on an unprogrammed way, a fact which has a negative bearing on their structure and characteristic features (Mathieson and Wall, 1981; Briassoulis, 1992). The tendency for territorial expansion of this kind of tourism and the use of culture as a mere spectacle or producer of souvenirs constitute the two important parameters in the process of degrading the quality and meaning of these resources.

It is certain that this situation is worse in tourist countries of the Third World, as well as in areas – irrespective of their development level – which tend towards 'single-course cultivation' of this model. As regards the way these two kinds of resources are developed, it is noted that the environment is used just as a 'host-space' for tourist infrastructure and tourists, while culture is only related to guided tours – usually of short duration – in monuments and folklore cultural activities for tourists, e.g. dance, music, theatre, etc. The tourist policy of the country tends to be, in most cases, identical with that of tour-operators, who play, directly or indirectly, a crucial role in development and management processes. In cases where this tourist development model is more balanced, either with the development of special and alternative forms of tourism or with the adoption of a tourist policy which favours sustainability, then the development and management of the environmental and cultural resources becomes more effective. This happens because there is a common 'central tendency' in supply and demand: infrastructure and services supplied follow a mild and balanced development, while at the same time, demand is characterised by respect towards local particularities and the search of cultural and environmental tourist activities.

Table 2. Factors affecting the development and management of cultural and environmental resources in tourist areas

1. Degree of growth of organised mass tourism 2. Specific characteristics of the resources (attractiveness, access, carrying capacity, local development characteristics) 3. Mass dimension of the growth of alternative tourism 4. Tourist marketing and advertising 5. Tourist policy in 'sustainable growth' programs 6. Conceptual definition of 'tourist resource' 7. The 'tourist dimension' in legislation concerning the protection and management of culture and the environment 8. Organisational development of the tourist sector in the country and its position in the international division of leisure 9. Local actors involved in tourist development	Development and management of environmental and cultural resources

3.2. SPECIAL CHARACTERISTICS OF RESOURCES

The existence of some, even quality, resources which are related to the environment or culture is not enough for their systematic management and development (Pearce, 1991; Pearce, 1992b). As demonstrated by international practice, the complex characteristics of these resources directly affect these processes. Attractiveness constitutes the first considerable parameter and is directly connected to the prospects for tourist exploitation of the particular resource, but also to the possibility to attract investors. A second crucial point concerns access and communication facilities offered by the area where these resources exist. Another important issue, which is also linked to the broader characteristics of the local social, economic, cultural and environmental structure is the carrying capacity of the area where the specific tourist resource is found (Coccossis and Parpairis, this volume). It is another crucial parameter, which influences the planning and management of the specific area and its resources. Finally, the characteristics of local development, especially the demographic features of the population and the composition of the labour force, have also to be taken into consideration in planning the development of these resources. If all these special features and parameters are taken into consideration, they lead to the strengthening of such a development program. On the other hand, an approach, which does not follow this course, may result in the development of a limited number of elements of the cultural or environmental resources, leading, therefore, to a less dynamic and shorter-term development of tourism in the specific area.

3.3. THE MASS DIMENSION OF ALTERNATIVE TOURISM DEVELOPMENT

In the modern world, there are many forms of tourism, which are connected to alternative tourism and their demand tends to acquire a considerably mass character (Weiler and Hall, 1990; Lane, 1993). All these forms are related to or are based on culture and the environment, which constitute the dominant parameters for their development. Some of the most important of these forms are: ecological tourism, agrotourism, cultural tourism, sports tourism, mountainous tourism, touring, religious tourism, health and natural life tourism, winter tourism, adventure tourism, educational – scientific tourism, tourism in the countryside, seaside tourism, etc. The gradual transformation of these forms from marginal activities to organised models of development in the past years has resulted in considerable complications and problems in the management of cultural and environmental resources, which are directly related with their development. Initially, the development of these forms of tourism followed the dynamic tendencies of demand, without always evaluating correctly the possibilities offered by local society and the environment in terms of management requirements set by tourist development. A typical example is that of agrotourism, which requires the parallel use of cultural and environmental resources found in the area. Its rapid development in many areas has resulted in the abandonment of productive occupations in agriculture and farming or in maintaining a similar occupation just as a 'window' in order to satisfy demand (CEMAGREF, 1986). At the same time, the rural culture, a binding element of the local social and cultural structure, loses its significance and is transformed into an organised spectacle which, as in the case of organised mass vacations, is offered to tourists. Similar examples also can be given for other alternative forms of tourism (cultural, ecological, winter tourism) and demonstrates the emergence of certain features, which have common characteristics with the consequences of mass tourism in seaside areas. A second issue has arisen from the mass development of infrastructure for alternative forms of tourism which lead to great management problems in environmentally sensitive areas. Another issue is connected with alternative tourist activities in mountains (sports, tours, ecological journeys, stay in the countryside, agrotourism, etc.), while the development of alternative tourism focusing on protected areas or natural parks also presents considerable problems. In both cases, the lack of planning has created problems, while the organised development of such activities has led to a 'complex' (Parpairis, 1984) of infrastructure and services in which certain forms with common characteristics dominate, e.g. agrotourism and cultural tourism or ecological tourism and tourism in the countryside, etc. Development in complexes ensures the organised management of local tourist resources while, at the same time, excludes phenomena of 'single-course cultivation' of alternative tourism which are observed in many cases, such as in sports tourism in mountainous areas (ski centres), agrotourism, religious tourism, and ecological tourism. A fourth issue concerns the social dimensions of the management of cultural and environmental resources at the local level and is linked to the special characteristics of the production and social structure of the countryside. The majority of alternative forms of tourism are developed in the natural environment or in rural communities, which brings the local population, mostly farmers, face to face with important dilemmas. The first is whether they will continue working in agriculture or they will move away from it and become 'bourgeois-farmers'. The second concerns the particular contribution of the environment and culture in the local productive and social structure. In the context of alternative tourism development, these resources become 'autonomous' and their management imposes new production and social relations on the local population. These relations bear many of the features of an urbanised area where emphasis is given on occupations in the service sector and not in productive occupations in the agricultural sector.

3.4. TOURISM MARKETING AND ADVERTISING

In this context, the issue of 'use' of cultural and environmental resources or activities in marketing and advertising becomes of great importance. Culture and the environment have always been two of the 'stereotypes' in tourist advertising and at the same time they constitute the basis of every organised marketing program at local or national level. The characteristics as well as the special role of these marketing programs differ depending on the model adopted for tourist development and according to whether agents involved in tourist development accept the significance of organised tourist marketing (Uzzel, 1984; Cazes, 1989). The most important of those characteristics are:

*   In organised mass tourism, where tourist marketing is mostly used, there are two main tendencies in the promotion of tourist resources related to culture and the environment. The first uses these resources as stereotypes, which make the tourist think of the history and the civilisation of the country, e.g. the Parthenon means Greece, Flamenco means Spain. In this tendency, advertising aims at reminding the specific 'stations' or 'stops' of tourist travel in the area. The second tendency promotes the environment and culture as products to be used by tourists, often giving the impression, especially when these resources are abundant, that they are parts of the infrastructure of an amusement park, where tourists, after paying a fee, may use them as they wish.
*   In alternative tourism, where tourist marketing is less common, two tendencies may be observed also. The first one specifies its messages and its sales policy,

aiming at approaching specific target groups. In this case, the promotion of cultural or environmental resources becomes more specialised: activities and services which respond to specific forms of tourism, e.g. ecological, cultural, touring, etc. The second tendency degrades, or depreciates, the significance of tourist marketing by considering it unnecessary as alternative tourism is not a 'product' or a 'package', or even that it has a 'given' clientele. In this case, both marketing and advertising are used in order to promote this very alternative dimension of tourism in the specific area.

In both development models, the role of tourist marketing and advertising is all the more essential and leads to new facts for local cultural and environmental resources. In the final analysis, their development and management, especially in the context of international competition, has to take into consideration the basic principle of marketing: that every destination or resource constitutes a potential tourist 'product' which is produced, priced, advertised and sold with all the consequences this could bring along for society or the environment.

3.5. TOURIST POLICY IN SUSTAINABLE DEVELOPMENT PROGRAMS

Another factor increasingly affecting the special characteristics of tourist regions or destinations is the search for sustainability in their development path. Sustainable tourist development (Komilis, 1994; De Kadt, 1992) is one of the most promising evolutions as concerns management and development of all sorts of tourist resources. Of course, this development was not accidental since the search for models of 'mild', 'balanced' or 'integrated' local socio-economic development had started a few decades ago. The special characteristics of sustainable tourist development, which are linked positively to the development and management of cultural and environmental resources, are the following:

- Sustainability, in the framework of tourist development, constitutes, directly or indirectly, a dominant element of tourist policy of many supranational organisations or entities, which deal with the various aspects of tourist development.
- The scientific discussion on this very issue links the model of sustainable tourist development to policies and initiatives for upgrading, promotion and protection of local culture and the environment.
- Management of tourist resources has to focus on local particularities and on the possibilities to feedback on these resources in order for them to remain attractive and qualitative.
- Alternative forms of tourism, which are developed in complexes, may constitute the basis of a local program of sustainable tourist development.
- The continuous strengthening of the local dimension of tourist development constitutes a leverage to governments and institutions, pushing them to adopt a tourist policy emphasising sustainability. This tendency is also encouraged by

the fact that sustainable tourist development responds to the motives of modern tourists.

A negative aspect is the fact that programs for sustainable tourist development, as well as scientific analysis of the issue, are still 'privileges' of developed countries. In most developing countries such programs are being implemented in 'pockets' without being linked to local know-how.

There is no doubt that the role of sustainable tourist development will become all the more crucial in future attempts for the management and development of environmental and cultural resources demand for which will keep on increasing as will be explained later.

3.6. DEFINITION OF 'TOURIST RESOURCE'

Determining the 'tourist resource' becomes crucial, in the perspective of modern facts of tourist development, especially as regards the issue of management and promotion of tourist activities related to culture and the environment. The gradual consolidation of organised tourist development internationally imposed certain considerable taxonomic distinctions of what was considered as a tourist resource. For several decades, and due to the identification of tourism with vacation, the most important tourist resources were: the environment, culture and tourist infrastructure (mainly accommodation facilities). The 'environment' mostly concerns the natural environment and climate, 'culture' concerns the monuments and cultural events, while 'tourist sector infrastructure' includes the variety of accommodation, dining and entertainment infrastructure. The modern motives of tourists and the development of special and alternative forms of tourism affect the semantic limits of the term 'tourist resource'. These changes concern the following issues:

- The turn towards tourist activities, characterised by the use or consumption of environmental and/or cultural resources, has shaped a new situation (Grolleau, 1988). Elements of the local handicraft or production tradition, rural landscapes, natural monuments, buildings related to rural life and activities connected with the local agricultural culture become 'tourist resources' demanding development and management.
- Areas of natural beauty, protected or difficult to access, are transformed into a field of development of tourist infrastructure and activities and are considered as tourist resources of high importance in modern tourism, e.g. national parks, mountainous areas, wetlands, etc.
- Settlements in the countryside or in urban areas become favourable (Dixon and Fountain, 1989) sites for the development of tourist cultural activities or for hosting tourists, e.g. stay in traditional settlements in the countryside, sightseeing or journeys in urban areas with a rich tradition.
- Specialised museums, entertainment areas or thematic parks are being constructed, constituting this way modern tourist resources which, aided by

technology, combine the environment with culture, creating tourist products of high demand (Urry, 1992; Eyssartel and Rochette, 1992).

Therefore, tourist resources, particularly those related to culture and the environment, increase in number and expand in terms of themes, posing considerable problems and dilemmas in their management. On the other hand, as far as their development is concerned, a new dimension emerges where almost anything can be considered as a tourist resource.

3.7. THE TOURISM DIMENSION IN LEGISLATION

This generalised use or consumption of different types of infrastructure, spatial entities, buildings and human activities as tourist resources may lead to considerable management problems. Such is the case of culture and, more recently, the environment, which constitute two 'sensitive' resources as regards their tourist use. Tourist development and the gradual transformation of many elements or sectors of culture as well as of the natural and built environment in national or local tourist resources of high importance led to new facts and institutional regulations (OECD, 1980) dealing with their protection, use or promotion. In several countries, legislation dealing with these issues ignores the tourist dimension, considering that both cultural and environmental resources need protection and, as a result, the tourist 'use' of these resources is drastically limited. In many of these cases, cultural particularities and religion play a very important role in shaping this legislation. In most countries, however, and especially the developed ones, which are rich in tourist resources, legislation takes into consideration the modern tendencies in tourism development. Usually, legislation allows, under strict conditions, agencies and organisations related to tourism to participate in the development of tourist resources. However, many are those who believe that this massive, and by now consolidated, use of tourist resources leads to their degradation. Undoubtedly, this situation is more problematic in countries or areas which host mass organised tourism. At the same time, the development of alternative forms of tourism, with the continuously broader use of similar resources, brought to the fore new institutional problems as regards their protection and management, especially in developed countries (Tsartas, 1993; Pearce, 1992a). In any case, this is one of the most complex issues in the study of tourist development, especially since the notion 'tourist resource' has changed, but also because it concerns socially sensitive issues, such as cultural heritage and the protection of the environment.

3.8. THE ORGANISATIONAL DEVELOPMENT OF THE COUNTRY'S TOURIST SECTOR

The countries that have been considered traditionally as tourist destinations are those that face the most serious problems as regards development and management of cultural and environmental resources. At the same time, these countries, due to the fact that the problems they face allow considering them as 'examples', are the ones that have made the most interesting efforts to solve such problems. To begin with, the rapid tourist development, especially during the 1950-1970 period, resulted in a series of negative consequences in the management and development of cultural and environmental resources. The lack of an institutional framework, the single-course cultivation of tourism, pressures by tour-operators, social expectations of locals from tourist development, tensions in international competition and the difficulties encountered in formulating tourist policy were the main reasons behind the problems, which appeared. The continuous effort to transform or promote any monument or area of natural beauty into a tourist resource ended up in generating weaknesses in the management, as well as problems in the protection of both the environment and the monuments. This situation, which was particularly obvious in the tourist countries of the Mediterranean, threatened tourist development, which is based on the presence of cultural and environmental resources. Tourist countries themselves searched for solutions either at the national or the local level. The same applied for supranational entities and organisations that are dealing with the management and development of these resources in a balanced way, in order to achieve long-term benefits for both the local population and the tourists. Organisations such as the United Nations, the World Tourism Organisation (WTO) and the European Union contributed by means of policies and specialised programs (e.g. the Blue Plan for the protection of the Mediterranean) in shaping a new framework for tourist development (WTO-UNEP, 1983; Gennon and Batisse, 1989; Fitzpatrick and Associates, 1989). In this way, 'tourist' areas become the field where all possibilities and scenarios for the management of tourist resources are tested in the context of both organised mass tourism and alternative forms of tourism. More recent (after 1980) tourist destinations benefit also as they become examples of proper management of tourist resources.

3.9. LOCAL ACTORS ENGAGING IN TOURISM DEVELOPMENT

Lastly, the participation of local agents in the process of tourist development has acquired a special bearing in tourist countries. The continuous decentralisation in decision-making has been a stable parameter of development, at least for the more affluent countries, during the post-war period. In the case of tourism, approaching the agents who affect local tourist development (WTO 1993a; Pearce, 1992; Tsartas, 1992, 1998) is a matter of high importance, since their actions usually determine the result of the procedures for the management of tourist resources. All these agents (businessmen, local authorities, representatives of professional and scientific associations, persons employed in the tourist sector, etc.) tend to support tourist development. In this particular phase of development, all local tourist resources are considered to be 'exploitable' and capable of producing income for the area, while expectations for quick development usually put aside problems related to the consequences this development may have on the cultural and environmental resources of the area. At this stage, those who are less related with tourism are more sceptical, considering that the natural and cultural resources of the area demand protection or a different way of exploitation. At the next stage, the local population should project the need for balanced management of local tourist resources. The more organised the local agents employed in the tourist sector are, the more dynamic they become in pushing for this cause. Younger persons, those working in tourism and those interested in the protection of the environment or the local cultural heritage are usually those who stand up to support the general turn towards forms of management and development which respect the local particularities. Needless to say that the position these agents hold on issues related to the management of tourist resources is not single or one-sided. However, the possibilities offered to them by a decentralised institutional framework as regards their dynamic participation in shaping local tourist development has turned them into a considerable factor influencing the management and development of local environmental and cultural resources.

In summary, the broad spectrum of cultural and environmental resources offered by tourist regions is the result of efforts made by the tourist sector to respond to tourist demand, which has all the more specialised motives. However, the procedures of developing these resources lead to considerable organisational and management problems, and, at the same time, they point to the direction of some extremely interesting developments in the course of 'production' and 'consumption' of these resources. These developments are examined below.

4. Conclusions

The preceding analysis leads to certain conclusions as regards the problems associated with environmental and cultural tourism resources and the implications for their management.

Firstly, the notion of 'tourist resource' needs to be conceptually clarified, emphasising resources connected with culture and the environment. The tendency of demand to become more specialised and the tendency of supply to 'produce' new resources has reversed the classic typology of tourist resources, while they also lead to a spatial expansion of tourist development. Since 'everything' can be considered as a tourist resource, could it be possible that we are approaching an era where all areas of the world could potentially be considered as 'tourist areas'?

Secondly, the continuous 'production' of tourist resources, usually with the help of modern technology, seems to aim at shaping new types of tourist travelling, where tourists will be able to choose the environmental or the cultural experience they wish. Recent developments indicate that such a prospect is possible in the foreseeable future.

Thirdly, the social significance of tourism in modern life is increasingly strengthened. The consumer dimension attributed to travelling and the general increase in the volume and types of travelling lead to a continuous expansion of the required tourist resources and services.

Fourthly, both culture and the environment, owing to many developments (socially responsible tourism, alternative tourism, sustainable tourist development), obtain a new dynamic, which turns them into tourist resources of great importance.

Fifthly, the management and development of cultural and environmental tourism resources in the modern world will continuously require more efforts in terms of planning and organisation. The reasons are the continuous expansion of organised tourism and the development of alternative tourism in many areas of the world.

Sixthly, the local dimension will become increasingly important in the management and development of tourist resources in the modern world. At the same time, dilemmas concerning the use of these resources for tourist exploitation, in contrast to their past uses, will grow (e.g. traditional agricultural settlements that are being transformed into potential tourist complexes).

Finally, the role of tourists in the process of development and management of environmental and cultural resources is enhanced; they become 'responsible' consumers and users of these resources.

References

BITS (Bureau Internationale du Tourisme Social) (1992) A charter on the ethics of tourism and the environment. *BITS Information* **110**, 12–13.

Briassoulis, H. (1992) Environmental impacts of tourism: a framework for analysis and evaluation. In H. Briassoulis and J. van der Straaten (eds) *Tourism and the Environment: Regional, Economic and Policy Issues.* Kluwer Academic Publishers, Dordrecht, pp. 11–22.

Bywater, M. (1992) *The European Tour Operator Industry.* The Economist Intelligence Unit, Report No 2141, London.

Cazes G. (1989) *Les Nouvelles Colonies de Vacances: le Tourisme Internationale a la Conquête du Tiers-Monde.* L' Harmattan, Paris.

CEMAGREF (1986) *Les Relations Agriculture–Tourisme.* CEMAGREF, Grenoble.

Cohen E. (1984) *A Phenomenology of Tourist Experiences.* Centre des Hautes Études Touristiques, Serie C, No 52, Aix-en-Provence.

D' Amore, L.O. (1993) Ethics and guidelines for socially and environmentally responsible tourism. *Journal of Travel Research* **31**, 64–66.

Dann G. (1981) Tourist motivation: an appraisal. *Annals of Tourism Research* **8**:2, 187–219.

De Kadt, Em. (1992) Making the alternative sustainable: lessons from development for tourism. In V.L. Smith and W.R. Eadington (eds), *Tourism Alternatives: Potentials and Problems in the Development of Tourism.* University of Pensylvania Press, Philadelphia, pp. 47–75.

Dixon, M. and K. Fountain (1989) *Contribution to the Drafting of a Charter for Cultural Tourism.* ECTARC – EEC, Wales.

Eyssartel A.M. and B. Rochette (1992) *Des Mondes Inventes: les Parcs à Theme.* Éditions de la Villette, Paris, 1992.

Fache W. (1992) Is the seasonal character of tourism beyond the immediate control of the business? In *Le Tourisme International entre Tradition et Modernité*. Actes du Colloque International, Université de Nice, Nice, pp. 239–265.

Faits et Opinions (1987) *Europeans and Their Holidays.* Commission of the European Communities, Paris.

Fitzpatrick J. and Associates (1989) Travel and Tourism in the Single European Market.The Economist Intelligence Unit, London, Special Report No 2014.

Gennon, M. and M. Batisse (eds), (1989) *The Futures for the Mediterranean Basin: the Blue Plan.* Oxford University Press, London.

Graburn N.H. (1977) Tourism: the sacred journey In V. Smith (ed.), *Hosts and Guests: the Anthropology of Tourism.* University of Pennsylvania Press, Philadelphia, pp. 33–48.

Grolleau H. (1988) *Rural Heritage and Tourism in the EEC.* Commission of the EEC Brussels.

Hultsman, J. (1995) Just tourism: an ethical framework. *Annals of Tourism Research* **22**:3, 553–567.

IUOTO (1975) *The Impact of International Tourism on the Economic Development of the Developing Countries.* IUOTO, Geneva.

Jafari, J. (1989) Socio–cultural dimensions of tourism: an English language literature review. In J. Byztrzanovski (ed.), *Tourism as a Factor of Change: A Socio–cultural Study.* Vienna, Vienna Center, pp. 26–31.

Komilis, P. (1994) Tourism and sustainable regional development. In A. Seaton, S.L. Jenkins, R.C. Wrd, P.U.C. Dieley, M.M. Bennet, L.R.M. Lellan, and R. Smith, (eds), *Tourism: The State of the Art.* Chichester, pp. 65–73.

KONSO (1987) The Journeys of the Europeans, Commission of the European Communities – Community of European Railways, Basel.

Krippendorf, J. (1989) *The Holiday Makers.* Heineman, London.

Lane, P. (1993) *Tourism Strategies and Rural Development: a Review for the OECD.* University of Bristol, Bristol.

Lanfant, M.F. and N.H. Graburn (1992) International tourism reconsidered: the principle of the alternative. In V.L. Smith and W.R. Eadington (eds), *Tourism Alternatives: Potentials and Problems in the Development of Tourism.* University of Pennsylvania Press, Philadelphia.

Lea, J.P. (1993) Tourism development ethics in the Third World. *Annals of Tourism Research* **20**:4, 701–715.

Mathieson, A. and G. Wall (1981) *Tourism: Economic, Physical and Social Impacts.* Longman, London.

OECD (1980) L' Impact du Tourisme sur l' Environnement: Rapport Géneral, Paris.

Parpairis, A. (1984) Eilogi periohon gia nees morfes tourismou: chorochroniki katanomi tis neas anaptixis (Selection of areas for new forms of tourism: Spatial division of the new development. *Tourismos ke Ikonomia,* November 1984, 60–62.

Pearce, D. (1991) *Tourism Today: A Geographical Analysis.* Longman, London.

Pearce, D. (1992a) Alternative tourism: concepts, classifications and questions. In V.L. Smith and W.R. Eadington, (eds) (1992) *Tourism Alternatives: Potentials and Problems in the Development of Tourism.* University of Pennsylvania Press, Philadelphia, pp. 15–30.

Pearce, D. (1992b) *Tourism Development.* Longman, London.

Tsartas, P. (1992) Local community and organisation of tourist resources: groups and intermediate social structure between the formal and informal: the Greek case. *Sociologia Urbana e Rurale,* No 38, pp. 199–204.

Tsartas, P. (1993) International trends of tourism development in the Mediterranean and its impacts on traditional and historic settlements: problems and dilemmas. Paper presented at the International Congress 'Tourism Impacts on Traditional and Historic Settlements of the Mediterranean and South European Countries'. Ministry of the Environment, Physical Planning, and Public Works, National Tourism Organisation, and UNESCO, Thessaloniki, Greece.

Tsartas, P. (1996) *Touristes, Taxidia, Topi: koinoniologikes proseggissi ston tourismo (Tourists; Voyages; Places: Sociological Approaches to Tourism).* Exantas, Athens.

Tsartas, P. (1998) *La Grèce: du Tourisme de Masse au Tourisme Alternatif.* L' Harmattan (Tourismes et Sociétes), Paris.

Turner, L. and J. Ash (1975) *The Golden Hordes: International Tourism and the Pleasure Periphery.* Constable, London.

Urry, J. (1992) *The Tourist Gaze: Leisure and Travel in Contemporary Societies.* Sage, London.

Uzzel, D. (1984) An alternative structuralist approach to the concept of tourism marketing. *Annals of Tourism Research* **11**:1, pp. 79–99.

Weiler, B. and C.M. Hall (eds) (1990) *Special Interest Tourism.* Belhaven Press, London.

World Travel and Tourism Council (1992) *Travel and Tourism: the World's Largest Industry.* New York.

WTO/UNEP (1983) Workshop on Environmental Aspects of Tourism. Madrid.

WTO (1993a) *Sustainable Tourism Development: Guide for Local Planners.* Madrid.

WTO (1993b) *Travel and Tourism Barometer.* Madrid.

EXPERIMENTAL ICONOLOGY: A TOOL FOR ANALYSIS FOR THE QUALITATIVE IMPROVEMENT AND TOURIST DEVELOPMENT OF PLACES

Joseph Stefanou
Department of Architecture
National Technical University of Athens
Athens
Greece

1. Introduction

Experimental iconology is a new method for image interpretation, which first appeared in the work of the Institute of Social Psychology and Communication at the University of Strasbourg. More specifically, research on the psychology of space, which, was pursued in the 1970-80 period particularly, gave rise to this methodology. By means of experimental interventions in the image of a place (landscape) – as this image is imprinted on a medium of broad circulation, such as the postcards – the interpretation of how this image is being perceived by the user of the place, the resident or the visitor is attempted on all levels: pre-pictorial, pictorial and iconologic.

It is believed that an analysis of the perception of the image of a place by visitors and tourists and its practical, emotional and ideological communication can contribute significantly to planning interventions for the design of tourist promotion and development of a place and the improvement of its aesthetic quality. The following exposition of the methodology purports to show its potential contribution towards these ends.

2. Basic assumptions

The definition, interpretation and clarification of the term 'landscape' which underlies the ensuing analysis is given first: "*Landscape* is the general impression a place offers to its permanent occupants, visitors, passers, and those who simply dream of it. Landscape is neither space itself nor the place but the 'image' of the place."

Space has an abstract meaning (it is a containing entity), and is measured by some quantitative measuring unit (e.g. the metre), but as a containing entity it is perceived to be empty. Our interest is in its 'capacity', its density, but not in its actual content. In contrast, place is a specific space; space enriched with something existing. The place exhibits many functions of a practical as well as an emotional and intellectual nature. A place has one or more purposes and, at the same time, it bears messages of a psychological or ideological content.

The landscape is considered by everyone – from artists who have struggled for centuries to depict it to scientists who have shown particular interest for its interpretation in the last century – to be a picture, a visual entity, a Gestalt (George, 1970). Through the term 'landscape' we perceive the place's image, the impression we are left with from the shapes that constitute the space and convert it to a place through the purpose each one of them conveys. Moreover, we do not deal with the visual image only but with the more global impression which is supplied from the other senses as well (smell, hearing, touch) and the knowledge of our emotional relation with the place. There are, for example, auditory landscapes (Moles, 1971) or thermal impression landscapes (Rimbert, 1973). It is clear that, since we refer to 'impression', subjectivity is inevitably involved.

At this point, we must draw attention to the fact that we talk about the 'collective' impression, the collective image, as Lynch (1960) defines the abstract structured image, which determines the perceptual structure of a place.

A last assumption, which is made concerns the kinds of shapes, which compose and articulate a place and, hence, affect its impression. Spatial shapes have been already mentioned. The behavioural patterns which build up a place's operational fabric (facts, deeds, happenings, etc.) as well as the historical moments and periods that define it are equally important (Stefanou, 1978b).

3. The need for a multidimensional treatment of urban space and its image

The aforementioned assumptions lead us to conclude that the urban landscape – as an image of urban space – as well as any kind of landscape, communicates with its occupants through a series of multi-natured messages and information. Following various procedures to interpret a city's image – by means of successive approaches of differing degrees of analytical depth – we can locate, interpret and evaluate this information.

In our contemporary times, an inability with respect to this interpretation is experienced, particularly regarding information and messages which do not belong to the class of physically quantifiable entities. This is due to the loss or, rather, de-activation of the old interpretation codes and dictionary (Hall, 1966). We see only images whose context and content we are unable to interpret. Slowly then, but surely, we create textless images since we have not acquired new codes to interpret this content.

Cities are recognised as possessing full scale polysemantics and, thus, we should be able to perceive the different shapes of a city from various points of view, to define the various aspects of the entity called 'city'. These aspects are of a material, psychic and intellectual nature and fulfil the corresponding material, psychic and intellectual needs of their inhabitants (Moles and Rohmer, 1978). If we accept the Aristotelian axiom that 'The city materialises the well-being and it is impossible to live in a city without virtue, freedom, wealth and nobility', we realise the importance that should be attached to the non-material, psycho-intellectual and social dimensions of the city, since virtue, freedom and nobility pertain purely to the interpretation of the shapes and behaviour from the points of view of aesthetics and semantics and belong exclusively to the psychological and ideological sphere.

4. An approach to the study of the psycho-social aspects of the urban landscape and its synthesis

Contemporary research has shown that it is possible to locate several notional configurations of the landscape with the aid of a medium belonging to the areas of mass communication, mass art and commercial production, like the postcard, which specifically deals with the depiction of places (landscapes) (Stefanou, 1978a). The first research attempts on the interpretation of the urban landscape with the aid of postcards have led to certain conclusions regarding:

1. The socio-dynamics of culture (Moles, 1966) and the ways in which the circulation circuit of postcards and a place's celebrity interact with each other.
2. How places with tourist interest can be located in order to lay out a tourist itinerary with a generous sightseeing content; i.e. an itinerary which contains a condensed repertoire of snapshots (assembled view units), which are touristically interesting, via the shortest possible route.
3. The significance attached to a place by the existence or not of some famous and trivial elements, namely a platitude core (centre de banalité), principally on what concerns a place's attributes and recognisability, as it has resulted from the method of experimentation and successive excisions (découpages).
4. The establishment of certain psychometric quantities, whose existence has been identified by the experimental researchers and whose mutations have direct consequences for the evaluation of the aesthetics of space.

A few of these psychometric quantities, which constitute the principal agents of interest of an urban landscape are:

1. The *originality* of the landscape. One of the most important pursuits of the occupants-consumers, on the one hand, and those who trade upon the place's sightseeing quality, on the other, is its rarity, the uniqueness of its character, its originality.

2. The *variety* of the elements of the landscape. In our days, the consumers of a place originate in different parts of the world. For them it is easier to search for and identify patterns through the variety of shapes rather than though their simplicity.

3. The *reality's specific density*, the credibility of a landscape, the degree to which it resembles reality or dream, the form or some glimpse of it, the specific situation or some projection of it. From this attribute follows

4. The *power of being imposing (pregnance)* through which a shape imposes itself on a spectator. This power is based on contrast and cultural addiction, i.e. on our familiarity with a shape and the ease with which we may recognise it as a cultural unit (Moles, 1971).

5. The *appeal to one's imaginative power (imageability)*; namely, the power of a landscape to produce strong impressions (Lynch, 1960).

6. The *legibility*. This is a quantity of a special visual nature and refers to the pure manifestation of an urban landscape; namely, the ease with which someone reads a place's plan and functional diagram.

7. The *connotation*. This is the hidden, subconscious landscape quality that relates the shapes to some deeper context that each one of us assigns to them.

8. The *symbolism*; namely, the semantic ability of space, the power of a picture to refer to things specific or abstract.

9. The *degree of exaltation*; the ability of a place to 'emit' grandeur or importance.

10. The *degree of sensibility* which refers mainly to the Dionysian and Narcissistic attributes of space (Bachelard, 1948; Barthes, 1970).

11. *Intricacy* which corresponds to the density of information and usually contrasts to

12. *Redundancy,* which may be thought of as the simplicity in conveying messages. All messages coming from a landscape are classified by a person into trivial and original. The ability to perceive the content of a message is associated with its redundancy which is maximum for a trivial message and minimum for an original one.

13. *Conveyance ability* is another purely semantic quantity directly related to the attitude of "it is like ...". In most cases, the reasoning behind a landscape's aesthetic appraisal coincides with this attitude: "it is nice because it looks like ...".

14. The *hierarchy of the constituents*. This quality refers to the structural properties of a landscape. Such a hierarchy is founded on geometric size criteria, existence or not of accentuated elements, their topology, etc. that define the classification order.

15. The *unity of the structure of perception*. This property is distinguished into tediousness and diversity. It ensures the adaptation of a landscape and its constituent elements into the surrounding environment.

16. The *degree of platitude* that bears a direct relation to the aforementioned power of being imposing.

17. The *degree of appropriation of a space* which also a function of the power of being imposing, the legibility and the power of orientation (De Lauwe, 1970).

All these psychometric quantities coexist, on a positive, negative or neutral basis, in every shape of the urban landscape and they define the space depending on their mode of participation. Before deepening into the use of these quantities in urban landscape analysis and evaluation, let us make some logical statements.

- An urban landscape is described by its occupants as tragic, sad, excellent, romantic, picturesque or, in general, it is classified in one or more aesthetic divisions that may describe an aesthetic object of such scale.
- This description is founded upon the participation in the shape of this landscape of a series of the previously mentioned psychometric quantities.
- These quantities are, in turn, based on specific shapes and functions of the landscape in question.

These statements imply that there are certain shapes, complexes of shapes and functions which, when invoked, signify the aforementioned quantities which, in turn, constitute the corresponding aesthetic division. It follows that every aesthetic division is based, in principle, on specific shapes. In this way, we can proceed to the stage of synthesis as follows. If we are aware of the shapes, which may be used as arguments to signify certain desired psychometric quantities, and then determine their degree of participation in order to achieve the aesthetic division sought for, we may set up, in a sense, the surrounding environment (aesthetic division) of the scene (landscape) as well as the script (function schedule). The ensuing discussion refers to the use of the postcard (as an agent of a place's picture with significant social appeal) to investigate the psychometric quantities cited above and their subsequent use to analyse the aesthetic evaluation of places. The approach utilised consisted of the following stages:

1. Decomposition of a landscape into units of space, time and behaviour.
2. Compilation of a morphological dictionary of the urban landscape and formalisation of its content.
3. Interpretation, using semiology and experimental iconology, of the basic types of shapes and extraction of the rhetoric and ideology of the pictures of the place under examination.
4. Interpretation of the relations among these shapes, their assembly and arrangement rules and drawing of conclusions concerning their composition.

In order to pursue such a multidimensional approach to the aesthetic study of the landscape complying with the requirements of contemporary scientific aesthetics, a basic prerequisite is the ability of the researcher for essential insight during interpretation at all the stages of perception which semiology and traditional iconology identify. These stages are:

1. The initial simple introductory stage which is the stage of the simple description of the shapes. It is called reality stage in semiology and pre-pictorial interpretation in iconology. The aforementioned dictionary and formalised shape classification ensure precision and scientific discipline during interpretation.
2. The function interpretation stage which gives the narrative record of the picture. In semiology it is called stage of denotation and, in iconology, stage of iconographic interpretation.
3. The deeper approach stage which may not be supported by the mere knowledge of the interpreter. This is the stage of intellectual connotation, which, for iconologists, is the precise iconological interpretation itself; namely, the deeper purpose and disguised ideological message of the picture.

At this point, it is noted that both theoretical semiology and traditional iconology fail to avoid the researcher's subjectivity and arbitrary judgement. However, there is another, even more serious, deficiency of these approaches. During any interpretation of an assembly of points (and landscape is such an assembly) the following are assessed and analysed:

1. The relations among all points. This analysis reveals their structure, i.e. the fashion in which they have been arranged in order for a rhetoric to be credited to them; this is the *structural analysis* (Barth, 1964).
2. The relations with the people. This assessment leads to their general meaning; namely, their *semantic interpretation*.
3. The relations with the specific occupants of space which cannot be assessed through the researcher's theoretical and subjective point of view since they refer to an objective social reality. This is the *real interpretation*.

Both approaches, semiology and iconology, gave no results on these points to date. Therefore, we formulated and applied an approach to the examination of the psychometric quantities of a landscape as follows:

1. Recourse was made to a specific aesthetic division whose historical past has been, on the one hand, theoretically analysed to ensure, without significant deviations, the structural and semantic interpretation of shapes on which the corresponding psychometric quantities are based and whose present, on the other hand, is vivid enough to enable us work essentially with the present occupants of the real interpretation stage. The class of romantic landscapes was chosen as the most relevant class since romanticism, as a psychological and ideological reaction to established entities and practices, is always more or less alive and since a new romantic climate has revived during the last years through the postcard with emphasis on the landscape.
2. The landscape, as an artistic and theoretical entity, the postcard as a means of approaching it and romanticism as an environment were thoroughly researched. Romanticism, in both its historical and present day manifestations, never ceased to employ a certain repertoire (the context of the myth), certain central characters

(the heroes of the myth), certain situations (the psychological framework of the myth) and certain space, time and behaviour patterns to express itself. Moreover, as a reaction to all academic and classical models, it rejects the established rules and does not cease to produce, through its own counter rules of synthesis, a characteristic compilation model of its own. The postcard, as a form of mass art has proven to be an excellent means for locating all these elements to such an extent that it has enabled the creation of a calibration matrix for the measurement of the degree of romanticism of a landscape.

3. Finally, experimentation was used in order to move to the deeper stage of interpretation of each image of the place and to study its correlation with the real interpretation; namely, the interpretation concerning the present occupants of space.

In this way, the significance of the romantic landscape (and this applies to every other aesthetic landscape class) was determined, the shapes bearing some of the most important psychometric quantities were established and, in addition, the transitional ability of a landscape to vary with its degree of romanticism – depending on the presence or not of new parameters or activities within it was realised.

Beyond this, experimentation helped to determine the significant vivid role of myth through which every practical, psychological or intellectual intervention in a place's image is implemented. It revealed also the imperishable value of the original and model shapes, which serve as an environment to satisfy the contemporary citizen's activities. Finally, it gave substance to the two fundamental aspects on which the concept of the landscape is based: visual whole and unity, on the one hand, and love of nature, on the other.

5. Concluding remarks

The presentation of certain basic precepts of experimental iconology and of the psychometric quantities employed within this context purported to expose this tool to a wider audience who may use it in various contexts for various purposes. One such context is tourist development and aesthetic enhancement of a place. Application of experimental iconology can help to pinpoint those elements of a place which make the greatest contribution to its tourist attractiveness and appeal, as well as of those which detract from it, those elements in need of enhancement, the qualitative aspects of a landscape most important to visitors, the psychological connotations of places and their relevance to the needs of visitors and tourists. The results obtained via this type of analysis provide the material necessary for synthesis, i.e. creation of tourist images and places which ensure the sustained attractiveness of an area, as well as for designing tourist policies to make this attractiveness pay.

References

Barthes, R. (1970) *Élements de Sémiologie*. Gauthier: Paris.

Bachelard, G. (1948) *L' Eau et les Rèves*. S. Corti: Paris.

George, P. (1970) *Les Méthodes de la Géographie*. P.U.F.: Paris.

Gibert, F. (1972) *Composition Urbaine*. Dunod: Paris.

Hall, E. (1966) *La Dimension Cachée*. SEVIL: Paris.

Lynch, K. (1960) *The Image of the City*. MIT Press: Cambridge, MA.

Lauwe, Ch. de (1976) *Appropriation de l'Espace et Changement Sociale*. I.P.S. SOCIAL: Paris

Moles, A. (1971) *La Communication*. Marabout Université: Paris.

Moles, A. and E. Rohmer (1978) *Psychologie de l'Espace*. Casterma: Paris.

Moles, A. (1966) *Socio-dynamique de la Culture*. Mouton: Paris.

Moles, A. (1971) *Art et Ordinateur*.: Casterma: Paris.

Panofsky, E. (1969) *L'Oeuvre de l'Art et ses Significations*. Gallimard, N.R.F.: Paris.

Rimbert, S. (1973) *Les Paysages Urbains*. A. Colin: Paris.

Stefanou, J. (1977) La Tarjeta Postal y el Paisaje Touristico. *Guadermos de Communication*. No. 27, Mexico.

Stefanou, J. (1978a) Les filtrei de la lecture. *Messages*, No. 10, Paris.

Stefanou, J. (1978b) *Dimensions psycho-sociales du paysage urbain*. U.L.P.:Strasbourg.

Stefanou, J. (1979) *Introduction to Semiology*. National Technical University of Athens, Athens (in Greek).

Stefanou, J. (1980) *Vers une Méthodologie de Protection du Patrimoine Architectural et Urbain*. Paris.

Stefanou, J. (1981) *Études des Paysages – Vers une Iconologie Experimentale de l' Image*. Strasbourg.

Stefanou, J. (1985) *Notes on Urban Synthesis*. National Technical University of Athens, Athens (in Greek).

THE CONTRIBUTION OF THE ANALYSIS OF THE IMAGE OF A PLACE TO THE FORMULATION OF TOURISM POLICY

JOSEPH STEFANOU
Department of Architecture
National Technical University of Athens
Athens
Greece

1. Introduction

The image of a place, its landscape, constitutes the principal means by which the tourist development of this place is attempted. Posters, prospectuses, advertising pamphlets, tourist guides, advertising spots, video-films or even films and documentaries are all mobilised for the tourist promotion of an area and the basic ingredient they have in common is the image of the area.

This portrayal, the imprinting of the landscape on the cellulose of a film, the paper of a prospectus or a postcard, the cloth of a shirt or a bag, leads to an indirect contact with the place itself, to an indirect knowledge and, by extension, to a first appropriation of this place.

The evaluation as well as the aesthetic and semantic interpretation of a landscape depends on the degree and mode by which the image is perceived and formed and the degree of mental, psychological and practical appropriation of the landscape by an individual. However, individual perception and appropriation are influenced by, and several times depend upon, a number of social factors. As Fulchignoni (1969) shows very successfully, constraints, or, more precisely, complexes of constraints (biological, psychological and social) of differing intensities determine, to a large extent, human behaviour. Psycho-social trends are the most intense and most complicated of these constraints. These trends manifest themselves in individual behaviour, of course, but also in the behaviour of organised groups (mass behaviour) and, more importantly, in the relationships between the two.

Because a series of criteria is necessary for the evaluation of a landscape, an effort is made in the present study (Stefanou, 1978) to elicit them by means of a medium which captures the image of a place on a large scale. This medium is the postcard. The next section outlines and justifies an analytical approach proposed towards this purpose. Applications of the approach are presented in the third section while the last section offers some concluding remarks.

2. An approach for eliciting landscape evaluation criteria on the basis of postcards

Communication through a given medium means that somebody knows the codes of the system this medium uses to transmit messages. Communication by means of the portrayal of a landscape, sometimes disengaged from several other explanatory signs, presupposes the use of an information system equipped with a code which is known among the communicators. In other words, it presupposes that the receiver is in a position to interpret, to decodify the message sent by the sender. In the case of the postcard system all factors, which participate in the creation of the signifying signs of the system are assumed to be known. At this point, it must be emphasised that this communication mode, i.e. the postcard, offers great services in the development of a visual education, as Arnheim (1967) has conceptualised it. In this respect, the postcard is a real mass art whose codes are easy and popularised and, thus, fit to everybody's perception. Parallel to this, however, the postcard system is an commodified system whose object, the landscape, is used as a consumer product.

Therefore, the postcard is a medium, which depends, on the one hand, on the laws of the mass art and, on the other, on the laws of consumption either direct (applied to the landscape), or indirect (visual-cultural consumption of the image of a place). In order to obtain information or to draw some inferences on the criteria of the psycho-social evaluation of the landscape on the basis of the postcard system, it is necessary to analyse (1) the behaviour and stance towards this medium of those who communicate by means of the postcard, and (2) the behaviour and actions of those who produce this product and of all those who participate in the whole circuit "conception, materialization, production, sale, selection, use of the postcard", i.e. of the producers, publishers, photographers, etc.

Thus, the process of analysis should consist of three phases. The first phase includes the collection of information on the activities and behaviour, with respect to the postcard, of all the groups involved in the circuit mentioned before on the basis of representative samples drawn from these groups. The second phase includes the semiotic analysis and interpretation of the information collected on the basis of the postcards of various places. Finally, the third phase is directed to the analysis of the places themselves depicted on the postcards in order to compare the elements which the postcard shows (indirect approach of the place) and the direct contact with a place. The analysis of the places whose images are reproduced by the postcards could also help in better identifying those collective images of a place which its visitors, the tourists, form (Rimbert, 1973).

Following this approach, the relationships among the place (experienced reality), its landscape (perceived image of the place) and the postcard (portrayal of the landscape to be consumed) are clarified. In this way, it is possible, to observe the

phenomenon of the socio-dynamic of culture (Moles, 1966) by means of the feedback:
- the postcards choose a landscape because it is famous, and
- landscapes are famous because postcards diffuse their image throughout the world.

The last point is supported by the fact that when the postcard distribution system changes, tourist movements change as well. The same phenomenon is evident in places of a great aesthetic interest which are preserved and protected exactly because they have been established by means of the great diffusion of the image (postcard) to which these places offer content. Characteristically, the case of the 'broad postcards' is mentioned here as Korosev-Serfaty (1976) and her group of the Psychology of Space Research Team of the University of Strasbourg call them. The image of these 'broad postcards' solidifies as a picture and as a decoration, it is protected by the law, and it is carefully preserved. Such is the case of many village squares of Alsace.

The study of the postcards becomes of interest once they are considered as images of a given place and inferences are sought about the behaviour of the users towards this landscape. The term 'landscape' here is defined as the image-perception of a place (Stefanou, 1981); i.e. it is the collective image, as Lynch (1960) puts it, as it is shaped as a whole by the real ideological and psychological perception of a place. When the postcard is considered as the image of a place, it must be accepted that it is essentially the picture of the landscape of a place, i.e. the image of the image of the place or, more precisely, image of the whole collective perception of the place. This view became very clear during the conduct of a special research by the author (Stefanou, 1977). The creators of the postcards seek very carefully to place their pictures against a background or within a frame, to accompany them with legends, advertising slogans, emblems, symbols, etc. in order to complete, in this way, the impression of the image of the perception of a place. On the other hand, it is equally obvious how the buyers of the postcards pay special care, during the selection phase, to buy an image which is full of references to the real givens (characteristics) of the place as well as to its ideological, emotional and symbolic dimensions.

The process of analysis suggested before permits the identification of the patterns of behaviour towards the image of a landscape and the drawing of inferences as regards the criteria by which this landscape is judged and evaluated by its users. In the following section, some empirical applications of this approach are given, the focus being especially on the production of the postcards, i.e. on the treatment of the landscape by the producers and creators of the postcards. These examples are characteristic of the potential for analysis offered by this medium.

3. Some empirical applications of the approach

A first application concerned the determination of short travel itineraries with a high density of visits. The study started with an examination of the most popular postcard publishers in Greece (Stefanou, 1978) which resulted in the drawing of a map showing the distribution of the themes of the postcards by publishers all over the country in relation to the interesting aspects of the places.

The next step was to develop a histogram of the Greek cities according to the number of postcards in which they appear (at the time of the research). On the basis of this histogram, a map showing the first 35 cities appearing in the greater number of postcards (minimum 20) was developed (Figure 1). These steps led to the formulation of several itineraries, which exhibit the largest density in terms of places with landscape and, consequently, tourist interest. An example of such an itinerary appears in Figure 2.

These itineraries are based on the selection of the shortest and simultaneously more dense, based on the number of landscapes included, route leading to several destinations. In other words, their laying out includes more and bigger circles which represent, depending on the scale, the number of postcards dedicated to a certain place

A second application concerned the examination of the evolution of the ideological direction of tourism policy in a given place over time. In Greece, for example, depending on the themes and the way they are represented in postcards, five phases are distinguished (Figure 3).

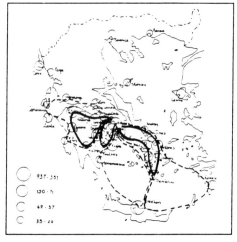

Figure 1. The first 35 cities appearing in the greater number of postcards

Figure 2. An example of a dense and quick tourist itinerary

During the *first phase*, the ancient monuments, the sea, the sun and the picturesqueness of the country are brought forward. This phase can be called the phase of the 'innocent intents'.

A *second phase* follows during which elements of traditional architecture and culture are added to those of the first phase in the postcards; e.g. neo-classical buildings, statues, small local monuments, etc.

The *third phase* characterises the period of rapid tourist growth and the element, which is projected strongly, is the consumption of luxury. Large hotel complexes, swimming pools, luxurious beaches, cosmopolitan environment occupy most of the surface of the postcards.

During the *fourth phase*, the element of luxury passes on the consumption of the picturesque and romantic as well. The impression of the 'picturesque happy poverty' is set aside by the enjoyment of the person who experiences it as a picture for his/her pleasure in comfort and luxury.

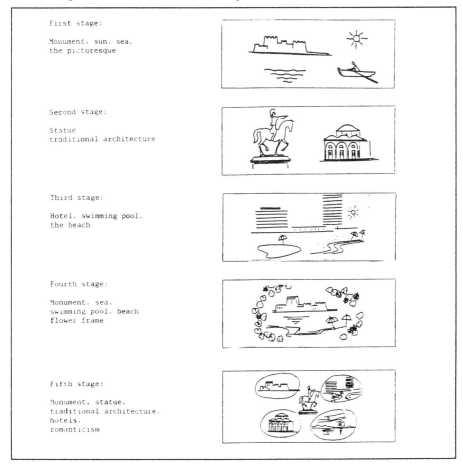

Figure 3. The five stages of the evolution of the ideology of tourist policy in Greece

Finally, in the *fifth phase*, which corresponds to the phase of tourism recession, the picture of the 'sell out' is intense. Historic monuments, traditional architecture, cultural elements, luxury hotels, romanticism, picturesqueness and everything else that is demanded at the given moment are squeezed in a picture and they are sold at the price of a single postcard.

A third application concerned the identification of the number of times the most interesting places within a settlement are portrayed in postcards. This is an application of special interest at the level of community planning when planning interventions are directed either to the preservation of the physiognomy of a community or to its tourist development. The city of Nafplion is used as an example of this application, which shows, at first glance, the degree of importance each element holds in building the tourist image of the city. So, in the 44 postcards in circulation in 1976-1977 (time of the research) depicting the city of Nafplion, the island of Bourtzi appears 63 times, the castle Palamidi 44 times, Akronafplia 28 times while various tourist facilities 38 times (Figure 4). It is worth noting that the tourist image resulting from recording these representations with a line on the map is completely identical to the image resulting from other experimentations or questionnaires to tourists.

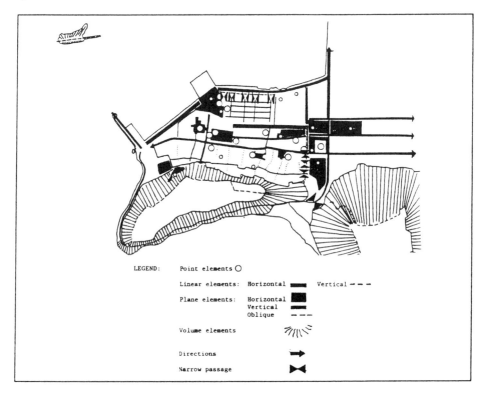

Figure 4. The residents' image of Nafplion (using Lynch's notation)

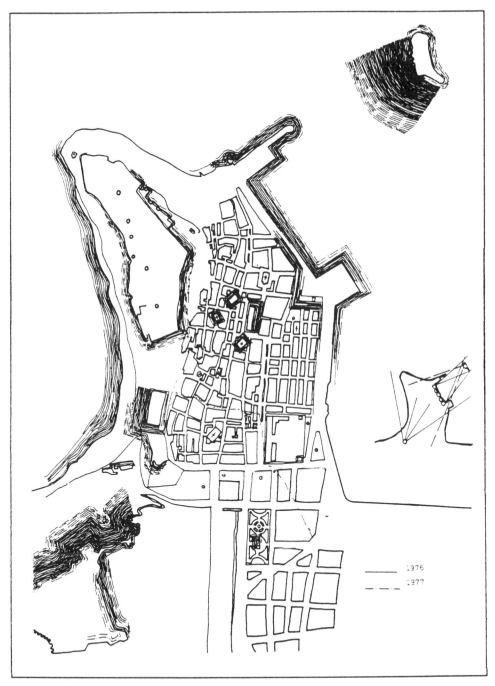

Figure 5. Number of times the various elements of the city of Nafplion appear in the postcards in circulation in 1976 and 1977. Each line represents one appearance.

It is also characteristic that the tourist image of the city differs significantly and uses different elements and hierarchies from the resident's image (Figure 5). For example, while the primary role in building the residents' image is held by the central square, or the square of St. Spyridon where the first Greek governor Kapodistrias was assassinated, these elements are totally absent in the tourist image. On the contrary, in the tourist image the most important element is the castle Bourtzi that has just a decorative meaning for the residents. In fact, Bourtzi, according to another set of experiments (Stefanou, 1984), constitutes a centre of banality in the perception of this place. This was shown by means of three sets of experiments:

(a) Bourtzi appeared on a postcard of Nafplion but other elements of the city were replaced by elements of another city.
(b) Bourtzi appeared on a postcard in which the rest of the city of Nafplion was replaced by another city.
(c) Bourtzi was placed on a postcard of the port of another city. In all cases, the persons questioned 'identified' the city appearing on the postcard as Nafplion!

Moreover, the way in which the map of the tourist image is constructed on the basis of the number of times the various elements of a city appear on the postcards offers the opportunity to identify the 'points of banal view' within a settlement (Stefanou, 1984); in other words, these viewpoints from which the various elements of the settlement present their most well-known, their most 'banal' view. This explains the fact that not only a large number of postcards present a square, for example of Mykonos, from the same vantage point, but also that every day all amateur photographers visiting the island as tourists crowd these points to take the same so well known shot of their own. The identification of these centres of banality as well as the points of banal view help greatly in the selection of the proper interventions for the tourist promotion and development of a place.

A last application of the approach suggested here, which may contribute to the efforts for improving and enhancing the aesthetic quality of a settlement and, thus, to its tourist promotion, is the identification, with the use of postcards, of a series of psychometric magnitudes, which characterise qualitatively a place. These psychometric magnitudes lead to the aesthetic characterisation of a place depending on the degree of their participation in the construction of the image of a place. This topic is analysed in another section in this volume.

4. Concluding remarks

The postcard is usually considered as a communication medium, a consumer product and as mass art. The approach proposed in this study and its applications briefly described show the utility and value of the postcard as a tool for the scientific analysis of the landscape of a place to aid, among other things, in designing more informed tourism policy and relevant planning interventions.

References

Arnheim, R. (1967) La Pensée Visuelle. In *Éducation de la Vision*. Bibliothèque de Synthese et la Connaissance. S.A.: Bruxelles.

Fulchignoni, E. (1969) *La Civilisation de l'Image*. Payot: Paris.

Korosec-Serfaty, P. (1976) *Appropriation Vecue, Revée Impossible*. Third International Conference of the Psychology of the Built Space, Strasbourg.

Lynch, K. (1960) *The Image of the City*. MIT Press: Cambridge, MA:.

Moles, A. (1966) *Socio-dynamique de la Culture*. Mouton: Paris.

Rimbert, S. (1973) *Les Paysages Urbains*. Armand Colin: Paris.

Stefanou, J. (1977) La Tarjeta Postal y el Paisaje Turistico. *Cuadernos de Communication*, No. 27.

Stefanou, J. (1978) *Dimensions Psychosociales du Paysage Urbain – Critères d'Analyse du Paysage par la Méthode des Cartes Postales*. U.L.P.: Strasbourg.

Stefanou, J. (1981) *Études de Paysage vers une Iconologie Experimentale de l'Image. Institute de Psychologie Sociale*. U.L.P.: Strasbourg.

Stefanou, J. (1984) The Perceptual Structure of Space. *Archaeology* **12** (in Greek).

YVON. French Publishing Company Using Ten Thousand Themes and Producing 120 Million Postcards Annually. In general, postcard production is a major industry producing millions of pictures. In Greece every postcard publisher produces about 10 million postcards annually with about two thousands themes.

COMMUNICATION STRATEGIES, TOURISM AND THE NATURAL ENVIRONMENT

CEES VAN WOERKUM,
NOELLE AARTS
and
CEES LEEUWIS
Agricultural University
Wageningen
The Netherlands

1. Introduction

Several contributions in the present volume show that the relationship between nature and tourism is not without tension. In some cases, tourism and nature are largely separate, but, more often than not, tourists prefer to spend their leisure time in beautiful natural areas, where serving tourists facilities are available. From the perspective of improving the biodiversity of natural areas, tourism is an opportunity as well as a threat. By experiencing the beauty of nature, tourists can learn and appreciate its richness and develop an awareness of the importance of nature as part of a national heritage. But by travelling around and living in these beautiful areas, tourists can spoil elements of nature and contribute to the everlasting destructive influence of mankind on the surrounding natural resources.

In this chapter, we argue that professional communication can play a role in bringing about a sustainable balance between tourism and nature. Evidently, communicative intervention is but one of the strategies that can be employed to this end. After all, human activities are not only a result of the interpretative processes induced by communication, but also of economic incentives, power relations, laws and regulations, etc. At the same time, no meaningful social change will come about without human communication. Communication is a very important mechanism through which human beings process reality, develop perceptions and thoughts, and (re)produce practices (see, e.g.. Giddens, 1984). We look at communication in relation to tourism and nature from a sustainability perspective. In other words, communication must function to make a certain region capable of preserving its 'natural capital' *vis-à-vis* the claims made by tourism and other activities such as agriculture, housing, traffic, and industry. Regional development is becoming increasingly more than a summation of all kinds of separate sectoral plans, whereby solutions in one sector tend to bring about problems for others. Instead, we are witnessing integrated efforts to manage and combine diverse sectoral interests within planning processes (Tatenhove and Leroy, 1995; Koppenjan *et al.*, 1993; Glasbergen, 1995). Tourism is only one of these interests.

In this paper, we will not only consider communication as an instrument of governments or NGOs to promote respect for nature and nature conservation by means of public awareness raising. We will also discuss professional communication as a means of bringing various stakeholders together in order to negotiate better policy plans, that is, integrated regional plans, in which nature values and other interests are combined and preserved in a creative and sustainable manner (for an overview of both approaches, see Van Woerkum, 1999). First we discuss 'instrumental communication' as an element in a wider policy context to enhance nature conservation; we then move to what we call 'interactive policy making' where communication possesses a constructive role.

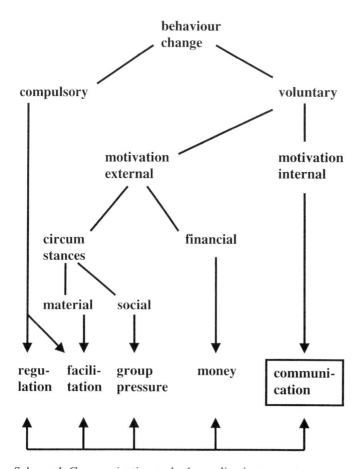

Scheme 1. Communication and other policy instruments

2. Communication as a policy instrument

Policy is meant to influence (patterns of) human behaviour. Building on policy theory and social psychological models (Fishbein and Ajzen, 1975; Petty and Cacioppo, 1986) we can construct a model which gives an overview of the different 'policy instruments' that are available to this end (Scheme 1).

In relation to nature development and nature conservation these 'policy instruments' can be operationalised in many ways. We will give a few examples and then we shall elaborate on the communication instrument.

The instrument of regulation is often operationalised in the form of laws or other rules of conduct; one can forbid tourists to enter vulnerable sites or oblige farmers to take measures in order to preserve the environment. Regulation has an image of being a 'strong' policy instrument. But in practice, we often see that rules are not adhered to. This may relate not only to a limited motivation to conform, but also to a lack of visibility and enforcement opportunities. Especially in the field of the environment, human behaviour is very difficult to control: "You can't put a policeman behind every tree."

'Provisions' are all those facilities and forms of organisation that help to preserve and enhance natural resources, and/or prevent their being abused. The institution of natural parks is one of them. High fences around them is another.

In European democracies, social pressure is not an instrument that can be easily mobilised as a separate policy instrument (unlike, for example, in Mao's China). However, social pressure it is to be kept in mind, for much human behaviour is influenced by the pressures of others. Thus, in a more indirect manner, social pressure can play an important role in the 'policy instrument mix', especially in combination with the instrument of communication. In fact, many behaviour change models are based upon combining these two instruments (see, e.g., Fishbein and Ajzen, 1975).

Money can be relatively easily used as an incentive to promote certain types of behaviour (if you want farmers to plant and maintain trees, pay them for it). Of course, it can also be used to discourage undesirable phenomena (if you want to decrease the number of tourists in a forest: ask a considerable entrance fee).

Lastly, we arrive at communication as a policy instrument. First, communication can be considered an instrument on its own (Weiss and Tschirhart, 1994). With the help of professional communication, policy makers can attempt to change people's behaviour by influencing their knowledge and attitudes. For instance, we can explain to tourists the importance of undisturbed breeding and nesting zones for birds, persuade them to go to certain parts of a nature reserve and to avoid others. This strategy, however, has certain limitations, to which we will return shortly. Secondly, communication can function in support of other policy instruments and help improve their effectiveness.

Communication as a 'stand-alone' instrument has a rather positive image as well as an optimistic flavour. Why not influence people by telling them the truth and effect internally motivated, voluntary change? Of course, we know that such behavioural change is possible. Indeed, some people have learned to love and understand nature: those who look after their behaviour from an environmental perspective, who do not litter, do not use their cars to 'enjoy nature' and refrain from picking rare flowers. However, in most people such inclinations have to compete with other motives: to choose the most comfortable way, to maximise short-term profits, and so on.

One of the big problems is that many people would, in principle, be willing to behave in an ecologically sustainable manner, if only they were convinced that everybody would do so (Baron *et al.*, 1992; Messick and Brewer, 1983). In the absence of trustworthy institutional arrangements to this effect (Ostrom, 1990), everybody is waiting for the others to act first and nobody is doing anything. This is, in a nutshell, the social dilemma problem, which refers to the conflict between individual and collective interests (Messick and McLelland, 1983; Orbell and Dawes, 1981). In these situations, individuals must choose between giving priority to their own immediate interests or serving the longer-term interests of a larger community. The case of nature preservation offers a good example of such a social dilemma situation.

From the literature on behaviour change, we know that there is a complex relation between cognition and human action (Fishbein and Ajzen, 1975; Petty and Cacioppo, 1986; Koelen, 1988). For one thing, it is clear that cognition is not a sufficient condition for sustained behaviour change. In many cases, people are cognitively aware of the negative impact of certain types of behaviour and the positive consequences of alternatives, but they have insufficient motivation and/or trust in others to follow suit. For this reason, it is important to include other, additional instruments in a policy plan, and devise other means to involve and motivate people. One of these means is to make them more responsible for the plans that are formulated. We discuss this latter issue in the second part of this chapter. First, we take a closer look at the role of communication in support of other policy instruments.

3. Communication in support of other policy instruments

The main function of communication in a 'policy mix' is, possibly, to make other instruments more effective. Regulation, provisions or money can be much more effective if careful attention is paid to communication and the communicative dimensions of the various policy instruments. In essence, this support function of communication can be twofold: (1) to inform the public and (2) to promote and/or increase acceptance. We will illustrate this function in relation to specific policy instruments.

Informing tourists about regulations regarding nature has an important role. In France, for example, some rivers and lakes are not open for motorboats and tourists have to know that. However, tourists do not usually read local newspapers; they are 'moving targets' for informative messages. Hence, one has to think creatively about an effective way of reaching these 'hard to reach' groups. In a similar vein, tourists need to know how they can reach a natural park, where they can park their car, etc. Many provisions run the risk of under-utilisation if nobody knows of their existence.

Promoting the use of natural parks or other provisions is another function of communication. If we want tourists to have contact with nature, we have to plan campaigns. An effective campaign requires systematic thinking and balanced decision-making with regard to the five basic elements of a communication plan (see Windahl and Signitzer, 1992). Thus, important questions are:

1. What are the objective we wish to pursue with the help of communication?
2. What is the target group we want to reach? What relevant diversity exists within the target group?
3. What communication strategy and approach do we wish to follow?
4. What mix of media do we need to use in order to be effective?
5. What organisational arrangements are required to run this campaign? What roles do different people play and at what point in time?

In contrast to the promotion of provisions, a fourth possibility is to use communication in order to make restricting measures more acceptable. Here, the idea is that the monitoring and sanctioning of deviant behaviour can be minimised if tourists understand the rationale of rules and regulations. Such understanding, then, may lead to acceptance and co-operation. Regulations regarding littering, for instance, are notoriously difficult to enforce. People often love the absence of other people in forests and nature reserves, which inherently goes along with an absence of controlling eyes. In such cases, a thorough and convincing explanation of the consequences of littering is one of the few options one has to prevent such behaviour. Here, too, a systematic campaign approach (see above) can yield good results (see also, Geller et al., 1982).

We have seen that communication can serve as an important ingredient in a so-called policy mix. As a stand-alone instrument, communication has a number of limitations in inducing sustained behavioural change. Nevertheless, no other instrument can be effective without proper communication.

4. Shortcomings of instrumental communication

The instrumental use of communication as an element in a policy mix has long been the dominant mode of thinking on the role of communication in policy processes. However, this approach is associated with a number of conceptual and practical problems. At the conceptual level, the approach tends to be rooted in somewhat mechanistic, deterministic, and rationalist modes of thinking about the nature of human action. Moreover, the approach seems to overemphasise the individual dimensions of human action (for a conceptual criticism, see, Leeuwis, 1993). As a result, human activity is seen as rather predictable and amenable to quite a bit of social engineering, if only one pulls the right strings. Many authors have pointed to the shortcomings of such modes of thinking (Giddens, 1984; Bourdieu, 1990) especially in the context of policy and intervention processes (Long, 1990; Long and Van der Ploeg, 1989). These critics argue that human action can be more fruitfully understood as an outcome of active social negotiation and construction processes through time and space. In practical terms, this means first and foremost that one cannot look at the acceptance or non-acceptance of policies in isolation from the way such policies have been constructed over time. Here we arrive at the main practical shortcoming of the instrumental approach; namely, that it is positioned mainly as a tool to implement an already formulated policy. In other words, the philosophy is that first there is a plan and then communication follows in order to guarantee public acceptance. Below, we will illustrate that acceptance itself must be seen as a multi-dimensional process and cannot be properly understood without reference to a wider social and historical context (Aarts, 1998). When considering the acceptance of nature policy by tourists, for example, a number of dimensions play a role that tend to be disregarded within a conventional instrumental approach.

A first dimension of acceptance relates to the policy motive as connected with a particular problem definition. If tourists do not perceive ecological problems and reduction of biodiversity as serious and urgent threats, it is likely that they do not see the need for a nature policy in the first place. Secondly, even if tourists recognise that a problem exists, this does not automatically mean that they accept the intervention of particular agencies and/or authorities. It may well be that tourists feel that it is not legitimate for particular institutions to interfere with their autonomy, especially when it comes to how they spend their free time. In relation to this, tourists can be very creative in obstructing and undermining particular policies. Clearly, relational factors between citizens and institutions, as created through time (for example, on the basis of previous experiences and/or pressure from others) are important in shaping tourists' responses. Thirdly, it is important that tourists accept that certain measures are really effective in reaching a particular goal. Hikers in the Rocky Mountains, for example, are not allowed to leave the trails, whereas all kind of large animals can wander around freely. It is often not immediately evident why such regulations exist and how they contribute to solving a particular problem.

But still, accepting effectiveness is not enough. A fourth dimension of acceptance is that measures must be accepted as realistic and applicable. For many inexperienced hikers, making a large detour in order to avoid vulnerable areas is not realistic. Finally, measures must not only be effective and applicable, they must also be considered fair. On some lakes, for example, windsurfers are not allowed to sail, though all sorts of small boats have access. These windsurfers protest because they feel that they are not treated fairly.

The above points imply that acceptance emerges as a socially negotiated and complex process. It is therefore not realistic to expect that it can be ensured by instrumental communication as the final step in the policy process, in which those who are expected to accept the policy play no further role. Many policies are designed only within a small circle of policy experts, politicians, and specific societal groups who go through a learning process, which results in certain measures, based on a particular set of arguments. These arguments are the product of a lot of thinking and discussion. However, these arguments mean very little to citizens who are outside this small circle and who have not gone through the learning process from which the measures resulted. Here it is important to recall that "arguments are worthless unless people are looking for them."

The solution for these types of problems must be found in the policy process itself. Our conclusion is that stakeholders need to be involved in one way or another in the process of policy formation. The role of communication, then, is not to 'sell' a particular policy once it has been designed, but rather to contribute to the formation of the policy itself. In other words, the policy becomes the result of communication instead of the other way around.

We want to make it clear from the outset that this approach is not the ultimate solution to problems of policy acceptance. 'Interactive policy making' has its virtues, but, as we will see there are also a number of pitfalls. Still, it is a new line of thinking, which is worthwhile exploring further.

5. Interactive policy making: communication and negotiation

The essence of interactive policy making is to organise and facilitate a communication process in which stakeholders negotiate and work their way towards viable policies (Aarts, 1998). In relation to policies concerning nature and tourism, the stakeholders that immediately come to mind are NGOs in the field of nature conservation, representatives of the tourist industry, farmers' unions, governmental authorities at various levels, and NGOs interested in preserving the cultural history of landscapes and regions. Here we encounter an initial difficulty: who represents the interests of tourists in all this? The problem is that 'tourists' constitute a rather diffuse and highly segmented category that is often poorly organised. Moreover, tourism usually is only a minor element in a person's existence and identity (unlike farming), so that few are really motivated to join policy processes in their capacity as 'tourists'. This means that somehow tourist representation must be organised deliberately. On the one hand, we can point to very special segments, like mountain climbers or hikers, cyclists, hunters, bird watchers, sailing enthusiasts, etc. Sometimes these groups have their own organisations and/or journals that may play a role as representatives. On the other hand, there is also the more general public. In the Netherlands, for example, interest groups of automobilists also feel that they have a mandate to represent the manifold interests of their members in their role as tourists. Similarly, general consumer organisations often can and do function as representatives in this respect. Both these general and specific organisations can serve as a link between the policy process and a variety of societal interests, desires, and expectations.

Involving these groups in policy making is a challenging task. Yet, we feel it can be worthwhile in specific situations. If, for example, a governmental body wanted to make a regional plan for sustainable tourism, it would be quite feasible to invite representatives of different groups to co-operate in an effort to design an acceptable plan. Clearly, certain basic conditions must be met within the regional government for such an effort to succeed at all. For example, there must be a degree of political and bureaucratic commitment; stakeholders need to be allowed a certain space for manoeuvre, and bureaucrats must be willing to transcend sectoral boundaries (Aarts, 1998; Leeuwis, 1995). From a tourist-and-ecology perspective, the need to pay attention to interactive policy making is reinforced strongly by the tendency towards integrated policy making. Regional governments, especially, no longer tend to produce separate policy plans that deal with just one domain or sector (be it nature, agriculture, tourism, water management, etc.). Instead, they tend to make combined plans, in which various domains are integrated. This development stems from an increased recognition of the 'multi-functionality' of natural resources. For instance, water resources have several functions: to deliver drinking water for the local population and tourists, irrigation water for agricultural activities, water for the preservation of nature in general. Therefore, thinking people and organisations that deal with tourism and ecology-related activities 'condemn' each other. Hence, the need to reflect on interactive policy-making.

It would be a mistake to regard interactive policy making as a smooth and friction-free process in which stakeholders are automatically willing to learn from and understand each other, and care for other stakeholders' interests. Rather, interactive policy making must be seen as a negotiation process in which conflicting interests are explicated, recognised, and eventually translated into a constructive plan. In this process, facilitators can play an important role in organising constructive communication (Pruitt and Carnevale, 1993; Fisher and Ury, 1981; Rubin, 1994). Here it is important to realise that two completely different modes of negotiation exist, which can be labelled as 'distributive' and 'integrative'. In Scheme 2, the differences between these two styles are summarised. In essence, a distributive negotiation style is based on a cake that has to be divided, whereas an integrative negotiation style is based on the baking process, in which the cake's consumers are directly involved in producing the cake they want (Pruitt and Carnevale, 1993). The differences between the two styles will be clear enough from Scheme 2. Only 'joint fact finding' perhaps deserves some explanation. In distributive negotiations, actors mostly use their knowledge to construct arguments that can serve as 'ammunition' to attack the other party. It will be clear that in such cases the reliability of the knowledge is likely to be disputed by the others. The result is a struggle for 'the truth'. These struggles seldom have winners. Joint fact finding refers to the creation of a common knowledge base that offers facts that are accepted by everybody. This can be realised by making all parties responsible for knowledge creation.

Scheme 2. Negotiation styles

Distributive	*Integrative*
* starting from fixed positions	* starting from interests and/or visions
* closed about motives and background	* open
* no joint fact finding	* joint fact finding
* overcharging	* no overcharging
* threats	* no threats
* no relationship building	* relationship building
* no learning effects	* learning effects
* no concern about the other parties	* concern about other parties

The aim of organising and facilitating a negotiation process is to produce a better plan that is widely supported by different stakeholders. If this is achieved, implementation of the plan is likely to become much easier. However, the feasibility of the potential plan depends on the quality and extent of the learning processes that stakeholders have gone through during the negotiation process. That is, the tourist industry, tourist representatives, farmers, etc. need to have learned what natural values are at stake and why certain measures are necessary. But nature conservationists must also have come to understand the motives and backgrounds of tourists and other stakeholders. This is what we call social learning: learning about other social groups. Without this process negotiations often remain distributive in nature. Here, stakeholders stick to their own views and perceptions and fight to maximise their share of the cake and/or minimise negative consequences. In this case, the outcomes of the negotiation process tend to be rather fragile, as there is little 'real' recognition of the perspectives and interests of other stakeholders (Aarts, 1998). Thus, distributive agreements tend to be supported only temporarily. With every new problem, new negotiations are necessary and the chances that existing agreements will be disputed are relatively high.

For a viable regional plan on sustainable tourism, 'real acceptance' based on social learning processes is a prerequisite. In other words, we need an integrative rather than a distributive negotiation process; that is, a creative communication process in which new perspectives, problem definitions, and win-win solutions are presented. A key prerequisite for an integrative negotiation process is that the stakeholders involved feel interdependent (Holling, 1995; Aarts, 1998). That is, they must be convinced that they cannot achieve their interests in a satisfactory manner without the co-operation of others. This means that a certain balance of power exists among stakeholders, whereby each party has sufficient options and resources at their disposal to effectively hinder or frustrate others in achieving their aims. We often see that, at a certain point in time, certain organisations are not willing to seriously negotiate with other stakeholders; in other words, they do not feel interdependent. In these circumstances, integrative negotiations are not yet an option. However, in practice, we see that stakeholders can employ various strategies to influence balances of power and feelings of interdependence, and thus help to create the conditions for integrative negotiations. For example, a tourist industry that is not concerned about the impact of their activities on nature, and is not willing to engage in negotiations, can be attacked by environmental organisations using the mass media as their outlet. By doing so they can damage the image of these industries (i.e., the Greenpeace strategy), making them realise that they have to come to the negotiation table.

In the present case, the results of these negotiation processes must be that the actors involved in tourism will learn to think in an ecologically responsible way. This may lead them to identify and/or accept new solutions as well as the restrictions imposed by certain policy instruments (such as regulations). A successful negotiation process may produce several beneficial impacts. First, it may result in more informed decisions concerning, for example, the conditions under which their activities are possible, the number of tourists that can be accommodated according to the carrying capacity of a natural resource, or how vulnerable sites are protected by zoning. Secondly, it may also lead to a re-conceptualisation of what 'recreational products' are. For instance, in a Dutch province, the initiative to create a new golf course was contested initially because it was supposedly at odds with plans to improve natural values in the region. Eventually, the initiators constructed a new concept for a golf course: the ecological golf course, whereby ecological aspects were expressly integrated into the design.

6. Facilitating integrative negotiations

Building on a variety of literature on communication and negotiation (Susskind and Cruikshank, 1987; Van der Veen and Glasbergen, 1992; Huguenin, 1994), Van Meegeren and Leeuwis (1998) have identified a number of tasks that need to be addressed when facilitating integrative negotiations (Scheme 3) (Van Meegeren and Leeuwis, 1998; Leeuwis, 1999).

Thus, communication serves different goals at different points in the negotiation process. Communication strategies and skills are especially required in tasks 3 to 6. Some of the tasks mentioned are especially important during the early stages of a negotiation process, whereas others become important during the process. All tasks remain relevant throughout the process as many iterations are likely to occur. In principle, all parties involved bear responsibility for the distribution of these tasks. However, in the early stages, the initiators of a process have a special responsibility. Moreover, specific persons may be attracted whose job is to facilitate and monitor the process. It may be worthwhile to also consider the appointment and/or training of such facilitators in the field of tourism and nature. Eventually, they can act as network managers to bring the parties together and to create platforms for constructive discussions and negotiations.

Scheme 3. Tasks in integrative negotiation processes

Task 1: Preparation
  - exploratory analysis of conflicts, problems, relations, practices, etc. in a historical perspective
  - selecting participants who feel interdependent
  - securing participation by stakeholders
  - establishing relations with the wider policy environment
Task 2: Preliminary protocol
  - creating a code of conduct and a provisional agenda
Task 3: Joint exploration and situation analysis
  - group formation
  - exchanging perspectives, interests, goals
  - analysing problems and interrelations
  - integration of visions into new problem definitions
  - preliminary identification of alternative solutions and 'win-win' strategies
  - identification of gaps in knowledge and insight
Task 4: Joint fact-finding
  - developing and implementing an action-plan to fill knowledge gaps
Task 5: Forging agreement
  - spelling: clarifying positions, making claims, use of pressure to secure concessions, create and resolve impasses
  - securing an agreement on a coherent package of measures and action plans
Task 6: Communication of representatives with constituencies
  - transferring the learning process
  - 'ratification' of agreement by constituencies
Task 7: Implementation
  - implementing the agreements made
  - monitoring implementation
  - creating contexts of re-negotiation

For each of these (sub)tasks, a large number of concrete recommendations can be made based on the negotiation literature. Authors like Fisher and Ury (1981) and Susskind and Cruikshank (1987), for example, developed a number of simple and crucial guidelines for facilitating negotiations in relation to various tasks. For instance:

- distinguish between substantive and relational conflicts and deal with the relational problems first (related to task 3)
- encourage people to communicate in terms of interests and fears, not in terms of positions and 'facts' (task 3 and 5)
- encourage people to communicate in terms of new alternative proposals rather than in terms of their positions *vis-à-vis* existing proposals (task 5)

This is not the place to elaborate on these practicalities. However, there is one important concern within integrative negotiations on which we would like to elaborate: the relation between representatives and their constituencies (task 6). The representatives tend to go through a learning process, while their constituents, on whom acceptance the result ultimately depends, do not. Thus, it is crucial that stakeholders make efforts to communicate the social learning process itself, for example, by means of regular feedback meetings with constituents, articles in journals, etc. Thus, tourist magazines may start to write about nature, not only as a commodity that can be used, but also as a value that needs to be respected on good grounds. In a similar vein, journals on nature conservation may write about tourism in a different, more inviting way than is currently the case. In this manner, integrative negotiations can lead to integrative thinking on a wider scale and can encourage tourists to reflect more continuously on the ecological implications of their activities. Of course, transferring learning processes by intensive and well-organised communication is a slow and gradual process, but eventually it may help to influence the behaviour of stakeholders and in a much more sustained manner than can be achieved through distributive negotiations and/or instrumental forms of communication. In the end, changing the communication network in the field of tourism can be more effective in raising ecological awareness than an effort to reach the public directly.

7. Hindering factors

As we have shown, the instrumental approach to communication has severe limitations. However, the question remains whether interactive policy making on the basis of integrative negotiation is just theory and wishful thinking or a practical option. There are still many lessons to be learned. There are indications, however, that the approach can work. In several countries, breakthroughs have been achieved along these lines in the relationship between farmers and nature conservationists. Both parties increasingly co-operate in bringing about a better balance between nature and food production. In the Netherlands, for example, the concept of 'agricultural nature management' is nowadays widely supported in agricultural circles. This is the result of a shift from distributive to towards integrative negotiations between farmers and nature conservationists (Van Woerkum and Aarts, 1998). However, the Dutch case also shows that the inclination among stakeholders to stick to their positions and to fight only for their own interests is strong. Also, we see that well-intentioned representatives underestimate the necessity to communicate the learning process to their members. In many cases, they tend to keep silent until they can offer 'concrete results'. This tends to get hem into trouble when defending the results to their members, as the 'learning gap' cannot be resolved in just one meeting at the end of the process. This creates the danger that the representatives end up in a 'boundary role conflict' (Turner, 1992) – different parties tend to confront them with pressures and demands which are difficult to agree upon without seriously jeopardising the opportunity to reach compromises with the others. The essence here is that representatives should consider their constituencies as stakeholders with whom they have to negotiate as well. These experiences show that facilitators can indeed play an important role in safeguarding an integrative process and ensuring communication with constituents.

As mentioned above, a serious problem can arise from the large imbalance of power that often exists between stakeholders. If one party is able to influence the situation by using its dominant position, the others will be inclined to move away from the negotiating table. Industrial firms can challenge the others by threatening to remove their plants if environmental measures are taken, which means a reduction of local employment. Environmental groups can use legal power to ensure that certain norms will be taken seriously. Often, they have media power: by influencing the media they can damage the image of commercial organisations. Similarly, stakeholders can also decide to retire if they have the feeling that they lack the power to 'win'. Since the starting point for integrative negotiations is perceived interdependency, a serious analysis of power relationships is needed. If insufficient perceived interdependency exists, it may be necessary to first explore options and strategies for changing power balances (Leeuwis, 1995; Leeuwis, 1999). Clearly, there may not always be sufficient scope to achieve this.

Another problem is the diversity in interests of those involved. Many negotiations between societal groups suffer from the problem of 'mixed interests' (Pruitt and Carnevale, 1993). It is not easy to find 'objective' criteria for weighing different values and interests (e.g. balancing aesthetic values against economic benefits) within negotiations. For tourists, visiting a particular scenic area is just one out of many options. For farmers, big restrictions can mean the end of their business. The interest of nature conservation should be articulated fully in order to gave it the necessary weight.

Finally, we point to the complexity of the issues under discussion. Biodiversity is a question that is difficult to understand for those people with little ecological awareness. As long as they see green trees and lots of birds, they do not perceive an immediate problem. This situation calls for conscious and creative efforts to communicate ecological problems in an insightful manner during negotiations, for example, with the help of visualisation, simulation and understandable analogies.

8. Conclusion

We have described two approaches towards professional communication aimed at achieving a balance between nature and tourism functions. In the instrumental approach, the function of communication is to influence behaviour directly or, perhaps even more importantly, to support other policy instruments, given an already existing policy plan. The second approach we have labelled interactive policy making. Herein, professional communication plays a role in the process of policy formation itself and is geared mainly towards facilitating learning and negotiation processes. We have argued that this second approach is a more viable option for bringing about an effective balance between nature and tourism as, for example, in moulding sustainable tourism solutions.

The two approaches do not exclude each other. Interactive policy making rarely achieves everything policy makers strive for. Moreover, the results of an integrative negotiation process will have to be communicated effectively once they have been achieved. Hence, instrumental campaigns remain relevant. But the interactive approach and the broad communication it provokes can lay the ground for wider public awareness and understanding. Moreover, it may help bring about effective co-ordination within the communication network around tourism, and prevent tourists getting contradictory messages from different instrumental campaigns organised by a variety of separate stakeholders. One may wonder which messages will be most attractive to them.

References

Aarts, M.N.C. (1998) *Een kwestie van natuur; een studie naar de aard en het verloop van communicatie over natuur en natuurbeleid* (A matter of nature; a study of communication about nature and nature policies). Wageningen: Agricultural University, group Communication and Innovation Studies (dissertation).

Bourdieu, P. (1990) *The Logic of Practice.* Polity Press: Cambridge

Fishbein, M. and I. Ajzen (1975) *Belief, Attitude, Intention and Behaviour.* Addison-Wesley: Reading, Ma.

Fisher, R. and W. Ury (1981) *Getting to Yes: Negotiating Agreement Without Giving in.* Penguin Books: Harmondsworth.

Geller, E.S., R.A. Winett and P.B. Everett (1982) *Preserving the Environment; New Strategies for Behaviour Change.* Pergamon Press: New York.

Giddens, A. (1984) *The Constitution of Society: Outline of the Theory of Structuration.* Polity Press: Cambridge.

Glasbergen, P. (1995) *Managing Environmental Disputes; Network Management as an Alternative.* Kluwer Academic Publishers: Dordrecht.

Holling, C.S. (1995) What barriers? What bridges? In L.H.Gunderson, C.S. Holling and S.S. Light (eds) *Barriers and Bridges; to the Renewal of Ecosystems and Institutions,* Columbia University Press: New York, pp. 15–33.

Huguenin, P. (1994) *Zakboek voor Onderhandelaars: Vuistregels, Vaardigheden en valkuilen* (A Pocket Book for Negotiators: Rules, Skills and Challenges) Bohn Stafleu Van Lochem: Houten.

Koelen, M.A. (1988) *Tales of logic: A Self-presentational View on Health-related Behaviour.* Landbouwuniversiteit: Department of Extension Science, Wageningen (dissertation).

Koppenjan, J.F.M., J.A. de Bruijn, and W.J.M. Kickert (eds) (1993) *Netwerkmanagement in het Openbaar Bestuur. Over de Mogelijkheden van Overheidssturing in Beleidsnetwerken.* (Management of Networks in Public Policy. About the Possibilities of Steering by the Government in Policy Networks), VUGA: The Hague.

Leeuwis, C. (1993) *Of computers, myths and modelling: The social construction of diversity, knowledge, information and communication technologies in Dutch horticulture and agricultural extension.* Wageningen Studies in Sociology, Nr. 36.

Leeuwis, C. (1995) The stimulation of development and innovation: reflections on projects, planning, participation and platforms. *European Journal of Agricultural Education and Extension: International journal on changes in agricultural knowledge and action systems* 2:3, pp. 15–27.

Leeuwis, C. (1999, forthcoming) Metaphors of participation for sustainable rural development. Beyond planning, decision-making and social learning, Submitted to *Development and Change.*

Long, N. (1990) From paradigm lost to paradigm regained? The case for an actor-oriented sociology of development. *European Review of Latin American and Caribbean Studies* **49**, December, pp. 3–32.

Long, N., and J.D. van der Ploeg (1989) Demythologizing planned intervention. *Sociologia Ruralis* **24**:3/4, 226–249.

Messinck D.M. and M.B. Brewer (1983) Solving social dilemmas. A review. In L. Wheeler, and P. Shaver (1983) *Review of Personality and Social Psychology* 4, pp. 11–44.

Messinck, D.M. and C.L. Mcbelland (1983) Individual adaptations and structural change as solutions to social dilemmas. *Journal of Personality and Social Psychology* **44**, 294–309.

Orbell, J. and J. Dawes (1981) Social Dilemmas, *Progress in Social Psychology* **1**, 37–63.

Ostrom, E. (1990) *Governing the Commons. The Evolution of Institutions for Collective Action.* Cambridge University Press, Cambridge.

Petty, R.E. and J.T. Cacioppo (1986) The elaboration likelihood model of persuasion. In L. Berkowitz (ed.) *Advances in experimental social psychology,* Vol. 19, pp. 123–205, Academic Press, New York.

Pruitt, D.G. and P.J. Carnevale (1993) *Negotiation in Social Conflict,* Open University Press: Buckingham.

Raising Public Awareness for Nature Conservation in Hungary (1996). ECNC: Tilburg.

Rubin, J.Z. (1994) Models of conflict management. *Journal of Social Issues* **50**:1, 33–45.

Susskind, L. and J. Cruikshank (1987) *Breaking the Impasse; Consensual Approaches to Resolving Public Disputes.* Basic Books: New York.

Tatenhove, J. and P. Leroy (1995) Beleidsnetwerken: een kritische analyse (Policy networks: a critical analysis). *Beleidswetenschap* **8**, 128–145.

Turner, D.B. (1992) Negotiator-Constituent relationships. In L.L.Putnam and M.E. Roloff (eds) (1992) *Communication and Negotiation, Sage Annual Reviews of Communication Research, Vol 20.* Sage Publications: London, pp. 233–247.

Van der Veen, J. and P. Glasbergen (1992) De consensusbenadering; Verkenning van een innovatieve werkvorm om regionale milieuconflicten te doorbreken. (The consensus approach; Exploration of an innovative approach to solve regional environmental conflicts). *Bestuurskunde* **1**, 228–237.

Van Woerkum, C.M.J. and M.N.C. Aarts (1998). Communication between farmers and government over nature: a new approach to policy development. In N.G. Röling and M.A.E. Wagemakers (eds) *Facilitating Sustainable Agriculture.* Cambridge University Press: Cambridge, pp. 272–280.

Van Woerkum, C.M.J. (1999) *Communication and Interactive Policy Making.* Wageningen: Group Communication and Innovation Studies, Agricultural University.

Weiss, J.A. and M. Tschirhart (1994) Public information campaigns as policy instruments, *Journal of Policy Analysis and Management* **13**, 82–119.

Windhall, S. and B. Signitzer (1992) *Using Communication Theory.* Sage: London.

SUSTAINABLE TOURISM POLICY IMPLEMENTATION: AN EX ANTE CRITICAL EXAMINATION

HELEN BRIASSOULIS
Department of Geography
University of the Aegean
Karantoni 17, Mytilini, Lesvos 81100
Greece

1. Introduction

It is a rather simple truism that policy prescriptions divorced from reality are unlikely to be implemented. For sustainable development policies, including sustainable tourism development policies, this condition is critical as the existing policy formulation and implementation structures and apparatuses may not be fit to the demands of relatively new policy ideas and orientations. Several analyses of the implementation difficulties of sustainable development, including sustainable tourism development policies, have appeared since the mid-1980s (Redcliff, 1987; De Kadt, 1990; Dutton and Hall, 1990; Environment Canada, 1990; Opschoor and Van der Straaten, 1993; Van der Straaten, 1994; McKercher, 1993). This paper adds an ex ante critical examination of the implementation potential of sustainable tourism development policies (henceforth, sustainable tourism policies). It is premised on the idea that tourism is not a simple economic activity or sector, in the traditional sense, but a multifaceted activity complex, which presents multidimensional implementation particularities. As the formulation of truly sustainable tourism policies is at an early stage, an *ex* ante policy implementation analysis within the actual tourism implementation environment is expected to aid in the design of implementable, effective policies.

The paper elicits the characteristics of theoretical truly sustainable tourism policies from the pertinent literature and, then, poses the question: how will the current tourism policy implementation environment respond to the demands of these policies while it operates under a different mode currently? Figure 1 outlines the methodological schema adopted to address this question. The tourism system interacts with both the tourism policy formulation and implementation environments while it is influenced by external, non-policy factors. This interaction – proposed policies implemented under the current conditions – is evaluated by a set of criteria, specifically chosen to evaluate policy implementation. Implementation issues are identified and classified and actions to facilitate sustainable tourism policy implementation are proposed. It is noted that the present evaluation is general and not context-specific (e.g. country-specific) as most sustainable tourism policy prescriptions are not context-specific either. Moreover, the main intent of this paper is to show how to think about sustainable tourism policy implementation problems rather than to analyse concrete real world situations.

The paper devotes three sections to a theoretical analysis of the implementation potential of sustainable tourism policies. The second section describes the tourism system, clarifies the concept of sustainable tourism development and outlines the main elements of truly sustainable tourism policies as prescribed by theory.

The tourism policy implementation environment and the evaluation criteria chosen are presented in the third section. The fourth section analyses and classifies the potential implementation issues. The fifth section proposes ways to address these issues. A postscript suggests future research directions.

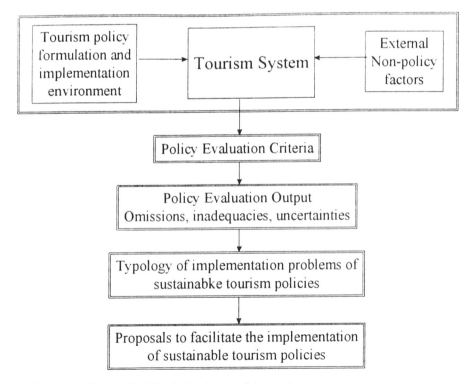

Figure 1. The methodological scheme of the study

2. The tourism system and sustainable tourism development

Figure 2 provides a simplified representation of the relationship between the tourism system, the state and the environment. Tourism is an activity complex with a variety of economic sectors contributing to the production of the tourist product and a variety of actors being involved in the consumption of this product. In addition, a number of private and public sector intermediaries mediate tourism production and consumption.

The supply-side of tourism, tourism production, is multiactivity and multisectoral. Mainly, it involves the accommodation, catering, travel, entertainment and services sectors (Baud-Bovy and Lawson, 1977; Inskeep, 1994). Numerous, mostly small, firms, of a variegated nature, sell different types of products each one of them making its particular contribution to and affecting the quality of the tourist product. These numerous, small firms constitute a large number of centres of decision generating conflicts frequently and hampering the control of the aggregate result of their actions – 'a tyranny of small decisions' (Kahn, 1966).

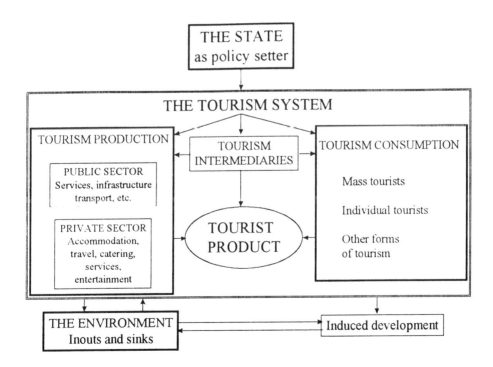

Figure 2. The tourism system, the state and the environment

The tourist product is multidimensional and multiattribute. It consists of material (tangible) – the various sectoral products and the natural and cultural resources of a tourist area – and immaterial (intangible) components – landscape aesthetics, socio-cultural values, tourist satisfaction. The tourist product does not have a *single* price as it is composed of quantifiable items which can be sold in the market and unquantifiable, unpriced ones. Hence, it cannot be subjected to uniform price controls and regulations, nor can it be treated with the traditional tools of economic analysis. Also, this product consists of reproducible (e.g. services offered by hotels, shops, etc.) and non-reproducible components (e.g. natural and cultural resources). If the latter undergo damages or irreversible alterations, the product looses its original quality and diminishes in quantity, since the original quality product becomes less. Finally, this product is a mix of private and public goods making difficult the application of purely public or purely private types of policies. Environmental inputs (public goods), in particular, are overused and eventually abused, because they are unpriced (Briassoulis, 1995).

On the demand side, tourism consumption and consumers take various forms depending on the purpose and organisational form of the activity - such as individual and mass tourism, vacation, religious, eco- and agro-tourism, etc. (Komilis, 1986). The tourist market has a domestic and an international segment implying different patterns of demand and pressure on resources. Because tourism is multiactivity, tourist demand is rather rarely spatially concentrated but spreads over a wide territory, the transportation network contributing critically in this respect. Also, tourist demand is seasonal and, viewed over the long run, it varies in space with changes in tastes, tourist preferences and marketing. Consequently, the impacts of tourism shift and spread over space and time (Briassoulis, 1995).

The production and consumption of the tourist product take place very frequently in the same place and at the same time. Thus, conflicting demands are placed on local resources, which serve also local activities. Moreover, once a component tourist activity suffers damages or the natural and/or cultural resources are destroyed, correction cannot be postponed to a later date nor the damage removed to another place without harming the competitiveness and profitability of tourism. Finally, tourism induces urban development in host areas. Activities benefiting from tourism or from tourism-related infrastructure concentrate in tourist areas increasing, thus, the burden on local resources and capacity (Briassoulis, 1995).

Tourism intermediaries (tour operators and travel agencies, mainly) play a crucial role in the tourism development process as they act as brokers coupling consumer demand with supply of particular types of tourist services. On the one hand, they manipulate tourist preferences, behaviour and expectations. On the other, they influence the supply of tourist services taking advantage, frequently, of the isolation and small scale of several tourist destinations, their uncertain economic environment, the volatility and proneness of tourism to unexpected socio-economic and political events, and so on. Finally, the state affects all three groups (producers, consumers, intermediaries) and the relationships among them, to various degrees, through direct tourism policies as well as through more general social, economic and spatial policies.

2.2. Sustainable tourism development

The Brundtland report defines sustainable development as development which meets the needs of the present without compromising the needs of the future generations (UNCED, 1987). This definition implies economically efficient use of resources, protection of the environment and the natural capital and equitable distribution of the costs and benefits of development within and across spatial units and over time. Whenever tourism represents a basic economic activity in an area, a two-way relationship develops where the sustainability of development of the host area secures the sustainability of tourism and vice versa. Hence, tourism development is sustainable when:

1. the complex of interlinked public and private activities and enterprises constituting the tourism activity complex operates profitably and efficiently over time, makes wise use of available resources and preserves (and enhances perhaps) the natural capital of the area, the principal constituents of the tourist product and chief determinants of the profitability of tourism
2. the same is true for the rest of economic activities taking place in the area
3. conflict is avoided and co-ordination is promoted among all activities in the area, including tourism, to ensure intra- and intergenerational fairness in resource use (e.g. in the use of local environmental resources and services; in responding to external and/or unexpected events and developments, etc.). It is reasonable to expect that, if at least one of these conditions is not met, tourism development will not be truly (or, completely) sustainable

Various policy prescriptions for sustainable tourism have appeared so far in the literature (see, among others, De Kadt, 1990; Dutton and Hall, 1990; Environment Canada, 1990; Eber, 1992; Sadler, 1992). These can be classified, according to their degree of specificity, into: goals (abstract, broad statements), objectives (concrete, operational statements), and targets (measurable outcomes) (Table 1). The table shows that most of the prescriptions are goals not adequately translated into objectives. Only certain goals have been translated into more practical objectives. Similarly, only certain objectives have been translated into specific targets, which, however, need further elaboration to become operational.

Table 1. Selected policy prescriptions for sustainable tourism development classified into goals, objectives and targets

	Goals	Objectives	Targets
	Time scale (Short-term / Long-term) · Spatial scale (Local / Global)	Time scale (Short-term / Long-term) · Spatial scale (Local / Global)	Time scale (Short-term / Long-term) · Spatial scale (Local / Global)
Protection of the environment	Sustainable use of energy and natural resources (all time and spatial scales)	Reduce tourist use of energy and natural resources (all time and spatial scales)	Small scale tourist dev. (Local)
	Ensure ecosystem stability and ecological balance (all time and spatial scales)	Impose limitations on growth (all time and spatial scales)	Less massive tourism (Local)
	Maintain biodiversity (all time and spatial scales)	Observe carrying capacity of tourist areas (all time and spatial scales)	Fits to local natural env. (Local)
	Pollution prevention and reduction (all time and spatial scales)	Avoid destructive land uses (all time and spatial scales)	Limits to the number of tourists per region
Economic efficiency	Ecological economy (all time and spatial scales)	Strong/weak sustainability criterion (all time and spatial scales)	Use local materials/resources (Local)
	Support local econ. (Local)	Adapt to local econ. potential / Small firms (Local)	
	Maintain productivity of resources (all time and spatial scales)	Integrate tourism development into planning	
Socio-cultural well-being	Social equity (full access to socio-cultural opportunity and continuity) (all time and spatial scales)	Equitable distribution of costs and benefits of tourism development	
	Distributive justice (all time and spatial scales)	Meet present needs / Meet future needs	
	Maintain cultural integrity and stability of host areas (all time and spatial scales)	Reduce economic inequalities	Fits local cult. envir. (Local)
	Quality of life (all time and spatial scales)	Priority to domestic/regional tourism development	
		Quality jobs/ Quality of local life	
	Consultation and participation of locals (all time and spatial scales)	Local control of developm. (Local)	Benefits and does not exploit locals (Local)
		Cooperation among parties/Non-violent conflict resolution	

Finally, most prescriptions refer to tourism development without putting it *explicitly* in the context of the broader development of the host area, missing, thus, the holistic nature of sustainable tourism development.Truly sustainable tourism development policies, the object of the present evaluation, possess characteristics with important implications for their implementation. Firstly, the core element of these policies is long-term, strategic tourism planning on a large scale, which constitutes the guiding framework for short-term, local level policy and planning actions. Secondly, these policies are holistic and broad in scope encompassing and impacting on many, diverse public and private sector parties, a fact giving rise frequently to goal and value conflicts. Finally, and more importantly, successful sustainable tourism policies should be integrated at all levels: the spatial, the temporal and the administrative/organisational. The continuous feedback from one level to the other is an essential requirement, which implies mutual interaction among all interests involved in a continuous learning and experience exchange process. For the present purposes, it is stressed that, although formulated tourism policies may satisfy all essential requirements of sustainable development, deviations occurring during implementation may lead to unsustainable outcomes; hence, the importance of an ex ante policy implementation analysis. For example, a particular system of incentives may prove inappropriate for economically efficient resource use; or, incomplete enforcement may violate the environmental protection principle; or, strong value conflicts and power imbalances may forestall the achievement of spatially and temporally fair allocations of the costs and benefits of tourism development, and so on.

3. The tourism policy implementation environment

This section proposes a general schema to analyse tourism policy implementation environments – namely, categories of policy implementers and their relationships – and suggests three criteria to evaluate the implementability of sustainable tourism policies. Furthermore, for the purposes of the ensuing analysis, it speculates on some features, which appear to be common to most tourism implementation environments currently. The discussion is inevitably procedural and general rather than substantive and specific as the exact nature of these environments depends critically on the socio-economic, political, and cultural circumstances of an area. To make valid generalisations, detailed analyses of particular cases are needed which are missing at the time of this writing. Policy implementation in general is the product of the actions and interactions of categories of implementers, positioned at various spatial and organisational levels. The pertinent policy implementation literature suggests a typology of implementers (Nakamura and Smallwood, 1980), which is adapted here to the case of tourism.

Table 2 is an indicative list of potential members of each category distinguished according to the spatial/organisational level to which they belong. Several broad observations are in order. Firstly, the exact members of each category will vary from country to country. Some categories may be even missing in some cases (e.g. this author could not identify formal implementers at the international level where certain policy decisions are made). In some other cases, certain categories may dominate. For example, in the case of tourism 'monocultures', especially in small isolated host areas, policy implementation may be essentially in the hands of lobbies and constituency groups.

Table 2. Indicative list of tourism policy implementers

Types of implementers	Spatial/organisational level		
	Local/regional (Loc/Reg.)	National (Nat.)	International/global (Int.)
Policy makers	Loc/Reg. councils/ governments	Nat. government tourism bodies	Int. organisations (e.g. UNEP)
Formal implementers	Loc/Reg. planning offices Loc/Reg. tourism management offices Local branches of state agencies Banks and state lending agencies	Nat. tourist organisations Ministries of Tourism, Transportation, Planning, N. Economy, Finance, etc.	Not known with certainty
Intermediaries	Various (see text for explanation)	Various (see text for explanation)	Various (see text for explanation)
Lobbies and constituency Groups	Loc/Reg. chambers of commerce, hotels, entertainment, etc. Loc/Reg. tourism and travel agents Loc/Reg. Environmental organisations Loc/Reg. NGOs	Nat. chambers of commerce, hoteliers, etc. Nat. tourism and travel agents Nat. environmental organisations Nat. NGOs	Int. tourism and travel agents and operators Int. tourism organisations (e.g. WTO) Int. environmental organizations (e.g. WWF, Greenpeace) Int. NGOs
Recipients/ consumers	Local hotels, banks, entertainment, restaurants, services	Domestic tourists, day visitors	Int. tourists (all types)
Media	Local radio, television, press	Nat. radio, television, press	Int. press, TV, radio

Secondly, intermediaries (not to be confused with tourism intermediaries) at all spatial levels take so diverse forms (some of which cannot be known with certainty a priori) that their identification is not easy. For example, banking interests (e.g. the World Bank), local building associations, employee unions, local property owners and many others may mediate and influence the tourism policy implementation process. Thirdly, there does not exist a unitary implementation authority. Instead, many public and private sector entities get involved in the implementation of particular policy aspects. Each of these entities possesses particular privileges, commands certain amounts of power and authority, and interacts variously with the other actors and the external environment. Thus, the ultimate implementation of ostensibly similar prescribed policies will differ both within and between countries.

Certain patterns of relationships seem to exist currently among implementers explained, in general, both by the variegated nature of tourism as well as by the diversity of implementers. Formal implementers interact with policy makers rather rarely. This is due to the heavy workload of both groups, the different conditions under which they operate as well as the spatial, temporal and organisational 'distance' between them. For example, a tourism police officer competent for issuing operations permits to tourist facilities may not even know the policy maker(s) who drafted the relevant legislation nor he can consult him or her in case of vague and disputed quality requirements.

The interaction between formal implementers and policy consumers and recipients depends on: (1) how clear and unambiguous the messages (policies) sent are, and (2) whether the transmission channels utilised (media, instruments, personnel) are appropriate. Sometimes, the channels distort the original message or they are absent between certain organisational/spatial levels of implementers. A case in point is the policy of tourism promotion. Depending on how promotion is carried out, tourists are attracted differentially to, perceive differently and behave variously in a particular tourist place.

Depending on local circumstances, policy makers interact with lobbies and constituency groups, especially when the latter demands favourable treatment or pushes for some other cause (e.g. environmental protection in host areas, which are habitats of endangered species). In this case, policies may reflect (or, anticipate) the desires and interests of lobbies and constituency groups, a critical determinant of their implementability.

Formal implementers may interact with intermediaries when necessary but this is probably a weak and random interaction, as the latter group is usually not known a priori. For example, archaeological services may demand modification or even cessation of touristic operations if they conflict with the character and conservation needs of archaeological areas. Also, formal implementers may interact necessarily with lobbies and constituency groups if policy success hinges on the co-operation of the latter. For example, chambers of hoteliers and/or commerce may assist national and/or local state tourism bodies to implement those policy aspects, which promote their interests.

Speculating about the rest of possible linkages between categories of implementers is risky as: (1) these are numerous given the multiplicity and diversity of implementers as well as the variety of organisational and spatial levels implicated; (2) certain linkages are not always visible (e.g. between intermediaries and the media or between policy makers at the national level and recipients at the local level); and (3) in some cases, they may not be active perhaps either because implementers do not exist at a certain level or because the available channels do not fit the particular transmission and communication needs of tourism policies (e.g. there may not exist procedures for national policy makers to interact with policy consumers at the international level/international tourists).

Tourism policy is put in effect by means of both formal and informal procedures. It is not uncommon for informal procedures to be more powerful and effective than the formal ones especially in heavily developed tourist areas where the stakes are high. This implies that the original policy is transformed during implementation and what is observed and assessed is perhaps not what policy makers had contemplated originally. In general, as it is the case with all policies, the particular implementers involved, the available means, compliance mechanisms and culture, bureaucratic and social norms and performance criteria, the degree of deviation from normal practices as well as external, non-policy factors (Sabatier and Mazmanian, 1980) determine the outcome of the actual implementation process. In fact, it is rarely possible to identify, assess, and evaluate with certainty the implementation outcomes of tourism policies and, consequently, to suggest corrective actions and measures.

The negative facets of the current tourism policy implementation environment are:

1. certain categories of implementers are missing at specific spatial/ organisational levels, a fact implying potential implementation 'gaps';
2. interaction among the extant categories of implementers is sometimes problematic if not remotely acceptable;
3. not all tourism policy implementers possess adequate power to influence proper actions at their respective level, a situation portending potential implementation failures;
4. neither formal, nor autonomous co-ordination among the various implementers does seem to exist to a satisfactory extent currently.

To evaluate ex ante the implementation potential of proposed sustainable tourism policies three criteria are suggested, which reflect the most important implementation concerns. Following the work of Mazmanian and Sabatier (1981), these criteria are: (1) tractability of the problem the policy addresses,

(2) potential for compliance with the policy, and (3) non-statutory factors affecting implementation.

4. Sustainable tourism development policies: analysis and classification of implementation issues

This section filters the main policy prescriptions for *truly* sustainable tourism development, as summarised in Table 1, through the three evaluation criteria with the purpose to: (1) assess their implementation potential, within the current tourism policy implementation environment described before, and (2) classify the potential implementation issues (Figure 3).

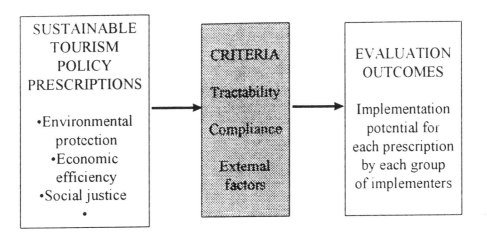

Figure 3. The analytical scheme of the evaluation

4.1 . TRACTABILITY OF THE POLICY PROBLEM

4.1.1. Protection of the environment
The scientific definition and analysis of environmental problems in tourist areas encounters several difficulties. Firstly, several concepts do not have

unambiguous definitions and operational measures (e.g. use of energy and natural resources at each spatial scale and by each tourist actor, ecological stability, balance and biodiversity). As regards the definition and measurement of environmental carrying capacity for tourism (Lindsay, 1986), the number of tourists an area can support is not the sole consideration. The local population and the indirect generation of users (e.g. employees of tourism related-services and other activities) are also important factors to be accounted for although these are often overlooked.

Secondly, information on several environmental parameters (e.g. stock of resources, ecosystem state and structure, current pollution conditions), the many and diverse actors and their relationships, and important macro-factors (e.g. autonomous environmental, socio-economic, political change, etc.) is rarely complete and certain. Consequently, operational ecological, economic/environmental, and policy impact models are difficult to build at the necessary level of detail required for implementation, to: (1) analyse the contribution of each tourism-related sector to the environmental problems of a region, (2) prescribe the proper values of the environmental parameters of interest and (3) conduct policy impact assessment and trade-off analysis between environmental quality and socio-economic goals. In addition, it is difficult to distinguish the role in and contribution to environmental protection of the public and private sector activities involved in tourism development.

4.1.2. Economic efficiency

Similarly, operational definitions and measurements of the components and structure of an ecological tourism economy as well as of 'economic efficiency' are not straightforward for several reasons. Tourism is multi-activity; it coexists with other economic sectors; informal activities and tourism-induced urban development demand on unknown amounts of resources; more than one spatial levels are involved; and so on. Hence, information on economic parameters is difficult to obtain, thorny assessments and valuations are required, and, consequently, operational policy impact model building is impeded. For example, to specify sustainable tourism economy patterns, it is necessary to know the preferences of tourists and non tourists, the stock of resources under consideration, to assess the costs and benefits of natural resources use, the effects of macro-factors on local economies, and so on. Similarly, the strong and weak sustainability concepts (Barbier, Markandya and Pearce, 1990) require information on macro variables such as investment, consumption, imports and exports not only for the tourist but also for the non-tourist sectors.

4.1.3. Socio-cultural well-being

Concepts such as social equity, distributive justice, cultural stability and integrity and quality of life have diverse meanings and their operational definition and measurement depend on the socio-cultural context of their use. In the case of tourism, these problems are more complex. Definitions should, among others, include public and private sector interests and values, account for the autonomous socio-cultural dynamics in host areas, view distributional issues broadly as tourism encompasses numerous, diverse, small and big, local and supralocal, current and prospective stakeholders, account for illegal tourist operations, and so on. Actual socio-cultural conditions are rarely known with certainty to serve as reliable baselines for setting policy targets and prescribing proper policy actions. Needless to say, that the systematic assessment and evaluation of the socio-cultural impacts of tourism policies does not seem to be feasible at present.

In summary, in all respects – environmental, economic, socio-cultural – the sustainable tourism development problems are not easily tractable. This may be one of the reasons why policy makers formulate imprecise and abstract policies (goal statements) which do not concern directly the real problem being addressed at the actual level of implementation (absence of essential targets). Thus, policy implementers are, provided with little guidance and are allowed ample discretion to decide on policy issues during actual application.

4.2. COMPLIANCE

The most important determinants of compliance with a policy are considered here to be the availability of proper implementation procedures and resources, the interest and compliance culture of the implementers and their co-ordination (Nakamura and Smallwood, 1980). The potential for compliance with sustainable tourism policies along these determinants for the various tourism policy implementers is broadly assessed in Table 3 and the main points are summarised below.

4.2.1. Implementation procedures
Implementation procedures are usually sparse, incomplete and not available for all types of implementers. The many and diverse tourism policy makers, although a potential type of implementers, engage normally in more than one policy areas. Hence, they cannot oversee continuously the implementation of tourism policies especially if this is not within their formal assignments. Formal implementers are supposedly obliged to put in effect the policies. However, they rarely possess clear and unambiguous policy implementation procedures and instructions, performance criteria, enforcement mechanisms and other details needed to guide their actions due to the tractability problems discussed before.

Thus, they have ample discretion to prioritise objectives, to value costs and damages, to construe the sustainability principles and criteria, and so on. For tourism policy recipients and consumers, explicit implementation procedures for sustainable tourism policies are very difficult to institute, communicate and enforce. Because they act, more or less, at the most remote from the top policy making and implementation centres, the local level, they do not always receive the policy messages, or they do not receive them in their original form. Thus, accommodation, catering, entertainment, travel and other service sector firms will have to find by themselves how to comply with environmental protection measures, to adapt to new modes of economic functioning and accounting and to adopt longer term instead of short term planning outlook. The issue of available implementation procedures does not apply to intermediaries, lobbies/constituency groups and the media as these are, essentially, informal implementers and policies do not provide procedures for their actions. Intermediaries, especially, hold at present an unrecognised place in the formulation and implementation of tourism policies, acting within a lax environment without explicit applicable rules, procedures and compliance mechanisms.

4.2.2. Resources

Resources of all kinds are limited, insufficient and ill-distributed among the various types of implementers for effective implementation. For policy makers this issue is not very critical as, more or less, they are not involved directly and systematically in implementation. Formal implementers must allocate, usually limited, financial and human resources to the public and the private sectors and this task is not always straightforward especially in mixed economies.

Implementing agencies usually lack the scientific and technical expertise and resources needed to apply new policy instruments and performance criteria (e.g. tourism green taxes, environmental accounting). The current administrative apparatus is compartmentalised, insufficient, and structured to serve different types of relations among tourism-related activities. Hence, it cannot support the holistic management of tourism as an activity complex, on the one hand, and, on the other, as one of the many economic activities taking place in an area. Intermediaries, when they decide to get involved in tourism policy implementation, have usually enough resources of all kinds to support their actions (with the exception of administrative resources, which does not apply to them). The resources of lobbies and constituency groups are variable depending mainly on their economic power and the technical support they can avail. Policy recipients and consumers, with rare exceptions (e.g. large hotel chains and tour operators), have rather limited resources which, in addition, are shared by other tasks.

4.2.3. Interest/compliance culture.

These are highly variable among the various types of policy implementers. Formal implementers may not be always interested in environmental protection or the redistributive economic and social policies for tourism. This may be especially true in tourism 'monocultures' due to the heavy dependence of the local economy on tourism. Moreover, their compliance culture varies with the spatial/organisational level they belong to and the norms of the broader cultural context. At the lower levels of the hierarchy, co-optation by local interests and corruption are probable while, at higher levels, interest in compliance may be diffuse and low.

For lobbies and constituency groups, their outlook, 'green consciousness', aspirations, long-range view and the extent to which sustainable tourism policies harm their interests exert important influences on compliance.

Table 3. Evaluation of the compliance potential with sustainable tourism policies

Type of imple-menter	Available procedu-res	Resources				Interest/ Compli-ance culture	Co-ordi-nation	Overall evalua-tion
		Finan-cial	Tech-nical	Administ-rative	Human			
Policy makers	?	N/A	N/A	N/A	N/A	N/A	?	None to Low
Formal imple-menters	Inappropriate definition tourism Incomplete procedures	Variable and limited	Limited	Problematic	Limited, un-trained staff	Variable	Problematic	Low
Interme diaries	N/A	It depends Usually enough	It depends ?	N/A	It depends Usually enough	Variable	Low or none	None to Low
Lobbies and constituency groups	N/A	It depends	It depends ?	N/A	It depends ?	Variable	Low or none	None to Low
Recip-ients and consum-ers	Vague, incomplete procedures and instruc-tions	Usually very limited	Limited Lack of support	N/A	Usually limited	Diverse Depends on green consciousness	Inadequ-ate (Tyranny of small decisions)	None to Very Low
Media	N/A	Enough	Enough	N/A	Enough	Depends on clientele	?	Low

Chambers of commerce, for example, may not welcome strict environmental policies and may feel threatened by the 'ecological style' of economic functioning with which they are not familiar. They may comply if there is overt and pressing tourist demand for 'green services'. Similarly, the potential for compliance of international tour and travel agents and operators and tourism developers may be low if their interests are at variance with certain requirements of sustainable tourism (e.g. small scale development adapted to the local natural environment). Recently, however, the idea of sustainable tourism is gaining ground among international economic tourist interests and efforts are made to employ several of its principles (CEMP, 1994). Local and national NGOs and environmental organisations are already active in informally supporting sustainable tourism policies. Finally, recipients and consumers will respond positively depending on their 'green values' (which are gaining ground currently) and the degree of behavioural change required on their part. Tourists may have to change their resource use patterns and travel budgets (with expenses reflecting the total cost of their using the environment of a host area), a process which is influenced by their economic status as well as by complex socio-cultural factors.

4.2.4. Co-ordination

Co-ordination among and within all types of implementers appears to be problematic overall. This can be attributed to the compartmentalised administrative structure, the multiplicity and diversity of implementers, the poor communications among them and the lack of proper legislative and institutional frames of action. Especially for policy recipients and consumers, lack of co-ordination results in a 'tyranny of small decisions' situation, as conflicting demands and actions lead to undesirable environmental, economic and social outcomes (such as waste of resources, excessive pollution, land use conflicts, and the like).

In conclusion, as judged from the examination of the main determinants of compliance, compliance patterns and strength vary considerably among and within the various types of implementers. Under the present conditions, compliance with sustainable tourism policies will possibly be practically limited.

4.3. External, non-policy factors

A host of external, non-policy factors facilitate or prohibit the adoption and implementation of sustainable tourism policies. The concurrent use of resources by both tourists and locals, especially in mixed tourist areas, interferes with environmental protection efforts in the tourist sector as: (1) moderation in tourist resource use may not be matched with moderation in non-tourist resource use and (2) nontourist uses affect the quality and quantity of the tourist product during the off-tourist season.

Developments within and in the periphery of tourist areas (e.g. housing, agriculture, industry, transportation) as well as more global changes reinforce or weaken the effects of the environmental conduct of tourist actors. Property rights and administrative boundaries (dissecting otherwise integral natural areas) may obstruct the implementation of environmental protection tourism policies. Cultural values and the institutional context affect significantly implementation success. Localities traditionally living by environmental values, possessing suitable legislative instruments and established fair justice cultural norms and mechanisms will exhibit greater compliance with the requirements of sustainable tourism.

A broader implementation culture favouring compliance with the law will be a positive factor also than would be the case otherwise. Economic vested interests influence the application and adherence to prescribed policies as they are strategically positioned within the implementation system (WTO, 1980; Kasperkovitz, 1993; Dutton and Hall, 1990; Opschoor and Van der Straaten, 1993). Localities subject to the exploitative influence of powerful interests will face greater difficulties to comply with sustainable tourism policies. Finally, macro-economic, social and technological factors (e.g. changes in exchange and interest rates, the economic structure of tourism origin and destination areas, the influence of international environmental organisations, political restructuring, new energy technologies, etc.) influence significantly the implementation potential of sustainable tourism policies.

Table 4. Implementation issues - sustainability 'ingredients' 'lost' and relevant spatial scales

Type of issue	Sustainability 'ingredients' 'lost'	Relevant spatial scales
Conceptual	Integration	All scales
Scientific-technical	All ingredients	All scales
Economic	Economic efficiency	National to local
Socio-cultural	Social fairness/choice	Regional, local
Institutional	Coordination, social choice	All scales
Legislative	All ingredients	All scales
Political	Integration, social choice	All scales
Administrative	Coordination	National to local
Physical	Environmental protection, economic efficiency	Regional, local
Communication	All ingredients	All scales

Based on the preceding analysis, ten broad types of implementation issues are identified and discussed separately below although in practice these are interrelated and interdependent. For each type, Table 4 shows which 'ingredient' of sustainability is most probable to be 'lost' during implementation as well as the spatial scale at which the issue will probably be important.

4.3.1. Conceptual issues

The current conception and definition of tourism is incomplete and inadequate as it ignores the fact that tourism is an activity complex as well as that it is but one of the many extant and/or potential activities and 'users' of the resources of an area. Resulting policies are imprecise and vague and they are directed to a few of the many components of the tourism complex. Naturally, they disregard the sectoral and economy-environment linkages and they do not introduce the required co-ordination mechanisms, leading, thus, to compliance problems.

4.3.2. Scientific-technical issues

The scientific uncertainty, the difficulties to develop integrated and policy impact models for tourism, and the related information needs (data collection and monitoring) for documented policy impact assessment and evaluation at all spatial/organisational levels delay the development of a rational, systematic basis for policy making and implementation. This situation is more serious in countries with underdeveloped information and monitoring systems. In addition, as technological developments in the various sectors do not proceed in parallel, there are no uniform technological solutions to environmental pollution problems for tourism considered as an activity complex.

4.3.3. Economic issues

Many implementation problems relate to the perceived costs and the redistribution of costs and benefits implied by sustainable tourism policies. Additionally, several environmental protection measures cannot be implemented as financial resources are insufficient and ill distributed among formal implementers across different spatial/ organisational levels. Macroeconomic factors add to the uncertainty associated with tourist demand. Strong vested economic interests distort the functioning of markets and state policies. The impact of these issues will be probably greater in small isolated economies.

4.4.4. Socio-cultural issues

The socio-cultural conditions in origin and destination areas influence the implementation potential of sustainable tourism policies. Lax and permissive compliance cultures foster weak enforcement of policies. Host communities may favour short-term economic benefits over longer-term environmental integrity. The many and diverse tourism actors each apply their particular value system and cost-benefit calculus leading collectively frequently to socially undesirable, economically inefficient and environmentally harmful outcomes.

4.4.5. Institutional issues

Gaps between policy formulation and policy implementation exist for several reasons. Sustainable tourism policies do not exist for all spatial/organisational levels currently. Policy makers and implementers at these levels do not have any role to play. Even at the macro-level, which these policies address mostly, formal policy implementers lack guidance in terms of environmental, economic and social standards for sustainable tourism development and, more generally, operational sustainability criteria. Unless these are provided to them, it is questionable if practical, day-to-day decisions for sustainable tourism can be made. Moreover, at certain levels (e.g. the regional or the multiregional) no institutions exist at present to implement the respective policies.

The current distributive policy making mode hinders the implementation of sustainable tourism policies, which are inherently redistributive. Moreover, the regulatory tradition of state policy making blocks the application of more flexible action schemes adapted to local conditions. Frequently, regulatory requirements are not matched with adequate resources leaving, thus, ultimately the problems unsolved.

Property rights represent another serious implementation barrier. On the one hand, policies implying developments different than those which landowners desire will prove inapplicable (especially, where the tendency to put land to tourist use for instant profits is strong). On the other, tourism-induced urban development is difficult to control and distorts whatever positive effects sustainable tourism policies may have.

4.4.6. Legislative issues

Proper legislative frameworks for sustainable tourism, which integrate the environmental with the economic, sectoral, tourist and planning (land use, regional) legislation and which include the proper executive means for policy implementation, do not exist yet. Moreover, critical value judgements are not made in legislation but are left for resolution at the local level where the interests of the numerous and diverse local and supralocal tourist actors may be in conflict. Long delays in implementation ensue and policies may be rarely implemented as formulated originally.

4.4.7. Political issues

The structure of power relations from the local to the international level determines where and to what extent sustainable tourism policies will be adopted and implemented. Local tourist interests are usually not well organised into a coherent and permanent interest group while supralocal interests have the resources and ability to do so (Pleumarom, 1994). Unless there is strong tourist demand for environmental quality, vested interests will continue to dominate and exploit local resources undermining, thus, the materialisation of sustainable tourism.

4.4.8. Administrative issues
The current administrative system operates along sectoral lines. On the one hand, it ignores the interrelatedness and inseparability of many tourism production sectors as well as their contribution and linkages to non-tourism sectors and activities. On the other, it ignores the natural continuity of host areas. It suffers from lack of co-ordination, co-operation and sharing of authority and responsibility over issues of managing activity complexes such as tourism. At higher spatial/organisational levels, where co-ordination is much more needed and problematic at present, administrative bodies either do not exist or they do not have adequate power to carry out the required actions. At lower levels, administrative mechanisms and provisions for local integrated management are mostly absent. Where tourist areas belong to more than one jurisdictions, multilateral agreements, which facilitate real solutions to management problems, are difficult to achieve.

4.4.9. Physical issues
A number of physical issues render problematic the implementation of sustainable tourism policies. The broader issue is the sustainability of spatial development of the host area/region, which is the backdrop of tourism. If this is not pursued systematically, it will nullify practically all efforts directed to tourism. The spatial structure of many host areas, both in the developed and the developing world, is plagued already by problems of unplanned and haphazard development, environmental and physical degradation, congestion, and lack of co-ordination between infrastructure and superstructure development. Land use conflicts between tourism and adjacent uses of land, such as agricultural, commercial, industrial and residential, are frequent and resolved in ad hoc fashion. Tourism-induced urban development generates an additional complex of activities and land use interests, pressing for resources, which must be controlled for making spatial development and tourism sustainable. Naturally, the severity of the physical issues depends critically on the size and state of development of tourist areas.

4.4.10. Communication issues
Several of the issues mentioned before inevitably lead to communication problems between tourism policy formulation and policy implementation – imprecise and vague policy statements which, in addition, are not co-ordinated among tourism and non-tourism sectors. During transmission from the higher to the lower levels, policies get on different meanings and importance depending

on local conditions and interests. Sometimes they may be 'lost' rendering implementation practically null 'by definition'. Policy implementers in separate tourism sectors have ample discretion to construe them as they like independently of one another especially when co-ordination is costly and not officially mandated. Policy recipients are also unaware of the policies and, if they are, they may not always have incentives or the resources needed to comply with them.

Table 5. Correspondence between types of implementation issues and suggested corrective actions

Types of issues	Type of action considered		
	Retain and improve positive aspects of current policy implementation environment	Introduce new forms of implementation tools/procedures	Feedback to policy formulation environment
Conceptual			X
Scientific-technical		X	X
Economic	X	X	X
Socio-cultural			X
Institutional		X	X
Legislative	X	X	X
Political		X	X
Administrative	X	X	X
Physical		X	X
Communication	X	X	X

5. Proposed measures for sustainable tourism policy implementation

The unsustainable tourism problems will persist for a long time in the future as these are difficult problems to cope with and solve. On the one hand, deep socio-economic changes are needed to provide the proper context within which sustainable development will be accepted and practised. On the other, state legislative and administrative apparatuses will take a long time to adapt to the philosophy and practice of sustainable development so as to provide the required policy implementation mechanisms. Nevertheless, measures can be taken to tackle certain of the implementation issues identified before.

Consequently, the three evaluation criteria will be met, to some extent at least, and the policy implementation potential of the proposed policies will increase (read Figure 1 from bottom up). This section discusses three possible general action avenues: (1) retain and possibly improve certain positive aspects of the current tourism policy implementation environment; (2) introduce new forms of policy implementation instruments; (3) feedback to the policy formulation environment to design more implementable policies. Table 5 suggests a correspondence between types of implementation problems and types of action to be considered. The exact policy measures for particular tourist areas will, naturally, be context-specific and socio-culturally determined.

5.1. RETAINING AND IMPROVING THE POSITIVE ASPECTS OF THE CURRENT TOURISM POLICY IMPLEMENTATION ENVIRONMENT

Although most problems draw from the ills of the current policy implementation system while some of them cannot be handled by however perfect this system is, certain issues can be addressed, at least partially, by exploiting specific features of this system. These concern mostly tourism production in the public and private sectors as current policies target mainly the supply side of tourism. In the public sector, extant economic, fiscal and administrative instruments (general and green taxes, charges, interest rates, decision making authority, etc.) can be used to finance the costs of environmental protection in host areas as well as to correct the maldistribution of costs and benefits of tourism development. Another possibility is to activate and enforce strictly the existing environmental protection and land use planning legislation in host areas and their surroundings and to impose environmental standards in all tourism-related activities and not only to the accommodation sector. Rating systems for tourist destinations can be generalised to spur progress towards sustainability in tourist areas. Improvement of existing data collection and monitoring systems to evaluate policy implementation performance can be encouraged also.

In the semi-public domain, existing municipal, non-governmental and tourism development bodies can be given broader scope and powers to undertake environmental protection actions. In the private sector, adoption of holistic environmental management in large hotels should be encouraged and extended to all tourism-related activities. Initiatives by local entrepreneurs to improve the environmental quality of their localities can be supported and publicised to serve as models for others. Finally, current spatial development mechanisms and procedures may provide minimal coordination among existing tourism implementing bodies both within the same territory and across spatial and/or temporal levels.

5.2. INTRODUCING NEW FORMS OF SUSTAINABLE TOURISM POLICY IMPLEMENTATION INSTRUMENTS

Selected new forms of implementation instruments are sketched briefly below grouped into: supply-side measures, demand-side measure and measures to influence tourism intermediaries.

5.2.1. Supply-side measures

Formal implementation bodies should be established at those levels where they do not exist at present (mainly the regional level within countries and the multiregional among countries) with adequate authority to support the continuous and consistent implementation of sustainable tourism development policies. However, it is more important that these bodies act in close co-operation, or, are integrated with implementation units competent in other policy areas; i.e. joint implementation of tourism and other policies should be the goal as this seems to be a more promising successful avenue than would be the case otherwise.

Clear and unambiguous communication between all types and levels of implementers may be facilitated by:

1. the establishment of information clearinghouses
2. the creation of state consulting and technical assistance offices to assist policy recipients to comply with the requirements of sustainable tourism policies in accordance with their possibilities and contextual conditions, as well as to construe ambiguous and uncertain policy prescriptions
3. the training of the staff of formal implementing agencies to assist policy recipients effectively
4. the staffing of formal implementing agencies with environmental professionals to provide the scientific and technical resources needed to improve sustainable tourism policy implementation

The last two measures are critical, as sustainable tourism is a new concept for most implementers who, naturally, cannot respond effectively to the requirements of their new task without proper training and education.

Non-adversarial conflict resolution by means of environmental negotiation and mediation must be introduced and legislated to avoid stalemates over issues of environmental and land use management and, consequently, further degradation and devaluation of the tourist product. Flexible implementation mechanisms to facilitate the implementation of negotiation outcomes include systems of economic incentives, compensation in money or in kind, transfer of development rights, land redistribution, etc.

The private sector needs to adopt innovative tourism management practices, too. Self-regulation within the tourism complex may aid in the resolution of conflicts among tourism-related enterprises and strengthen their position vis-a-vis the non-tourist sectors (Briassoulis, 1995). Public-private partnerships may help promote tourism activities not easily undertaken by each sector separately.

Finally, penalties for illegal and incentives for legal operations should be devised to control illegal tourism development. The exact nature of these measures has to be determined locally as illegal activities vary with socio-cultural and economic organisation, conditions and norms.

5.2.2. Measures to influence tourism intermediaries

The aim of such measures will be to control the activities of tourism intermediaries – basically to break their monopoly and monopsony powers – and to co-ordinate their activities with those of other implementers. Several issues must be resolved, however, such as: which country (origin or destination) will be responsible for their control (e.g. where conflicts should be adjudicated and the proceeds from charges should go), how they will be controlled (e.g. by means of regulatory measures, market-based approaches, etc.), what kinds of penalties should be imposed for non-compliance, how changes in their status will be recorded, etc.

5.2.3. Demand-side measures

The demand side of tourism is the least controlled at the moment and measures are needed to couple tourist demand with tourist supply policies. Tourist demand should be manipulated positively towards more environment-respecting behaviour and away from resource-consumptive to resource-preserving uses. Education, general and specific, is a principal means to effect such behavioural changes. Promotion and consumer awareness of available alternative forms of tourism (e.g. agrotourism, ecotourism, etc.) and means to pursue them (e.g. information on destinations, facilities) will increase tourist demand for more sustainable forms of tourism.

5.3. FEEDING BACK TO THE POLICY FORMULATION ENVIRONMENT

General and specific changes in the policy formulation environment will provide the enabling framework for the design of implementable sustainable tourism policies. The first and foremost change concerns the adoption and active use of the concept and principle of sustainability in policy making. This implies changes in:

1. the policy making mode (from distributive to redistributive)
2. the policy making approach (from incremental to strategic)
3. the decision making tradition (from top-down to participatory)
4. the policy future outlook (from short- to long term)
5. the basis of planning and management (from sectoral to integrated), with integrated spatial planning playing the co-ordinative role as land mediates all economy-environment-society interactions

Within the tourism policy formulation environment, a first priority change is the adoption of a holistic definition of tourism as an activity complex, where both public and private sector actors participate, and which encompasses the natural and cultural resources of an area as well as its broader physical, socio-cultural and economic context. In other words, tourism should not be treated as a separate but as one of the economic activities which needs to be managed properly to contribute to sustainable development.

To support the rational, scientifically documented and technically correct design of policies, a systematic analytic basis must be established with main elements: (1) the concept of total (i.e. not only tourism) carrying capacity, (2) economic-environmental policy models, even in elementary form, and (3) general and tourism-specific sustainable development indicators and performance criteria to guide the actions of policy implementers.

Integrated planning and management and total development control should be instituted (where this is not already done) and should constitute the basis for policy design. Spatial planning, specifically, will suggest the most sustainable combinations of economic activities for the development of an area's resources, will provide infrastructure congruent with the planned superstructure and will co-ordinate all activities operating within and among tourist areas. Participatory integrated tourism planning, if properly practised, is expected to lead to policies tailored to the needs and nature of host areas. This will happen because it may accommodate the demands of competing groups, aid in conflict resolution, determine exact performance criteria and, thus, reduce the incidence of conflict-laden developments.

Certain complementary changes in the tourism policy formulation environment are necessary also. At the institutional level, a rearrangement of the responsibilities and competencies of the public and private sectors is needed as the private sector should bear its burden of environmental degradation for the heavy use of environmental resources. At the legislative level, environmental and economic legislation on tourism and other socio-economic issues should be co-ordinated. The environmental dimension should be introduced in all sectoral legislation with environmental standards mandated for all tourism-related sectors and operations. Similarly, environmental impact assessment in tourist areas should apply at two levels: (1) the level of individual major operations and activities as usual in EIA practice and (2) the level of the whole tourist area (areal impact assessment) to anticipate and manage effectively the cumulative impacts arising out of the many small tourist and non-tourist activities not subject to individual EIAs.

State (local or regional, at least) planning will assume the responsibility to monitor and manage the cumulative effects of other sectors on tourism. Policies should include systems of incentives and disincentives for all parties involved – producers, consumers and intermediaries – to induce environment-respecting and resource-conserving behaviour and discourage the proliferation of illegal tourism-related activities. Finally, at the administrative level, co-ordination among related administrative bodies, streamlining of administrative procedures, reduction of overlaps and elimination of conflicting competencies should be provided, in the short-term. For the longer term, co-ordinating, strategic bodies, at all spatial scales, should be the basic units for the formulation, co-ordination and implementation of sustainable development, including tourism, policies.

The expected outcome of the proposed changes is the design of more understandable, precise and unambiguous sustainable tourism policies. If the links and co-ordination among policies at different spatial levels and over time are designed and not left to chance or at the discretion of the numerous policy implementers, it will be easier to communicate, explain and implement them at the lower implementation levels.

6. Postscript: the need for further research

The preceding ex ante critical examination of the implementation potential of sustainable tourism policies offered a broad brush assessment of the main issues involved. To prove helpful in actual tourism policy making, this assessment must be supported by research conducted on two main fronts. Firstly, detailed case studies must be undertaken covering a variety of socio-economic, political, and cultural settings with a three-fold goal:

1. to analyse the actual structure and functioning of various implementation environments
2. to assess the implementation potential of sustainable tourism policies within these environments
3. to identify particular local aspects which must be taken into account in the design and implementation of sustainable policies. In addition, in a comparative analysis perspective, these studies will be invaluable in drawing valid generalisations over the broad patterns and dynamics of sustainable tourism policy implementation

The second front concerns the organisation and systematic pursuit of scientific research on the particular environmental impacts of tourism policies borrowing, of course, from the broader environmental research field. Development of operational measures of total, and not only tourism, carrying capacity as well as development of general and tourism-specific sustainable performance indicators are much needed contributions from science to serve the purposes of sustainable tourism policy formulation and implementation. In parallel, inquiry into technologies suited to the socio-cultural milieu of particular host areas must be undertaken to provide practical means for handling the resource and environmental impacts of tourism and, thus, facilitate the implementation of sustainable tourism policies.

References

Barbier, E.B., A. Markandya and D.W. Pearce (1990) Environmental sustainability and cost-benefit analysis. *Environment and Planning A* **22**:9, 1259–1266.

Baud-Bovy, M. and F. Lawson (1977) *Tourism and Recreation Development.* The Architectural Press, London.

Briassoulis, H. (1995) The environmental internalities of tourism: Theoretical analysis and policy implications. In P. Nijkamp and H. Coccossis (eds) *Sustainable Tourism.* Avebury, London, pp. 25–39.

CEMP (1994) *Sustainable Tourism for the 21st Century: Practical Strategies and Future Directions, An International Think Tank.* Lagos, Algarve, 20–26 November 1994. Organised by the Centre for Environmental Management and Planning, Aberdeen, Scotland.

Dutton, M. and C.M. Hall (1990) Making tourism sustainable: The policy/practice conundrum. In *Environment, Tourism and Development: An Agenda for Action?* Workshop Papers, 4–10 March, 1990, Valetta, Malta. Centre for Environmental Management and Planning, Aberdeen, Scotland.

Eber, S. (1992) *Beyond the Green Horizon: Principles for Sustainable Development.* Discussion paper by Tourism Concern, WWF, U.K.

Environment Canada (1990) *Implementing Sustainable Development.* Report of the Interdepartmental Workshop on Sustainable Development in Federal Natural Resource Departments, Mont Ste Marie, Quebec, Canada, June 1990.

Inskeep, E. (1994) *National and Regional Tourism Planning: Methodologies and Case Studies.* Routledge, London.

Kadt, E. de (1990) Making the alternative sustainable: lessons from development for tourism. In *Environment, Tourism and Development: An Agenda for Action?* Workshop Papers, 4–10 March, 1990, Valetta, Malta. Centre for Environmental Management and Planning, Aberdeen, Scotland.

Kahn, A. E. (1966) The tyranny of small decisions: market failures, imperfections, and the limits of economics. *Kyklos* **19**:1, 23–47.

Kasperkovitz, J.M. (1993) Sustainable development against vested interests. In F.J. Dietz, U. Simonis, and J. van der Straaten (eds) *Sustainability and Environmental Policy: Restraints and Advances.* Sigma Verlag, Berlin, pp. 138–149.

Komilis, P. (1986) *Tourist Activities.* Athens, Center for Economic Research and Planning. (In Greek).

Lindsay, J.J. (1986) Carrying capacity for tourism development in national parks of the United States. *Industry and Environment* **9**:1, 17–20.

Mazmanian, D. and P. Sabatier (eds) (1981) *Effective Policy Implementation.* Lexington Books, Lexington Ma.

McKercher, B. (1993) The unrecognized threat to tourism; Can tourism survive 'sustainability'? *Tourism Management* April, 131–136.

Nakamura, R.T. and F. Smallwood (1980) *The Politics of Policy Implementation.* St. Martin's Press, New York.

Opschoor, H. and J. van der Straaten (1993) Sustainable development: An institutional approach. *Ecological Economics* **7**, 203–222.

Pleumarom, A. (1994) The political economy of tourism. *The Ecologist* **24**:4, 142–148.

Redcliff, M. (1987) *Sustainable Development: Exploring the Contradictions.* Methuen, London.

Sabatier, P. and Mazmanian, D. (1980) The implementation of public policy: A framework for analysis. *Policy Studies Journal* (special issue), 538–558.

Sadler, B. (1992) Sustaining tomorrow and endless summer: On linking tourism and environment in the Caribbean. In *Environment, Tourism and Development: An Agenda for Action?* Workshop Papers, 4–10 March, 1990, Valetta, Malta. Centre for Environmental Management and Planning, Aberdeen, Scotland.

Straaten, J. van der (1994) Sustainable development and public policy, Paper presented at the International Symposium *Models of Sustainable Development. Exclusive or Complementary Approaches to Sustainability?* Université Pantheon-Sorbonne, March 16–18, 1994, Paris.

UNCED (1987) *Our Common Future.* Oxford University Press, Oxford.

WTO (1980) *The Manila Declaration on World Tourism.* World Tourism Organisation, Madrid.

HERITAGE TOURISM AND URBAN ENVIRONMENTS: CONFLICT OR HARMONY?

G.J.ASHWORTH
Urban and Regional Planning, Faculty of Spatial Sciences
University of Groningen
The Netherlands

1. Cities are also environments

Most studies of the impact of human activities upon the environment concentrate upon rural areas and define environmental quality and the threats to it in terms of aspects of the natural physical environment. The very terminology of the environmental debate is 'green' whether applied to political and ideological movements or 'books' of planning policy (Commission of the European Communities, 1990; Government of Canada, 1990; Minister of Housing, Planning and Environment, 1989). The obvious assertion still needs to be made that the city is also an environment in the sense of being more than just a location in abstract space. Also the juxtaposition of natural and built, rural and urban, green and grey as two separate systems with dichotomous problems and planning solutions is not only unrealistic but also unhelpful, not least in considering the relations of the activity tourism with its environmental setting.

This symbiosis is accepted here in three senses. First, although that cities possess natural attributes of site, vegetational cover, building materials and the like while rural areas are also, to an extent, built environments. Both, therefore, have been restructured by deliberate intervention and design. The distinction between the so-called natural and the built-environment is one of the degrees of such intervention. Secondly, environmental management, in both cases, requires the selection of desirable or preservable elements as well as the choice of the pace and degree of acceptable change. The philosophical dilemmas stemming from the necessity for these choices are remarkably similar and give rise to similar practical consequences. For example, the use of authenticity in both the natural and the historic built environments as a defining concept, selection criterion and measure of policy success, raises similar questions about 'what is natural?' or 'what is historic?' together with the related questions of the process of authentication and the identification and motivation of the authenticator. Consequently, the inclusion of responsibilities for both natural and built environments in the same management organisations, governed by the same legislative frameworks and operated through the same working practices is globally more the rule than the exception. Thirdly, the distinction between rural and urban areas is similarly more a continuum than bipolarity. This is especially evident in tourism settings when the situation is approached from the side of the tourist use of place attributes rather than the supply of tourism resources. Tourists are rarely distinguishable by their rural or urban motives, spatial behaviour or consumption of environmental experiences nor are tourist places easily divisible into two distinct categories. Most, as for example the US tourist places in Figure 1, are located along the two spectra of natural/man-made and rural /urban.

This simple, and perhaps self-evident, argument needs to be restated here as an explanation of the serious imbalance in attention within current concerns for environmental quality and of the serious consequences that stem of this relative neglect, especially when tourism is considered. Although cities occupy a relatively small proportion of the earth's surface, they actually provide the working, living and, most relevant here, recreational environment of most of its inhabitants most of the time.

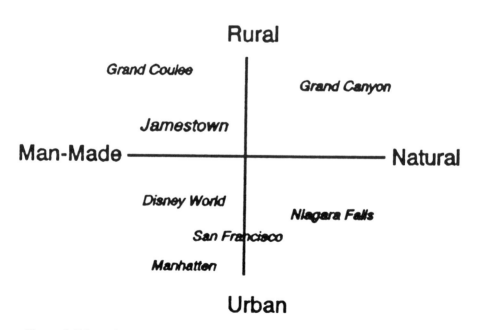

Figure 1. Dimensions in a classification of environments

2. Tourism: an opportunity or a threat?

Similarly, recent concerns about the relationship of tourism to environments originate from the perceived threat of the former to the latter. Not surprisingly, therefore, most studies of the impacts of tourism upon the environment have focused upon its undesirable effects and the policy reaction of public organisations and later, somewhat reluctantly, the commercial tourism industry has been defensive. Public policy has increasingly conceived of tourism as being essentially hostile to the maintenance of environmental quality and if it is to be tolerated as unavoidable for economic reasons, then the tourism flows must be contained and channelled and the inevitable ravages repaired (English Tourist Board, Countryside Commission, Rural development Commission, 1991; Tourism and the Environment Task Force, 1991). The commercial industry has, similarly, largely accepted the premises of the argument and countered its consequences by the promotion of a sustainable or eco-tourism alternative (for a broadly sympathetic account of these see, Coccossis and Nijkamp, 1996 or Hunter and Green, 1995 and for more sceptical critiques, see Wheeler, 1991 or Ashworth, 1992).

This chapter begins with three assertions contrary to the above arguments, namely:

- Tourism is as much an urban as a rural phenomenon and is as much dependent upon man-made as well as upon natural resources. Both the primary and the secondary tourism attractions are mainly located in cities and even when rural areas are used as primary landscape attractions, the supporting facilities are usually urban, and thus the impacts of tourism are likely to be more intense in cities rather than the countryside (Ashworth, 1988).

- Tourism and environmental quality can be positively as well as negatively related. The quality of the urban environment can have a direct and positive effect upon tourism but, equally, tourism can provide the justification and financial support for the creation and maintenance of quality in the environment.

- From these two assertions, the third can easily be deduced. It is possible to devise and implement mutually supportive policies with aims acceptable to both tourism development and environmental quality, in place of the contradictory policies of tourism stimulation and environmental protection that are frequently encountered currently.

Thus, tourism can be an opportunity for improving environmental quality as much as a threat to it and a high quality environment, in turn, presents opportunities for tourism development rather than a constraint upon it. The discovery of the conditions fostering the realisation of this possibility is the purpose of the following discussion.

It should be stressed immediately, however, that these assertions, and the arguments upon which they are based, do not imply any automatic symbiosis between urban tourism and the quality of the urban environment, even less that a harmonious coexistence is generally the case; only that this can be so, in a significant minority of cases, if suitable policies are devised and implemented. Nor is it necessary to accept that tourism is the only, or even the most important, user of quality urban settings: there are many other possible users and only rarely can tourism alone provide the stimulus or support for its maintenance. Nor is the use of a quality environment the only, or, again, necessarily the most important, form of tourism: to many tourist activities the quality of the environment, whether rural or urban, remains at best marginal and, at worst, irrelevant. Similarly, it is easy to document the role that some forms of tourism in some places have played in damaging the quality of the environment in towns or in the country; it is no part of this argument to deny that this has occurred, only to deny that it is inevitable.

The purpose of this chapter is to explore the necessary preconditions for the creation of symbiotic, mutually supportive policies. To that end, the various roles played by a high quality urban environment for particular forms of urban tourism must be understood. Then, the experience of a number of cases is drawn upon to provide a set of guidelines for suitable policies and allow conclusions to be drawn.

3. The urban environment as a tourist resource

There are, of course, many possible relationships between the broad collection of diverse activities that can be loosely bundled together as tourism and the physical environment of cities. For the environment, tourism is just one of a large number of many possible functions to be accommodated in various competitive or compatible combinations. This approach to tourism, as a varied and continually changing mix of functional associations and spatial clusters, is developed in Jansen-Verbeke and Ashworth (1990). For tourism, however, an approach borrowed from marketing science, which treats the environment as a resource, may provide a usable framework for analysis.

At its simplest level, buyer-benefit analysis would provide a means of focusing on the actual patterns of use. Buyer-benefit analysis classifies products according to the utility benefits obtained by the consumer not the characteristics produced by the producer; thus, the same physical product becomes different products to different consumer groups and, applied specifically to place products in Ashworth and Voogd (1990). In the argument here the question would be, 'which consumer benefits are actually being sought and enjoyed through the various uses of the urban environment. The main types of relationship to emerge in this way can be classified as:

- Irrelevant, where it is space itself rather than any particular attribute of that space that is being consumed.
- Incidental, where the attributes of the environment fulfil an additional but subsidiary function, most usually as a backcloth or context for tourism activities principally dependent upon other features.

- Intrinsic, where the quality attributes of the environment are the principal benefit being sought by consumers. This intrinsic use, although probably describing a minority of consumers, is of most interest here as it describes the most active and reciprocal relationship between tourism and the urban environment.

There are, of course, many forms of tourism pursued in many types of cities and the general argument must be made more specific if only for convenience and clarity. The focus here will be upon heritage tourism in tourist-historic cities. Each of these categories relates, in the first case, to various overlapping types of tourists activity (Figure 2) and, in the second, to similarly overlapping categories of tourist city (Figure 3). However the focus is especially apposite as the heritage function makes particular demands, many of which are derived from tourism, upon the quality of urban environments, specifically their built components.

3.1. THE URBAN ENVIRONMENT AS TOURISM COMMODITY

If the environment, due to its intrinsic qualities, has a function as a resource for an activity, then a process of commodification, i.e. the treatment of an attribute as if it were a marketable commodity, has occurred. An understanding of this process is, therefore, critical. The simple analogy with manufacturing industry can be used to demonstrate the nature of the relationships between the component parts and, thus, the role of the environment as an industrial resource. This can be briefly elaborated for the case of urban heritage with the help of Figure 4 and the five components of such a model.

The urban environmental resource is a reflection and an evocation of the events, personalities, historical, literary and mythological associations of the past, expressed through the consciously conserved structures and forms that compose the city's contemporary self-image as well as view of the past. Selection is central and has been performed in part by the vagaries of time and human memory, but, principally, by the deliberate choice of those who have preserved, enhanced, rebuilt and re-created. Resources are not a fortuitous endowment determining their use but are activated by potential uses for them. The urban environment is, thus, best viewed as a quarry of possibilities, only a small proportion of which will ever be activated for use in the production of tourist products.

The assembly process by which resources are converted into products is interpretation and packaging. These are not merely the bringing together of a given set of resources but a means of selecting and transforming resources, in this case, the urban environment, into products, in this instance for environmental industries, of which heritage tourism is one. Selection, interpretation and packaging involve a series of choices about which product is being produced and, thus, which resources are to be used in which ways.

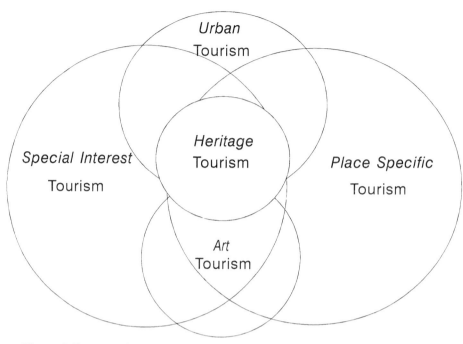

Figure 2. Some tourisms

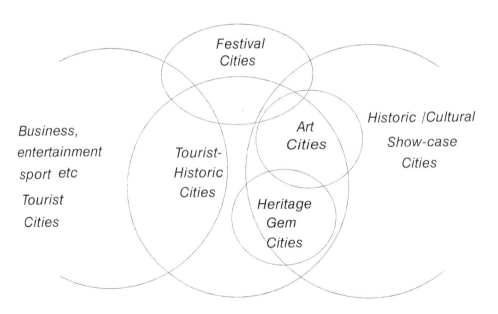

Figure 3. Some tourism cities

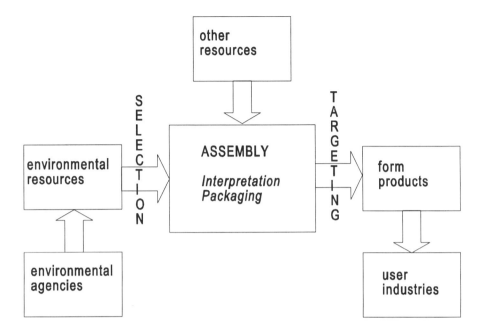

Figure 4. The commodification of the environment

The end product, in this case various heritage tourist packages, has a specific meaning which is not synonymous with the existing built-environment or even an aggregated set of buildings, spaces and cityscapes. It is a product defined by, and intended to be consumed by, defined markets. Thus, quite different products intended for quite different markets can be created from the same sets of environmental resources by varying the assembly process and especially the interpretation. In addition, it is worth reiterating that although the resources may in part be derived from the landscapes of the historic past, the industry being described here is a modern one, satisfying contemporary demands for contemporary products. Its effectiveness can, therefore, only be assessed in those terms.

The consumer defines the tourist product and this has a number of implications. First, quality in the urban environment is defined by the market for it, not the producer of it. One of the clearest variants of this is the 'whose heritage?' question (Tunbridge, 1984) and its rider ''who is thereby disinherited?' (Tunbridge and Ashworth, 1996). The answer is the heritage of the customer not of the producer. This has numerous implications for the selection of what environment is conserved as well as how it is presented for consumption. Secondly, quite different markets can, thus, create quite different products from the same environment, which are then sold to different consumers at the same physical locations. This is an aspect of the 'multiselling problem' that is intrinsic to all place-products, which raises numerous planning and management issues (Ashworth and Voogd, 1990).

The producer performs this critical transformation process and it is clear from each of the stages outlined above that deliberate choices about resources, products and markets are being made. The existence of choice raises the possibilities of using the commodification process in the pursuit of particular urban planning and management goals, in this case the creation and maintenance of physical environmental quality (Ashworth, 1991). The critical point that must be stressed here is that no producer exists in the same sense as in the production of most commercial marketable products. The manufacturing analogy breaks down as generally quite different producers are responsible for each of the separate stages described above: the management choices that are so essential in resource creation and maintenance, interpretative selection and packaging, and market identification and targeting are being made by different organisations for quite different objectives. In this case, urban heritage tourism, itself a collection of various public and private organisations at diverse spatial scales, is selling a product composed of resources it largely neither owns nor manages, which have been created, again by a variety of different organisations, to serve quite different goals. This central question of 'who produces environmental products?' will be returned to later.

3.2. THE QUALITY URBAN ENVIRONMENT AND HERITAGE TOURISM

An important question, is how important is the commodified built-environment as a heritage tourist product: the answers to this will largely determine the extent to which tourism can be regarded as a universally applicable and powerful means of maintaining and enhancing its quality environmental resource.

This is not the place to reiterate the many studies of this particular type of tourism (see Ashworth and Tunbridge, 1990; Boniface and Fowler, 1993; Herbert, 1995) and, in any event, it is only three principal points that are of relevance here. First, that the heritage environment is a major motive for a number of different types of tourism activity; secondly, that it is utilised in a number of different ways; and, thirdly, that this form of tourism can be associated with specific visitor characteristics important for the urban management arguments implicit in this chapter.

The origins of modern tourism in socially approved self-improvement can be traced (Towner, 1985; Towner, 1988) from the medieval pilgrimage through the eighteenth century 'Grand Tour', to what could be termed the Baedeker (or equally Michelin) Tourism of today, with the European cities, in particular, being 'the great museum' (Horne, 1984) or the 'cities of art' (Costa and Van der Berg, 1993), within which urban heritage plays a major, if often only vaguely defined, role, as the main motive for intercontinental tourism travel. Similarly, in many European countries, where studies of tourism motivation have been undertaken, urban heritage is generally the second most important main motive for intracontinental and domestic tourism as well as an important subsidiary motive for holidays together with other main motivations (Ashworth, 1993; Ashworth, 1995; Page, 1995). Finally, the important, if not dominant, role of urban heritage resources in short break, weekend and day excursions has been revealed in numerous local studies.

The failure to produce a definitive numerical measure of the strength of the market for heritage tourism at this point in the argument, or even to refer to such a statistic in the literature, reflects not the lack of importance of the urban environment in world tourism, but, quite the reverse, its all pervasive ubiquity, that renders impossible and pointless any attempt to separate out and measure a specific homogeneous category of urban heritage tourist. Both 'special interest tourism' and 'place specific tourism' are vaguely defined but strongly increasing attributes which would include urban heritage tourism (Figure 2) which is intimately related to 'art tourism', 'festival tourism' and many other specialist niche markets as to make its isolation difficult. What is certain is that the conserved built environment of cities is serving as a major motive for entire international and intercontinental holiday packages, as a background amenity factor for much international business and conference tourism, as a secondary attraction destination for excursions of holiday-makers primarily attracted to beaches or rural areas, as a brief cultural interlude for those also using the city for its many other commercial, shopping, entertainment and such mundane facilities. The task is not to seek to construct a definition of the urban heritage user from an aggregate of these and many other diverse categories of users but to understand the role being played by the conserved urban environment within such multi-resourced packages created for multi-motivated consumers in the multifunctional city.

The characteristics and behaviour of urban heritage consumers has a more direct relevance to the argument here than attempts to segregate specific categories of users. It has long been understood that the urban heritage consumer, especially when compared to the beach tourist, can be generalised as being child-free, belonging to the higher income and educational attainment groups, is relatively high spending, hotel-based, and less seasonally restricted (Prentice (1993) summarises much of this research into the character of the heritage tourist). The potential significance of these attributes in the context of this argument lies in the resulting ratio of local benefits to costs, in the spatial and economic consequences of potential product diversification and in the implications for management.

4. Lessons from some urban cases

All of the above analysis is a necessary preliminary to the development of deliberate strategy. It must be reiterated that, at many points in the above models, choice is not only possible, it is required. There is, however, nothing inevitable in the supportive relationship between urban tourism, and particularly heritage tourism, and the quality of the built-environment as a resource for this activity. Indeed, quite the reverse of such inevitability is the case and a key word that emerges is selectivity. Heritage tourism is only one tourism option among many, and is itself highly selective in terms of types, interpretations and locations of the environmental resources used. Certain environments in some areas of some towns may be used in this way – and, conversely, many, therefore, will not. Equally, it must be stressed that the policy as a whole is optional not compulsory; no city, regardless of the potential richness of its resource endowment, is locked into a policy of heritage tourism development. This is only one of many possible policies. The nature of this selection process and how it can be influenced by active goal-determined intervention must now be considered and this is best undertaken by generalising from the experiences of actual cases.

There are, of course, a vast and varied number of possible cases to draw upon and each is necessarily a unique heritage product developed from a unique set of environmental resources for a particular market. However, the objective here is neither a comprehensive coverage nor a detailed description, but the drawing out of a few important policy difficulties from a range of categories of cases so as to arrive at some general guidelines helpful in the development of the desirable strategy of harmony. For this purpose, an initial fundamental distinction is drawn.

4.1. MAJOR TOURIST-HISTORIC CITIES

These are cities with an established international reputation for high quality in their built-environment supported by an international consensus that their architectural and historic importance constitutes a 'world heritage resource' both in the restricted-supply defined sense of a UNESCO 'World Heritage Site' but also, more importantly, as legitimated by visitor numbers. This has immediate implications as to how such cities are regarded and resulted in both the development of major heritage tourism industries drawing upon a world market and also a widespread 'option' demand manifested in international financing and concern, which narrows the range of alternative strategies available locally.

Two important provisos must be added. First, despite their dominance in both externally and internally projected urban images, quality heritage settings occupy only a small proportion of the total land areas of such cities. These cities may be sold as a whole but they are consumed in pieces. Secondly, tourism and the related economic activities that depend upon the commodified environmental resource, however important, are rarely the most important source of local employment and incomes.

Three implications for management result. First, the concern for the maintenance of environmental quality of relatively small parts of the urban area dominates urban policy making for the city as a whole and inevitable distorts both spatial and functional planning. Secondly, the main planning priorities are effectively established, and to a large extent externally imposed, leaving little room for local manoeuvre. Thirdly, the size and historical evolution of most of these cities implies that a degree of multifunctionality must exist, even in or near the highest quality areas, creating potential land use conflicts needing various solutions.

Two of the many possible cases are selected here to serve as archetypes and source of generalisation for the argument considered. Many more such cases are discussed at greater length and drawn from a wider variety of geographical locations in Ashworth and Tunbridge (1990), and Tiesdell *et al.* (1996).

VENICE. The expression of these general points in the much studied case of Venice can be summarised as a spatio-functional segregation whose completeness borders on the schizophrenic and a set of policy conflicts that are manifested through a hierarchy of spatial administrative scales. The quality settings and the tourism dependent upon them are exclusively and intensively concentrated into the 'lagoon' city and associated islands, within which they have developed a near monopolistic position. The decline of the residential related commercial activities and relict industrial functions (see Burtenshaw *et al.*, 1991), and the restrictions on motor vehicle circulation have resulted in the creation of what is effectively a heritage city, both in the sense that this has become the only important function of the area, and that this function is accorded an absolute priority both within and even more significantly outside the city.

A consequence of the functional separation within the urban region has been the growth of the 'mainland' industrial, commercial and residential settlements especially Marghera and Mestre, which, in turn, has created distinctly different environmental planning objectives within the wider urban region (Figure 5). A result is that the last thirty years have been characterised by two types of largely unresolved confrontation.

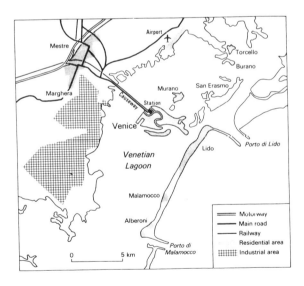

Figure 5. Venice

The first is between planning strategies for different parts of the urban region and at different levels in the governmental hierarchy. Simply stated, the maintenance, and, indeed, physical survival of the built environment of the lagoon city is dependent upon restrictions, especially on navigable access to the lagoon, water extraction and on atmospheric emissions, which, in turn, threaten the economic interests if not viability of the wider region. International concern, expertise and funding, especially since the 1966 floods, for what is perceived to be a threatened world environmental heritage, has been frustrated at the national and especially Veneto regional level, where quite different environmental and thus political priorities exist (for a fuller discussion of the threats see O'Riordan (1975) or, more graphically, Fay and Knightly (1976), and of the planning responses see Burtenshaw *et al.* (1991)).

Secondly, within the lagoon city the conflict has been between those concerned for the maintenance of the quality environmental resource and the principal user of it, namely heritage tourism, which is credited with damaging the resource that it is exploiting. The most usual type of policy response that has been suggested is the formulation of some form of carrying capacity followed by suggestions on restrictions on the number of tourists if these are exceeded. Although the concept of 'carrying capacity' has proved to be largely unworkable as a general instrument for the management of tourists in perceived fragile environments, at the highly specific local level where it can be almost continuously monitored in physical terms, it is attractive through its very simplicity. The physical lagoon setting of Venice is particularly appealing in this respect and Van der Borg (1990) has calculated such capacities and chronicled the many days on which they are exceeded. The result in Venice has been the consideration of a wide range of restrictions on various groups of visitors at various times which seem ultimately to lead to such controls on the number, types and timing of the visitor flow that Venice will become the world's first 'turnstile tourist city'. There is even discussion of a 'Venice card' allowing entry (to places or times) based on either economic contribution (e. g. expenditure in hotels) or even 'appropriateness of the intent' (presumably knowledge of or interest in, the historic resources).

In terms of the argument advanced earlier, there are many logical and practical difficulties with these approaches. If environmental resources can only be defined by their consumption, then the question to be posed to the 'Save Venice' (with the implied 'from tourists', or even 'from the wrong sort of tourists') lobby is always, 'for whom, then, is it to be saved?' Clearly, the confrontation here, as in other heritage gems at the same stage of development, is not between such fundamentally opposed positions as is sometimes implied. The question at issue is not use or no use, for heritage can only exist in terms of a legatee, but the way it is used. In so far as environmental resources are in fixed supply, and, at least, in some respects they may be at a specific time and place, then policy for their exploitation has many similarities with that for the use of a fossil fuel resource. The complication with this analogy is the option demand for heritage which leads to the position that Venice must be saved from present active users for the sake of a distant passive market that requires only the knowledge of its continued existence and intergenerational bequest.

An alternative more suited to the logic of the argument so far, than to either preserving the environment by severely restricting the demand or to sacrifice the environment to damage and destruction of over use is to respond to demand by creating more Venices elsewhere. The objection that Venice is unique is countered by pointing out that similarly unique heritage cities offering similar tourism experiences can quite easily be developed elsewhere. The 'supply' of Venice can even be extended south of the lagoon at Choggia, or even across the Adriatic in the string of Venetian heritage settlements from Capodistria (Koper), through Ragusa (Dubrovnik) to Crete and the Aegean.

BATH (Figure 6). A very similar set of spatial functional patterns can be seen in Bath but on a scale that is more detailed and is not manifested so sharply through spatial administrative hierarchies as in Venice. The 'Georgian City', as a result of its later, eighteenth century, construction, occupies sites which are mostly just peripheral to the older historic core, itself a mixture of Roman, Medieval, Georgian and modern styles. Although, there has been substantial residential suburbanisation, the spatial and functional polarisation between a conserved historic city and new 'cities' serving modern functions has not occurred

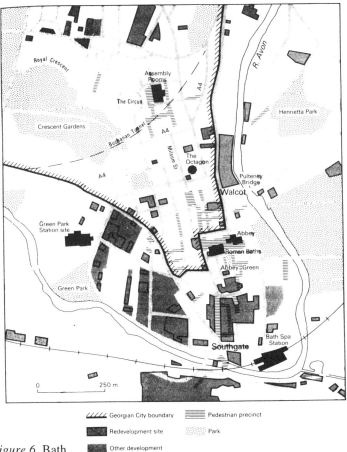

Figure 6. Bath

Redevelopment to accommodate modern transport, retailing, hotel and educational functions has been allowed to take place on the edge of the historic core, in areas of lower environmental quality, at least relative to the standards of this city.

There has been some attempt, therefore, to maintain a degree of multifunctionality, resulting in a modern shopping development and public transport interchange at Southgate and a multifunctional development including a major hotel at Walcot (see Burtenshaw *et al.*, 1991) both of which could be, and have been, challenged as a reduction in visual environmental quality.

The Bath case raises clearly two issues: firstly, the balance to be struck between the heritage function and other, quite legitimate, urban functions; secondly, the distribution of the benefits and costs of creating and maintaining quality in the built-environment. The former mainly accrue at the international level, through the psychic benefits of the continued existence of a 'world heritage city'; at the national level, through tourism foreign exchange balances; and, locally, to those economic activities dependent directly or indirectly upon tourism receipts. The costs, however, are both less easy to allocate and more diffuse, but include direct costs to the local authority and indirectly the negative costs incurred as a result of these priorities for local attention and expenditures. In addition, high environmental amenity is a positive attraction for certain functions as, for instance, specific residential and service functions but the restrictions necessary for its maintenance will equally repel others leaving the city with an unbalanced economic and social structure.

4.2. HERITAGE AREAS IN MULTIFUNCTIONAL CITIES

Medium sized, multifunctional cities, within which specific areas have been developed as high quality morphological settings as a matter of deliberate policy, are in many respects more typical in the sense of being the actual living and working environments of more people. They also offer, however, both more complexity and more opportunities for choice in policy creation, not least because no clear consensus exists, unlike in cities such as Venice or Bath, about the existence or location of high environmental quality areas. These cities can be characterised as:

- Multifunctional, by definition, thereby manifesting large variations in the quality of a built-environment which has evolved continuously over a long period in response to varied demands made upon it. The possibilities for the enhancement of such environments is complicated by their support of many activities other than those generated by historicity.

- A consequence of this long term multifunctional development is a variegated morphological pattern producing a palimpsest of ages, styles and qualities. The clearly spatially demarcated 'historic cities' of Venice or Bath are replaced by 'islands' of high environmental quality and historicity within a wider 'sea' of other areas. These are essentially micro-scale reconstructed stage-sets, and can only be considered at that scale.

- Having priorities and a range of strategic choices that are not only wider than those available in the major tourist-historic cities but are also to a large extent internally rather than externally determined.

A brief description of two such islands, which can be taken as typical of the many which have been created by planning actions in the last 25 years in most medium sized multifunctional European cities (see the cases in Tiesdell *et al.*, 1996) will exemplify these characteristics.

ELM HILL, Norwich. Elm Hill is a short street whose distinction lies in its role within the creation of heritage in Norwich as the centre of a multifunctional regional capital with an important tourism and cultural function. More widely, it served as an inspiration to many other cities who, consciously or not, have created their own 'Elm Hills' all over Europe, leading to the paradox that what originally was an expression of unique local identity through the physical environment, has become, through the imitation of 'catalogue heritagisation' (Ashworth, 1997), a standardised visual cliché of a historic conservation area.

It is the fourth most visited area of the city and, as can be seen in Figure 7. It occupies a strategic linking position between the city's three most important attractions, namely, the cathedral, the market place and the castle (Ashworth and De Haan, 1986). Yet, it contains no notable monument, recorded historic association or specific tourism attraction and, a little over 25 years ago, was narrowly saved from complete demolition and redevelopment. Although, it was a medieval street and is flanked by some buildings dating from the sixteenth to the eighteenth centuries, it was effectively reconstructed in the 1970s as a set of facades with attention being paid to the detailing of the street scene, including newly introduced cobblestones and street furniture with an antique appearance.

The reaction in land use was the growth of ground floor commercial activities that benefited from association with historicity and the presence of visitors (tea and coffee shops, antique and craft shops). In addition, rising rent levels effectively excluded other commercial users and the residential function, leaving many of the upper storeys under-used. The result has been to produce by deliberate planning action a high quality physical environment, whose location plays an important linking role between the cathedral and the market areas. It, thus, supported the creation, and subsequent maintenance (City of Norwich 1992), of a heritage city by the local authority in the service of a mix of local and tourist activities; (see Berkers (1986) for a description of the decision making that led to its rehabilitation).

The questions posed by the creation of Elm Hill, and many similar historic stage-sets created by planning action in most such cities in Europe (e.g. Bergkwartier, Deventer; Bottcherstraße, Bremen; Stonegate, York; Stokstraat Maastricht are among the best known) is not the relationship between historical authenticity and urban conservation policy. All environmental conservation policies involve so much conscious selection and subsequent enhancement, replacement and replication that they are more a process of creation of what is desired rather than of protection of a surviving relict environment. Elm Hill is, thus, a product of the commodification model described above. The implications of accepting this position are, however, that a series of opportunities are presented for the use of historicity in shaping a high quality urban environment for the achievement of specified planning objectives. Such an environment, then, can be judged only in terms of its effectiveness in meeting particular modern demands, including the authenticity of the experience rather than any intrinsic authenticity of the objects themselves, a concept that has little meaning, let alone relevance as a guide for planning decisions here.

Figure 7. Elm Hill, Norwich

MARTINIKERKHOF, Groningen. Groningen is a medium sized multifunctional provincial capital with an architectural heritage reflecting a millennium of existence as a regional ecclesiastical, government and commercial centre. The largest concentration of surviving historic buildings are clustered in the north-east quadrant of the inner city (the so-called Martinikwartier). Conservation policy from the 1960s concentrated upon the preservation and restoration of major monuments, such as the cathedral church, the Martinikerk, the provincial administrative offices, the Provinciehuis and the governor's palace and gardens, the Prinsenhof /Prinsentuin). The shift in emphasis in the course of the 1970s towards ensembles led to the whole area being designated a conservation area (beschermd stadsgezicht). As with Elm Hill, much of this area of Groningen was scheduled for large scale demolition and redevelopment, and this shift in policy is discussed in detail in Ashworth (1984).

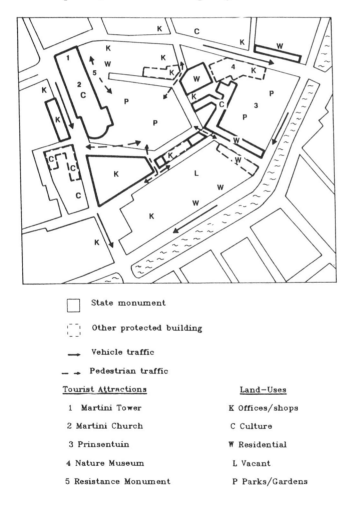

	State monument
	Other protected building
	Vehicle traffic
	Pedestrian traffic

Tourist Attractions	Land-Uses
1 Martini Tower	K Offices/shops
2 Martini Church	C Culture
3 Prinsentuin	W Residential
4 Nature Museum	L Vacant
5 Resistance Monument	P Parks/Gardens

Figure 8. Martinikerkhof, Groningen

The renovation and rehabilitation of many of the smaller domestic buildings was undertaken by the city rehabilitation corporation (stadsherstel) which purchases, renovates and resells on the free market. The deliberate creation of a high quality historic district centred on the Martinikerkhof (Figure 8) was supported by expensive public works to create the parked area, a new street pattern of granite setts, the closure of a through road, stringent restrictions on vehicle circulation and parking as well as attention to such detail as the removal of all conventional street furniture and the reintroduction of operating gas street lighting. The district was, therefore, planned as an historic area having a representative and symbolic function for the city, in sharp environmental contrast to the commercial and entertainment districts which it adjoins. It serves the recreational needs of city residents and visitors. The sort of tourism generated by the city favours the particular atmosphere of an historic district as ancillary to shopping, a setting for cultural attractions and as a heritage resource in itself (Ashworth, 1997).

Its creation by local authority planners and its exploitation by the tourist authority as heritage has produced a distinctive mix of uses of the buildings of the area. The parkland open spaces and the pedestrianised routeways are used for a wide range of informal recreational activities by residents, shoppers, workers and tourists. Five of the city's main tourism attractions are located in the district. There are, however, no supporting tourism facilities; catering establishments are found around and to the south of the Grote Markt which separates heritage from entertainment (Gemeente Groningen, 1992).

The four dominant land uses are culture, public offices, housing and commercial offices, with the first two occupying the larger and the last two the smaller buildings. The Martininkerk serves many public functions apart from the religious function; the governor's palace houses the radio station; a private art gallery and a dance school occupy other buildings. These cultural uses both gain from and contribute to the historicity of the area. The provincial administration occupies the partially restored and partly replicated Provinciehuis, seeking status, symbolism and recognition at the cost of difficult vehicular access and restricted parking for staff and clients. Housing is the dominant use of the smaller buildings and rented subdivisions have now given way to one-family owner-occupation leading to a distinct social change over the past 15 years.

Within a small area, therefore, there exists a mix of very varied uses, all of which have different relationships with the historic built-environment and with each other. As a self-consciously created historic district fashioned by the actions of local authority planners, the principal and continuous management task is the maintenance of that balance and the avoidance or resolution of conflict between users.

The spatial scale and the role played by local planners in using historicity are similar in the Groningen and Norwich cases. The differences lie in the mix of planning objectives and in the mix of functions using the high quality environment as a resource and ultimately major differences in the way city centres are used in the two countries (Dingsdale and Van Steen, 1997).

5. The lessons of experience

A difficulty apparent in drawing general lessons for the development of planning strategy from such cases is that the very selectivity stressed earlier emphasises the characteristics of the particular case. Each of the cities described above is special in two senses. First, the creation of particular heritage tourism products from particular environmental resources is essentially a unique reflection of local identity. However, this may be balanced, as argued above, by an erosion of that sense of locality by both its replication and the establishment of a national or even European conventional wisdom and working practice about how such areas should be treated and also by the expectations of the tourist about an historic area. Secondly, collectively, they represent only a minority of cities or areas within cities. They are the fortunate few, although this 'fortune', as has been argued above, has less to do with fortuitous endowment than with the conscious ability to exploit opportunity on the part of urban managers, which many more cities, and areas within them, could emulate. However, the preconditions necessary for such success can be summarised as integration, which can be elaborated in four different but related aspects.

- Organisational integration is the most obvious but also the most all embracing and most difficult to achieve. The whole commodification process described earlier depends upon an analogy with manufacturing production that ceases to apply at the organisational level. This is exacerbated but not solely caused by a distinction between the public sector, generally responsible for the shaping and maintenance of most environmental resources, and the private sector, which uses such resources in the production and marketing of products. Fragmentation of firms and agencies is typical of both sectors in tourism. In the Venice case, much of the problem is in conflict between the various levels in the jurisdictional hierarchy, while in both Bath and Norwich conflict has arisen between public bodies responsible for conservation and tourism promotion, respectively. This is only one manifestation of a much wider pattern of city (and especially given its multifunctional character, inner city) management tackled by many types of City Centre Management within which heritage and tourism normally play prominent roles (Ennen and Ashworth, 1995).

- Motivational integration is both a cause and a result of the above. Simply expressed, a high quality urban environment has been produced as a result of largely local demands in the Groningen and Norwich cases, and international demands in the Venice case, originally for a mix of symbolic, educational and cultural reasons. They now support, and, in return, are, to an extent, supported by major economic activities whose demands are imported.

- Financial integration is rendered necessary by the inherent maldistribution of costs and benefits. The economic benefits of activities, such as heritage tourism, accrue to specific and limited economic sectors within and outside the city. The urban environment being exploited is generally either a 'free' resource or supplied below cost as a result of the other markets it is serving. Taxation systems only imperfectly redress some of this maldistribution and, in any event, are far too indirect and insensitive to provide the feedback within the market assumed to exist by the model.

- Functional integration between a specific activity and a particular environment depends upon the nature of the functional association between them. This has numerous implications, some of which are relatively easily amenable to regulation. A variety of such associations, both positive and negative, between urban historic settings and tourism accommodation facilities, for example, are discussed in Ashworth (1990).
- Spatial integration stems, in part, from the functional associations and results in spatial clustering or dispersion. Both the historic city and the tourism city can be described as spatial entities and their spatial segregation, overlap, conflict and compatibility can be modelled (Ashworth and Tunbridge 1990; Tunbridge and Ashworth, 1992). Such spatial evolution is only a consequence of functional association but, generally, lends itself to management through the existing instruments of spatial planning.

There is no clear-cut blue-print for attaining such integration. Neither the size, antiquity, dominant political ideology, type of commodified heritage environment, nor particular mix of functions seems to offer clear guidance; nor does any particular organisational structure of partnerships, whether public-public or public-private. The relationship between this activity and this resource is complex, sensitive and multifaceted. An understanding of how they are related, whether through marketing models and planning cases as attempted here, or otherwise, is a precondition of management and the establishment of suitable organisational structures. A high quality urban environment and urban heritage tourism can be incompatible or mutually supporting opportunities; the choice between these alternatives is not predetermined by any particular set of conditions and thus remains open to deliberate decision.

References

Ashworth, G. J. (1984) The management of change: conservation policy in Groningen, *Cities*, 605-16.

Ashworth, G. J. (1988) Urban tourism: an imbalance in attention. In C. Cooper (ed.), *Progress in Tourism*. Belhaven, London.

Ashworth, G. J. (1990) Accommodation and the historic city. *Built Environment* **15**:2, 92–100.

Ashworth, G.J. (1991) *Heritage Planning; the Management of Urban Change*, Geopers, Groningen.

Ashworth, G. J. (1992) Planning for sustainable tourism. *Town Planning Review* **63**:2, 325–329.

Ashworth, G. J. (1993) Culture and tourism: conflict of symbiosis in Europe. In W. Pomple and P. Lavery (eds), *Tourism in Europe: Structures and Developments*. CAB

International, Wallingford.

Ashworth, G. J. (1995) Managing the cultural tourist. In G. J. Ashworth and A. G. J. Dietvorst (eds), *Tourism and Spatial Transformation: Implications for Policy and Planning.* CAB International, Wallingford.

Ashworth, G. J. (1997) Managing change in the city centre: the Groningen case. In A. Dingsdale and P. van Steen (eds), *The Management of Urban Change in Europe.* Groningen Studies 63, Groningen.

Ashworth, G. J. and T. Z. de Haan (1986) *Uses and Users of the Tourist-historic City: an Evolutionary Model in Norwich.* Field Studies 10, GIRUG, Groningen.

Ashworth, G. J. and J. E. Tunbridge (1990) *The Tourist-historic City.* Belhaven, London.

Ashworth, G. J. and H. Voogd, (1990) *Selling the City: Marketing Approaches in Public Sector Urban Planning.* Belhaven, London.

Berkers, M. (1986) *Norwich: Policy in a Tourist-historic City.* Field Studies 9, GIRUG, Goringen.

Boniface, P. and P. J. Fowler (1993) *Heritage and Tourism.* Routledge, London.

Borg J. van der (1990) *Tourism and Urban Development.* Faculty of Economics, Erasmus University, Rotterdam.

Burtenshaw D., M. Bateman and G. J. Ashworth (1991) *The European city: Western perspectives.* Fulton, London.

City of Norwich (1992) *Tourism Development Action Plan.* Planning Department, Norwich.

Commission of the European Communities (1990) *Green Book on the Urban Environment.* COM 90 218, Brussel.

Coccossis H. and P. Nijkamp (eds), (1996) *Sustainable Tourism Development.* Avebury, Aldershot.

Costa, P. and L. van der Berg (1993) *The Management of Tourism in the Cities of Art.* CISET 2, University of Venice.

Dingsdale, A. and P. Van Steen (eds), (1997) *The Managment of Urban Change in Europe.* Groningen Studies 63, Groningen.

English Tourist Board, Countryside Commission, Rural development Commission (1991) *The Green Light: a Guide to Sustainable Tourism.* London.

Ennen, E. and G. J. Ashworth (1995) *Centrummanagement: een Nieuwe Strategie voor Stedelijk Beleid? (Centre Management; a New Strategy for Urban Policy?).* Geopers, Groningen.

Fay, S. and P. Knightly (1976) *The Death of Venice.* Deutsch, London.

Gemeente Groningen (1992) *Binnenstad beter* (Inner City Better). Municipality of Groningen, RO/EZ, Groningen.

Government of Canada (1990) *Canada's Green Plan Environment Canada.* Ottawa.

Herbert, D. T. (1995) *Heritage, Tourism and Society.* Cassell, London.

Horne. D. (1984) *The Great Museum: the Re-presentation of History.* Pluto Press, London.

Hunter, C. and H. Green (1995) *Tourism and the Environment: a Sustainable Relationship?* Routledge, London.

Jansen-Verbeke, M.C. and G. J. Ashworth (1990) Functional association and spatial clustering: an analytical approach to tourism combinations. *Annals of Tourism Research.*

Minister of Housing, Planning and Environment (1989) *National Environmental Policy Plan: to Choose or to Lose.* SDU, The Hague.

O'Riordan, N. (1975) The Venetian ideal. *Geographical Magazine* **47**:7, 416–426.

Page, S. T. (1995) *Urban Tourism.* Routledge, London.

Prentice, R. (1993) *Tourism and Heritage Attractions.* Routledge, London.

Tiesdell, S., T. Oc and T. Heath (1996) *Revitalising Historic Urban Quarters.* Architectural Press, London.

Tourism and the Environment Task Force (1991) *Tourism and the Environment: Maintaining the Balance.* 5 Vols, English Tourist Board, London.

Towner, J. (1985) The Grand Tour: a key phase in the history of tourism. *Annals of Tourism Research* **12**:3, 297–333.

Towner, J. (1988) Approaches to the history of tourism. *Annals of Tourism Research* **15**:1, 47–62.

Tunbridge, J. E. (1984) Whose heritage to conserve. *Canadian Geographer* **26**, 171–80.

Tunbridge, J. E. and G. J. Ashworth (1992) Leisure resource development on cityport revitalisation: The tourist-historic dimension. In B. S. Hoyle and D. Pinder (eds), *European Port Cities in Transition.* Belhaven, London, pp. 176–200.

Tunbridge, J. E. and G. J. Ashworth (1996) *Dissonant Heritage: the Management of the Past as a Resource in Conflict.* Wiley, London.

Wheeler, B. (1991) Tourism's troubled times. *Tourism Management* **12**:2.

TOURISM AND THE CITY:
SOME STRATEGY GUIDELINES FOR
A SUSTAINABLE TOURISM DEVELOPMENT

JAN VAN DER BORG
EURICUR
Erasmus University
Postbox 1738, 3000 DR Rotterdam
The Netherlands

SET
University of Venice
Riviera S. Pietro 83, 30030 Oriago di Mira
Italy

1. Introduction

When analysing the impact of tourism on the environment, reference is usually made to the devastating effects mass tourism has on the natural environment. Only recently, the question whether or not cities, originally designed to host people, might have similar problems with tourism has arisen. A confirmative answer to this question implies that a city's policy for tourism development has to account for that city's limits to absorb visitor flows. In other words, urban tourism development strategy has to be compatible with the urban environment. The aim of this contribution is to discuss the principal characteristics of such a strategy, not only as far as environmental issues are concerned. It intends to give a comprehensive answer to the question whether, and under what circumstances, urban tourism may be worth developing, a crucial question for many cities that are at the moment considering to promote tourism development.

Recently, the market of urban tourism is rapidly expanding, and further growth may be foreseen for the 1990s. The favourable market conditions tempt many city planners to make tourism development an important ingredient of urban policy. They are apt to overlook, however, that the impact of tourism on the economy of the urban system is not under all circumstances stimulating.

The most pessimistic scenario of tourism development ends up with the conclusion that when the number of visitors of a tourist city exceeds a maximum related to its physical capacity of absorption, the long-term negative external effects of tourism, such as pollution and congestion, will readily become unacceptable. In the extreme case, the uncontrolled growth of tourism slowly consumes a city's heritage, the one major urban resource that unfortunately cannot be reproduced. The quality of urban life deteriorates fast; excessive pressures from tourism activity reduce the accessibility of the centre to a minimum. Most other economic activities have to find accessible business sites elsewhere. The city becomes unliveable not only for the residential population, but also for tourists and excursionists. The entire urban tourism system is about to loose even its last flourishing urban function, tourism.

But it is not only in such extreme situations that the development of tourism may be virtually ineffective as an instrument to stimulate local economic growth. We may encounter two other situations in which tourism development is only partially effective or even harmful. Firstly, as the market expands, competition between tourist destinations gets increasingly more intense. Most localities that aspire to promote tourism lack the basic requisites – that is, a highly diversified, appealing package of attractions and facilities – to obtain a large enough share in the market of urban tourism to guarantee their competitiveness in the long run. The modern tourist is very sophisticated. An abundant presence of natural and cultural resources is an absolutely necessary condition, but these assets cannot be marketed without the support of adequate facilities and a vast infrastructure. In fact, the costs of the initial, mostly public, investments and promotion are apt to be made in vain.

Secondly, tourism development may trigger a process known as "crowding out". The mechanism, first described for tourist resorts by Prud'homme (1986), tends to expel the less lucrative urban functions and replace them with tourist activities. The surpassing of the tourist carrying capacity of a city, interpreted as a social-economic limit to tourism (Canestrelli and Costa, 1990), releases this mechanism. The social-economic limit may be far more restrictive than the physical one, because it depends heavily on the health of the city's economic and social structure.

These scenarios, far from being hypothetical, suggest that the development of tourism may also have negative aspects. This certainly does not mean that fear of a negative outcome should keep cities from giving tourism a try. It does mean, however, that the development of urban tourism makes sense only if a set of preconditions has been satisfied and the strategy is rationally designed and implemented. Furthermore, the strategy for the development of tourism must be sustained by the choice of adequate policy instruments.

The aim of this chapter is to discuss the basic preconditions that make urban tourism development worthwhile, the alternative development scenarios, and the choice of the instruments that help to render the strategy effective. Some of the issues that will be addressed in the chapter will be illustrated by cases taken from a recent UNESCO study on urban tourism (Costa and Van der Borg, 1991). The chapter is structured as follows. The preconditions for successful tourism development and alternative development strategies are discussed in the second section while the implications for local tourist policies are discussed in the third section. The fourth section contains the conclusions of this study.

2. Tourism and urban development

2.1. TOURISM AND THE CITY

Several authors have argued that cities evolve in a cyclical manner. Periods of urban growth are necessarily followed by periods of urban decline. The mechanisms that cause the dynamics of the cycle have been described extensively by Van den Berg (1987). He has argued that periods of growth are characterised by a strong tendency towards spatial concentration. In these periods, families and firms are highly interested in a central location. During periods of decline, several groups of families and, perhaps with some time lag, firms, shift their orientation from the central city to the surrounding areas. Spatial deconcentration tendencies cause the economic base of the central city to shrink and the city falls into a persistent urban depression.

The urban life cycle consists of four development stages. After the urbanisation stage, which is characterised by an unconditioned growth of the central city, the city goes through the stages of suburbanisation and desurbanisation. During these stages, the municipalities around the central city are rapidly expanding, while the central city itself enters a period of social and economic stagnation. Many Western cities actually find themselves at a very delicate transition stage, slowly recovering from the crisis into which they had fallen. Tourism is seen by many cities as the most appropriate way to speed up the process of recovery.

The development process of tourism is also supposed to be cyclical. The dynamics of the life cycle of the tourist city can be sketched as follows (for a more complete description see Van der Borg, 1991). At the first development stage of urban tourism, the traditional excursionists, those who are visiting the city within 24 hours from their proper homes, are replaced by the residential tourists. Their visit takes more than 24 hours, which implies that they "consume" at least one overnight stay in the destination. The growing economic spill-off provokes the birth of new local tourist firms, mostly concentrated within the city centre itself.

If the interest in the city continues to increase, two new types of visitors emerge, the indirect and the false excursionists. The indirect excursionist visits the city within 24 hours from a holiday destination other than the destination of the excursion. The false excursionists, even if they have previously chosen the city in question as holiday destination, look for accommodation elsewhere. From their non-central location they will pay daily a visit to the city. The previously mentioned UNESCO study on tourism in cities of art has revealed that the share of excursionists in the total number of visitors in this development stage is far from negligible. It ranges from 85 percent in Heidelberg, via some 75 percent in Bath and Salzburg, to some 50 percent in Venice, Oxford and Sopron (Costa and Van der Borg, 1991).

At the same time, the growth in the residential segment of tourism stagnates, either because the hotel capacity cannot be expanded any further, or because the price level of accommodation has reached a level that repels additional tourists. The local market of tourism has become mature by then.

Since the indirect as well as the false excursionists visit the tourist city from a location other than the city itself, tourism will disperse across the territory. The term urban tourism system becomes appropriate then. The urban tourism system contains all localities that in one way or another depend on the central tourist city. The localities that generate false excursionism depend completely on the centre, while those supplying indirect excursionists only partly count on their own resources.

With the appearance of excursionism, almost automatically the risk of the number of visitors surpassing the social-economic or the physical limit to tourism increases exponentially. The violation of the social-economic limit feeds a process of crowding out, which afflicts especially the "normal", less profitable urban functions. It may be assumed that tourism is still able to compensate for these losses. The violation of the physical limit to tourism not only reduces the quality of life to the minimum, but also implies the destruction of non-reproducible natural and cultural urban resources.

In sum, the surpassing of these limits causes social costs, especially the local ones, to rise considerably, outproportioning the rise in local benefits. In such circumstances, efforts to develop tourism are apt to become ineffective, at least when the purpose of tourism development is to boost the process of urban recovery. That may not be evident in the short run, but in the long run it certainly does if the costs and benefits are considered over the entire tourism development cycle.

2.2. THE MODERN CITY AND SUCCESSFUL TOURISM DEVELOPMENT

2.2.1. *The basic conditions for successful tourism development.*

Despite the gloomy long-term prospects for the city that blindly strives to become a tourist attraction, the development of tourism can undoubtedly contribute to the social and economic health of a city, provided that three basic preconditions are satisfied as described below. The first two concern the city's image and the quality of its tourist product, while the third concerns the expected effectiveness of tourism development efforts in the long run.

First of all, the city must have an appealing image. There are cities of which people are convinced that they are pleasant to be in and there are cities of which people think it is better not to stay. The role of the image seems banal but its importance has been frequently confirmed. How far the image interferes with the choice of destination, and how far images correspond to the quality of the tourist product that is actually offered, is hard to assess, but examples of the clichés that circulate are not so difficult to give. Venice, for instance, is supposed to be romantic and pleasant; for most people it is hard to believe that there are people working in Venice, other than in the tourist industry. Rotterdam on the other hand, even if it actually possesses many unique attractions, is taken by many to be boring, while Amsterdam is regarded as a "convivial" and dynamic city. Cities in the North of France are by definition dull, while those in the South are supposed to be definitely sparkling. German cities are all the same, and New York is the heart of the world. Indeed, whether the images that are circulating are true or not is hardly relevant. What really matters is their persistence. They are almost impossible to change.

Secondly, a city must make sure that it can supply a range of easily accessible and highly competitive tourist products, attracting enough demand to make worth the efforts and investments needed for their launch. The competition on the world market of tourism is becoming so intense that the possession of natural and man-made resources alone is insufficient for a new attraction to join the ranks of established tourist destinations. Originality is a major strength, since curiosity is what drives most tourists; hence, copying the success stories of other cities is doomed to failure. Furthermore, the overall product offered needs to fit the image the city already possesses or wants to obtain.

The third precondition concerns the effectiveness of tourism development efforts. Only if the development of tourism promises to be effective through the whole cycle, the city can proceed with its stimulation efforts. The effectiveness of tourism development depend in its turn on:

- The present economic structure of the city. The openness of the local economy determines the amount of income and employment that leaks out through the already existing intersectoral and spatial linkages. The strength of the current central urban functions reduces the risk of an early violation of the tourist carrying capacity, which in the long run would reduce the centre to a monoculture. Paradoxically, the development of tourism seems more suited to (potentially) healthy and developed cities than to problem cities (see also Shaw and Williams, 1988).

- The existing and future possibilities to accommodate residential tourists, which determines the capacity to internalise a major part of the benefits from tourism.

- The proximity to already established tourist resorts which might profit from the neighbour's latest efforts, thus adding to the risk of the benefits from tourism being externalised.

- The possibility of the surrounding municipalities absorbing the spillover of unsatisfied residential tourist demand, which depends not only on the accessibility of the centre, but also on the alternatives these surrounding municipalities can develop themselves.

- The limits to tourism that can be foreseen. Special attention must be paid to the tourist carrying capacity in the social-economic sense, a limit that might easily be overlooked as its violation does not produce immediately visible costs, but affects the urban economy in the long run.

An analysis of the strengths and weaknesses of the city as far as tourist attractions and facilities are concerned, is indispensable in this context. If there are more weaknesses than strengths, the city would be well advised to abandon the idea of tourism development in the traditional sense immediately. Fortunately, there are some alternative strategies available to such cities, which will be discussed in the next section.

Summarising, if there are reasons to believe that both the city's image and its product are appealing enough, and the development of tourism is theoretically feasible, an estimation of the possible benefits and costs becomes urgent. The local tourism balance, a comprehensive overview of all the possible positive and negative effects tourism might have, can be used for a systematic assessment of the impact of tourism development. For more details the interested reader is referred to Van der Borg (1991).

2.2.2. Tourism development strategies

In addition to the basic conditions mentioned in the previous section, there are some supplementary ones that determine the outcomes of urban tourism development. These conditions differ from case to case and it is impossible to draw a complete list since they are closely related to a city's condition, its expectations with respect to efforts to launch tourism, and the strategy chosen to that effect. Some of the most important ones are discussed in the following.

As discussed before, the development cycle of tourism is mostly independent of the urban life cycle. Nevertheless, the choice of a development strategy does depend on the stage of urban development because the expectations regarding the effectiveness of tourism differ from one stage to the other. Obviously, different types of cities require different development strategies for tourism. There are, in other words, certain regularities that determine the choice of strategy. Mill and Morrison (1985) have proposed three strategies of tourism development:

1. Balanced growth: tourism is developed in a broad-based economy. Efforts are concentrated on producing locally as many tourist goods and services as possible, as well as the necessary intermediate goods. Policies are supply-oriented.

2. Unbalanced growth: tourism as a spark for the economy. The need to expand tourist demand is stressed. Policies are demand-oriented.

3. Co-ordinated growth: stimulating tourist demand in an already diversified economy.

The first strategy is the most appropriate for small and medium-sized urban economies – cities in the first stages of the urban life cycle – that are seeking to reinforce their economic structure by stimulating tourism, for instance. Such cities will be far less tense in their expectations from tourism than "problem"cities. The second strategy is the one followed most frequently, but least suited to urban attractions. Only if there are no external restrictions on the local economy and the environment can such a strategy be successful. The third strategy is to be preferred for already mature cities, that is, cities in the later stages of the urban life cycle.

In practice, the development of tourism is frequently perceived as a handy instrument for urban revitalisation. For cities at the transition stage between decline and recovery that cannot satisfy the conditions concerning their image, their tourist product, and the expected effectiveness of development efforts, a co-ordinated growth strategy – primarily stimulating latent tourist demand – is sufficient to help them effectively through the transition phase. However, the development of tourism would be particularly effective for the revitalisation of potentially strong and diversified urban economies, with a minimum of leakages because they are highly self-supporting, and whose tourist carrying capacity and physical limits can be hardly violated. Such cities can afford to choose the co-ordinated growth strategy. In this case, co-ordination and co-operation between the public and the private sector, as well as between the local and the regional or the national institutions, are additional necessary conditions for solving the problems of unevenly distributed costs and benefits ensuing from tourism. The private sector will be willing to cooperage only when it realises that its continuity is in jeopardy once the tourist resources are exhausted. Lastly, a frequently neglected condition for effective urban tourism development is the existence of appropriate tourism policy which, moreover, needs to be an integral part of a more general urban policy in order to avoid the clashes of interest between tourism and other urban functions.

A question that remains is what those cities that do not yet have an appealing image, nor a well structured, integrated tourist product, can do. It has been stressed that in such circumstances the chances of success of traditional tourism development strategies are poor. To build up a tourist product capable of conquering the tourist market from scratch requires massive investments in the city's infrastructure. The risk is high that the return on these investments will remain far below the usual real-estate standards. This does not mean that there are no possibilities at all for these cities to benefit from the growing tourist market. Three indirect strategies can be suggested.

Van den Berg *et al.* (1990) have outlined one possible strategy for the city of Rotterdam, an example of a city where traditional tourism development is doomed to failure. The authors suggest an indirect way to develop tourism, namely, stimulating congress tourism. Congress tourism is quickly gaining importance as a special branch of the incentive market. Because the congress visitor to a city is assumed to be attracted not so much by the image or the quality of its tourist product as by the congress itself, such a strategy might just help to overcome an initial threshold. The improvement of the tourist product and image will be gradual, and can partly be financed by the returns from congress tourism. For more details, the reader is referred to Van den Berg *et al.* (1990).

A second indirect strategy to launch local tourism is to stimulate recreation. The financial injection required to improve the recreation facilities of a city is relatively small, and, as the proposed improvements directly concern the quality of life of the city's inhabitants, easy to justify. Finally, a third alternative strategy is to capitalise on the attractions in the surrounding zones, and stress the accessibility the city offers to existing tourist areas.

A combination of these indirect strategies, if possible, may prove even more effective. In all the cases referred to, if the development of tourism has proved feasible, and an appropriate strategy has been chosen, an adequate tourism policy has to be developed to support the strategy. The most important features of such a policy are presented in the next section.

3. The instruments for sustainable urban tourism development

Traditionally, tourism policy has been identified with promotion. The policy for urban tourism was no exception. Only recently have some cities begun to realise that such a narrow interpretation does more harm than good. Although every city will have its own particular problems requiring specific solutions, the policy for urban tourism should at least consider three questions: how to market urban tourism; how to finance the development of urban tourism; and how to regulate the flow of visitors. As it will become clear later, the answers to these questions are not only interrelated, but also follow directly from the discussion of urban development and the launching of urban tourism in the previous sections.

To repeat that the policy for urban tourism necessarily forms an integral part of a more general urban policy seems almost superfluous. Accordingly, the principal objectives of a city's urban policy, namely, to promote and safeguard the interrelated interests of the families and firms living or working there, are reflected in the objectives of the urban tourism policy as well. In other words, the development of urban tourism has to contribute to the well being of both the local population and those directly interested: the local tourist industry and the tourists. This underlines once more the importance of embedding urban tourism policy within the broader body of urban policies taking into account the general criteria of efficiency, market orientation, and continuity.

The purpose of urban tourism policy, specifically, is to intervene in the life cycle of tourism in such a way as to reduce social costs to the minimum and boost the benefits as high as possible. In practice, this implies above all the promotion of residential tourism and the control of indirect and false excursionism, within the limits of tourist carrying capacity. In that context, the marketing of tourism can be looked upon as a "soft" instrument to guide demand and select segments, while the regulation of the number of visitors is a "hard" solution. According to Van den Berg, Klaassen, and Van der Meer (1990), strategic city marketing has to become a central element in urban policy. The marketing of tourism by a city aspiring to become an urban tourist attraction is an element of the overall strategy of city marketing addressing the potential visitor.

The basis of the marketing policy should be an analysis of the consumer market (market research) and of the strengths and weaknesses of the city's own supply in relation to that of competitors, if any. These exercises lead to a selection of those market segments that may be successfully approached, given the strengths and weaknesses of the products, the city's own included, in the relevant markets. Basic market research, described among others by Churchill (1987), also indicates how to fill in the marketing mix, which consists of four elements: the product; the price; the distribution; and the promotion.

As far as the product policy of an urban attraction is concerned, emphasis must be laid on the development of attractions, facilities, and infrastructure that serve the residential tourist segments. Hence, the local government must stimulate investments that help the development of residential tourism. In practice this means taking care, above all, that the total supply of tourist accommodation allows the city to accommodate a number of visitors almost equal to its tourist carrying capacity. As the occupancy rate of accommodation tends to be far below 100 per cent, a margin remains for the absorption of additional tourists *and* excursionists. This margin will shrink as interest in the city in question grows. Moreover, not only the capacity, but also the diversity in the supply of beds counts: the absence of modestly priced accommodation invites false excursionism.

The package of attractions and facilities should be such as to induce any visitor to stay the night. In other words, there should be too many attractions for visitors to "do the musts" in a single day. Furthermore, the locality should also offer nocturnal attractions and facilities (such as night restaurants, a casino, bars, and night clubs).

In many cities, experiments with a central price policy for goods and services used by tourists have produced satisfactory results. Actually, it is common practice to offer tourists special tariffs for public transport, or to sell residential tourists a package of attractions and facilities at a special price (including other spots in addition to the traditional interest spots already becoming congested), either directly, or indirectly in the form of a "tourist card", as has been proposed recently for Venice. The price discrimination between inhabitants and tourists can be extended to a distinction between tourists and excursionists, making the latter pay more for certain facilities. All these initiatives require co-operation and co-ordination between the public and private sectors, and between the local and regional or national tourist firms and institutions.

The distribution policy refers to the representation of the city by the network of tourist bureau's, tour operators and travel agencies for the purpose of reaching the market. Until now, most cities have been very reluctant to work with professional tourist intermediaries. The consequence is that tour operators and travel agencies that offer city packages to their clients loathe to take unnecessary risks, tend to stick to the traditional urban destinations, and continue selling congested destinations. A first step in the right direction might be to promote better co-operation between local tourist bureau's (in the Netherlands the "VVV"; in Italy the "APT") and tour operators and travel agencies, in order to reinforce the city's grip on the tourist market, as well as to make use of the vast know-how of these institutions. A similar approach has been suggested for Rotterdam by Van den Berg *et al.* (1990). A second, even more important, step might be to make local tourist bureau's function as tour operators. They can offer a variety of packages the city wants to sell and make for direct control of the visitor flows. Furthermore, they may permit the city to create incentives for tourists and excursionists to book their visits in advance. That would at the same time internalise part of the benefits which would otherwise leak out to the generating countries or localities. Again, co-operation and co-ordination are crucial for the ultimate success of these measures.

Last but not least, the promotion policy aims at attracting those segments of the market that are compatible with the specific requirements of the sustainable tourism strategy for the city in question. Positive promotion should begin by addressing residential tourism. The promotion policy might try to influence the share of foreign residential tourists, or the share of more specific market segments, in the total number of residential tourists. Negative promotion is also possible. Pointed publicity campaigns can be carried on to discourage excursionists, especially indirect and false excursionists, by convincing them that the only way to "taste" the city is to stay the night. In Venice, for example, indirect excursionists could be made aware of the fact that they have been taken in by some fancy tour operator. Overall, if the marketing policy of the tourist city is effective, more drastic measures can be avoided.

The sectoral and territorial anomalies in the distribution of the benefits and costs related to tourism are at the base of the excessive utilisation of urban and natural resources. In that context, an essential question is who is responsible for financing local tourism. In most cases, tourism development is largely financed by the local society, or, to be more precise, by the local government. It is the municipality that makes the major investments, especially in the initial phase, in such collective goods as infrastructure and additional facilities, as well as being responsible for the maintenance of the public natural, cultural, and man-made resources. The national fiscal systems do not always provide for municipal taxes to compensate such specific costs (that hold true, for example, for the Netherlands as well as Italy). The local private sector invests in its own structures and accounts for the associated operational costs. Tourist firms located outside the tourist city meet only the operational costs generated by excursionism. The inhabitants and firms who just have to accept tourism development in their locality suffer from externalities such as congestion and pollution.

A change in this situation seems overdue. A different way of financing local tourism might help to redress the skewness of the distribution of costs (and benefits, if any). Locally, forming public-private partnerships (PPPs) that take over responsibility for the management of the public tourist resources and infrastructure seems the most obvious solution. Experience in many countries and many parts of the urban society has shown that PPPs not only contribute to the solution of current fiscal and financial imbalances but may also help to overcome the lack of co-ordination between the public and the private sector.

To restore the financial balance between the centre and the rest of the urban tourism system, revision of the current system of tourist taxes might be considered. Usually, tourist taxes are levied on overnight stays, the revenues accruing to the host municipality. However, if a consistent flow of excursionists can be shown to exist between a locality and the tourist city, that locality should be obliged to use part of its revenues from tourist taxes to cover the costs incurred centrally. Alternatively or additionally, countries that collect taxes centrally for later redistribution among the municipalities should include an indicator of tourist pressure among the criteria by which contributions are determined. A system of price discrimination between inhabitants and tourists, applicable especially to public facilities not exclusively destined to the tourist market, such as local or interurban public transport, parking lots, museums, theatres, would allow for partial compensation of the costs borne by the residential population.

Hard measures to control the flow of incoming visitors come into view in cases of extreme pressure from tourism. Few cities have experience with such measures, mostly because they have not confronted a situation of emergency. The UNESCO study (Costa and Van der Borg, 1991) revealed that in art cities such as Bath, Salzburg, and Rothenburg limits to the circulation of traffic in the centre have already been implemented, while in other European cities of art similar measures are seriously studied. Only Venice is actually studying measures to control the flows by means of a booking and information system. Although both the Law and Constitution of all countries recognise that heritage needs to be protected and the quality of life preserved, measures regulating the flows of visitors to an urban attraction may easily be considered to go against the Constitution if they do not explicitly guarantee the rights of equality, freedom of circulation and sojourn, and economic freedom. In practice, all forms of control that do not materially impede access to the centre are legally acceptable.

4. Sustainable urban tourism development: some conclusions

Efforts to develop urban tourism are not under all circumstances productive. Three basic conditions have to be satisfied before such efforts can be considered seriously:
1. The city should possess an appealing image.
2. The city should possess at least some strong elements of what may become a competitive tourist product.
3. The expected long-term benefits from development should exceed the expected long-term costs.

Some external conditions, discussed in Section 2.2.1, may either favour or impede the development of urban tourism. The city's tourist strategy should be to anticipate and, if necessary, correct the effects of these external factors. Furthermore, it has to take into account the social-economic needs that are supposed to be satisfied by tourism, needs that depend, among other things, on the stage of urban development at which the city finds itself. Obviously, the development of tourism should be an integral part of the urban development strategy of the city.

Three, traditional direct, strategies for the development of tourism have been mentioned in this chapter. The first, the strategy of balanced growth, focuses on the perfection of the tourist product while the growth of the number of visitors is of secondary importance. This strategy is especially suited for smaller cities with a highly specialised economy, cities that are not so much looking for an additional economic stimulus as for ways to diversify their economy. The second, the strategy of unbalanced growth, is the one followed most frequently, and is oriented exclusively to enlarging demand. This strategy is not suited at all to urban tourism. The third strategy may be interesting, in particular, for major urban regions with an already diversified and stable economy, since it aims at co-ordinated growth, selecting specific segments of the tourist market and developing them gradually.

Naturally, there are cities that do not satisfy the fundamental conditions for successful tourism development. These cities may follow an indirect strategy, as was pointed out in Section 3. They may stimulate either congress tourism or recreation, and thus gradually improve their own tourist product, or cash in on their position with respect to already established attractions, at least if it is a central position. Van den Berg *et al.* (1990) have suggested that the city of Rotterdam opts for congress tourism. Cities that lack the industrial endowments Rotterdam has, may opt for the recreation scenario.

Finally, it goes without saying that any development strategy needs to be supported by carefully chosen policy instruments. The era in which tourism policy simply equalled promotion is definitively past. Tourism policy now has to find an answer to the questions of how to market tourism, how to finance its development, and how to keep visitor flows within the limits imposed by tourist carrying capacity.

Because the market of residential tourism is both economically and environmentally the most interesting, the marketing policy has to be oriented towards that market segment. What this means in practice for the contents of the tourism marketing mix has been discussed in Section 3. A well-designed marketing policy may render superfluous harder measures to regulate visitor flows.

Public-private partnerships (PPPs) take over responsibility for the development, financing, and exploitation of the tourist product. The creation of PPPs favours the matching of costs and benefits from tourism, which removes the incentive to exploit the attraction to excess, and ensures some indispensable sectoral co-ordination.

Tourism is usually considered to be some miracle cure for cities that require a positive economic impulse. Only recently, it has been accepted that unbalanced tourism development might unrestorably damage the urban environment. The benefits are not always worth the risk. This makes the design of a sustainable urban tourism development strategy of utmost importance.

References

Van den Berg, L. (1987) *Urban Systems in a Dynamic Society*. Aldershot, Gower

Van den Berg, L., J. van der Borg, J. van der Meer and I.A. Witmaar (1990) *Verblijfstoerisme in Rotterdam: Een Uitnodiging voor het Organiserend Vermogen van de Rotterdamse Regio*. EURICUR, Rotterdam.

Van den Berg, L., L.H. Klaassen. and J. van der Meer (1990) *Marketing Metropolitan Regions*. EURICUR, Rotterdam.

Van der Borg, J. (1991) *Tourism and Urban Development*. Thesis Publishers, Amsterdam.

Canestrelli, E. and P. Costa(1991) Determining Tourist Carrying Capacity: A Fuzzy Approach. *Annals of Tourism Research* **18**:2.

Churchill, G. A. (1987) *Marketing Research. Methodological Foundations*. Fourth edition. The Dryden Press, Chicago.

Costa, P. and J. van der Borg (1991) *European Art Cities and Flows of Visitors: A Framework for the Assessment of the Impact of Cultural Tourism*. UNESCO-ROSTE, Venice.

Mill, R. C. and A.M. Morrison (1985) *The Tourism System*. Prentice-Hall, Englewood Cliffs.

Prud'homme, R. (1986) *Le Tourisme et le Développement de Vénise*, mimeo. Université de Paris, Paris.

Shaw, G. and A.M. Williams (eds) (1988) *Tourism and Economic Development*. Pinter, London.

Smith, S. L. J. (1989) *Tourism Analysis*. Longman Scientific, London.

REGIONALISATION OF TOURISM ACTIVITY IN GREECE: PROBLEMS AND POLICIES

NICHOLAS KONSOLAS
and
GERASSIMOS ZACHARATOS
Regional Development Institute
Athens
Greece

1. Introduction

During the 1960s and 1970s, the opportunities which tourism development offered for the more general development process in tourist-host developing countries were a principal subject in political and scientific discussion. The main reason was the increase in foreign currency due to tourist receipts and, hence, the balancing of the almost continuous currency deficit of these countries. This balancing, expressed either as percentage of imports or as percentage of the deficit in the trade balance of payments covered by tourist currency, still constitutes the principal aim of tourist policy in most of the developing countries.

However, in spite of the fact that general discussion regarding the development of tourism still remains focused on the "currency question", another dimension of tourism development started to gain ground increasingly in the past decade. In particular, and compared with the autonomous character of policies for regional development during the past 15 years, the exclusive focus on the monetary aspects of international tourism is gradually abandoned and the regionalisation of tourism development is coming to be recognised as the second most basic aspect (after currency) of this development in host countries. In this perspective, tourism development is now promoted as one of the basic instruments of regional policy, especially for socio-economically depressed and problematic areas.

2. The evolution of policies on the regionalisation of tourism in Greece

Before proceeding to present in detail the regionalisation of tourism in Greece and the policies which support it, certain problems should be stressed which arise both from the nature of tourism itself and from the theoretical and practical approaches to it to date.

Although 15 years have passed since Gugg (1972) noted the exclusively descriptive nature of approaches to tourism, a generally accepted system of definition of tourism as a specific analytical category has not been established yet in the relevant literature internationally. In fact, in most tourism-receiving countries, tourism is not treated in a separate, distinct way, but on the basis of the existing system of criteria for the classifications of branches and sectors of the national economy. This weakness, reflected also in the general vagueness with which foreign tourism is treated in the national accounting systems of most host countries, has only recently begun to give way to the establishment of specific definitions as to what foreign tourism is and how it should be dealt with in a host economy. Briefly, independently of methodological differences in various countries (Stephen, 1989), the recent trend in international discussion and practice of foreign tourism, is to include tourism as a particular category of private consumption and consequently to specify it as such in national accounts. Moreover, only recently the distinction between domestic and national private tourist consumption has started to become clear.

We consider that this new trend, officially reflected at the international level in the OECD publication "Economic accounts for tourism" (Franz, 1990), will permit a significant redefinition, not only in theory but also in practice, of the role and impact of tourism in host countries. For it is one thing to discuss tourism as a "branch" of the "services sector" whose product could not until now be specified in terms of production characteristics, and quite another thing to discuss how to approach quantitatively, qualitatively and over time the development and growth of tourism as a particular category of private consumption, in a national economy or in some of its regions (Zacharatos, 1983). Obviously, the redefinition of the approach to the analysis of tourism and its effects on the balance of payments, income, employment and regional development will require, to some extent, the revision of the relevant policies of the state as well which up to now have supported tourism development; such policy revision will permit the simultaneous establishment of effective systems of assessment of this development, both at the national and the regional levels.

The first official use of tourism as a policy instrument for regional development in Greece dates back to the late 1960s. The attempt to use tourism as a means of regional development in Greece is shown in the various 5-year plans for economic and social development, on the one hand, and, on the other, in the rationale of credit policy to provide investment incentives aimed at the economic development of the country. The 5-year plans state the general goals for the national economy as well as the particular goals for the regions of the country, which must be fulfilled through the tourist development of the country. In legislation regarding economic development, the specific means of financing on which the achievement of such goals is based are defined. Consequently, the presentation of the effects of the regionalisation of tourism in Greece requires a comparative presentation both of the goals and the means of tourism policy to date and of the geographical areas where the formation of fixed capital for tourist superstructure facilities has developed. Tourist superstructure is considered here to be every kind and form of tourist accommodation. It is this functional form of capital, which, to a great extent, determines the geographical distribution of production and the annual actual tourist consumption, so that the regionalisation of tourism can be considered as identical with the regionalisation of tourist accommodation.

A close examination of the development of tourism in Greece during the post-war period leads to the following remarks as regards the phases of this development. The 1948-1952 period can be characterised by the direct intervention of the state in the creation of infrastructure facilities in tourist centres well known before the war (Athina, Delphi, Rhodos, Kerkyra), as well as by the introduction of a short-term credit policy aimed at the renewal and modernisation of hotel units which had suffered extensive damage during the war. In the thirteen years between 1953 and 1966, the newly founded National Tourist Organisation of Greece (NTOG), established in 1951 as the basic agency of tourism policy, began an extensive programme of public investment in tourist superstructure facilities (hotels, motels and organised beaches) in various regions of the country, with two main aims. First, with its direct investment intervention, the state (NTOG) attempted to create modern hotel and restaurant facilities in various parts of the country, in order for them to function as building and management models for further tourism development. Second, this direct state investment activity in profit-making tourist facilities was aimed at overcoming the reluctance of the private sector to invest in these areas by undertaking the cost of "setting up" or "first opening" various regions to tourism development.

Public investment in commercial facilities initially covered 100% of the total public funds available for tourism. In 1966 – i.e. at the end of this period – this percentage had dropped to 27% and it continued decreasing until the beginning of the 1980s, having in the meantime created a large number of tourist enterprises of various kinds in most areas of the country. This direct intervention on the part of the state in the construction and operation of tourist enterprises for the satisfaction of tourist demand, and especially their spread throughout the country by state decision, can be considered the first policy for the regionalisation of tourism in Greece, antedating any official economic-scientific arguments. In particular, from 1953, when the total public funds allocated to tourist growth were used for developing tourist superstructure facilities, until the beginning of the 1980s, when the state ceased all investment activity in this direction, the tourist enterprises given in Table 1 were planned and constructed, with the aim of "creating" tourist areas.

The regional allocation of the state hotel units by Prefecture is not characterised by particular choice as regards location, since 61 of the public enterprises of this kind now functioning are located in 37 of the country's 52 Prefectures, while the remaining 247 state tourist enterprises are scattered throughout the country's prefectural divisions.

Table 1. Tourist enterprises planned and constructed from 1953 until 1980

No.	Kind of enterprise	No. of units
1.	Hotels	54
2.	Traditional Villages	7
3.	Camp sites	15
4.	Beaches (bathing facilities)	9
5.	Spas	10
6.	Marinas	6
7.	Caves	3
8.	Winter Sports Centres	1
9.	Casinos	3
10.	Refreshment Stands, restaurants, kiosks	29
11.	Cafeterias	79
12.	Sea Shore areas	37
13.	Other tourist enterprises	55
	Total number of state tourist enterprises	308

Source: National Tourist Organisation of Greece (NTOG-EOT)

After reaching zero levels in the 1956–1957 period, private investments for new hotel facilities became active again from 1958 onwards and continue to grow without significant annual fluctuations. However, the regional distribution of private investment in hotel facilities does not follow the same pattern as the policy of state investment allocation in tourist superstructure; during this period 80% of total private investments were concentrated in three traditional tourist centres: Athina (52%), Kerkyra (12%) and Rhodos (16%).

The first official document in which the use of tourism as a regional policy tool is put forward on a theoretical economic basis was the 5-year plan of 1968-1972. "The organic inclusion of the tourism development programme within the context of the regional development of the country" (PLAN, 1968) constitutes one of its basic aims. In addition, within the context of the proposed tourist development of the country, the term "tourist zones" is used for the first time. The delineation and definition of these zones "should be made in accordance with the aims of regional development of the country", according to the same document. At this point, it should be noted that one of the implications of this proposal for tourist zones and the related tourism planning involved, was an attempt by the state to amass stocks of publicly owned land in which the proposed "tourist zones" would be established. Land for tourism would be zoned in order to control in this way the yield of private investment activity as well as the environmentally sound distribution of new hotels and other tourist facilities in space. Due to social reactions and pressures the attempt did not succeed in creating large areas of land, unified regarding ownership and suitable for characterisation as "zones" for further large scale tourist development.

Owing to the spasmodic manner in which expropriation orders were executed during this period, a large number of small pieces of land scattered throughout the country passed into public ownership. Thus the term "tourist zone", in spite of the fact that it is often used in general discussions of tourism in Greece, cannot be used as if it has a specific planning meaning.

During this period and parallel to the above efforts, the credit policy concerning the formation of a private fixed capital in hotel facilities introduced a new borrowing practice, the abuse of which created a large increase in investments with the following characteristics:

There has been an abolition of banking criteria for loans and their replacement with the direct and definitive authority of NTOG, as regards approval of the size and site of new hotel units. This permitted a more rapid formation of capital (see Table 2) in large capacity tourist superstructures, which, however, from the point of view of returns to capital are still problematic today.

Table 2. Geographical distribution of hotel capacity (in percentage of number of beds) in 14 Nomos in Greece

Nomos	1959	1971	1981	1988
Attikis	38.1	26.2	23.6	18.1
Dodekanisou	8.7	11.4	13.0	15.5
Irakliou	1.1	2.1	5.9	8.1
Kerkyras	5.0	4.0	6.4	6.9
Kykladon	1.5	3.1	3.6	4.8
Lassithiou	0.4	1.7	2.9	3.6
Evias	2.1	3.5	3.8	3.1
Chalkidikis	1.1	1.7	3.0	3.0
Argolidas	1.9	2.4	3.6	3.0
Magnisias	3.1	1.9	1.9	2.8
Rethimnis	0.7	0.3	1.4	2.5
Fthiotidas	1.5	2.8	3.2	2.4
Thessalonikis	8.3	6.1	3.2	2.2
Korinthias	1.2	4.1	2.4	2.1
Total of 14 nomos	75.0	71.3	77.9	78.1
Rest of nomos	25.0	28.7	22.1	21.9
National total	100	100	100	100

(*) Nomos: Main administrative division of the country. Greece has 52 nomos.

From the point of view of the regionalisation of new tourist units during this period, an even greater concentration, in terms of number of beds, can be seen in the traditional tourist centres (Attiki, Athina, Rhodos and Kerkyra), on the one hand, and, on the other hand, the creation of new tourist poles such as in Kriti (Iraklion and Lassithi areas) and in Chalkidiki with the construction of large hotel units. A further feature of the regionalisation of tourist units during this period is the distribution of a part of the hotel investments in areas where there is a total lack of infrastructure capable of meeting a sudden increase in tourist flows and demand. This fact has contributed to the creation of environmental and functional problems in these areas, the solution of which becomes increasingly expensive and administratively hard to implement as time goes by and the flows of tourists increases continuously. These areas (Ermioni, some small Aegean islands and certain parts of Kriti) constitute a classic example of complete lack of correspondence between investment in superstructure (an extremely rapid and large increase in capacity, and thus in total private consumption) and investment in infrastructure. Not only there was no increase in infrastructure but the constant overloading of the existing infrastructure resulted in the continuous degradation of the environment.

To summarise the assessment of the regionalisation policy as regards the basic tourist supply installations in the period referred to, the following remarks are made: The decisive, legally established intervention of the state (NTOG) to make administrative adjustments to credit policy as regards the size and location of hotel units for which loans had been given, without the existence of specific and enforced national spatial planning, and the purely indicative nature of the tourism development programme during this period contributed to: (1) an even greater concentration of tourist activity in the traditional tourist centres, (2) the creation of new large-scale tourist concentrations, and (3) the simultaneous creation of a number of new and small-scale tourist centres, all of them characterised by a lack of satisfactory social and environmental infrastructure.

In the middle of this period, which may be described as a phase of rapid development of tourism in Greece and one which laboured under the burden of both the general regional problems of the country and the spatial and functional problems of tourism development itself, a first attempt was made to harmonise tourism with regional development, by establishing spatial delineation criteria and adjusting the tourism-oriented credit policy. Thus, from 1971 on, when the first law on tourism development incentives was put into effect, up to the present, state intervention for the formation of private capital in tourism has been manifested in a series of five basic development laws (in 1972, 1976, 1978, 1981 and 1982). According to these laws, which constitute the basic framework within which the regionalisation of tourism in Greece functions today, the country is divided into 4 to 5 main regions and investment incentives are graduated in increasing order from the more to the less developed regions.

The role of regionalisation of tourism in regional development policies received greater emphasis in the 5-year plans which followed the 1968-1972 plan, as the spatial and environmental problems arising from and being exacerbated by tourism development became more pressing. Specifically, in the 1973-1977 5-year plan, after the increase in currency goal, the goal second in order of importance was defined to be "the use of tourism to a major extent for the promotion of the general development policy and the parallel protection of the natural environment and of tourist values" (PLAN, 1973). In addition, the same plan stresses that it is through tourism that "reinforcement of the demographic and social goals of the problematic areas of the country" (PLAN, 1973) should be attempted. In the 1976–1980 plan, after the marked investment activity of the 1968–1975 period and the spatial planning problems that were created in the old and new tourist centres, the problematic of tourist capacity in given areas came into question for the first time, and it was suggested that tourism planning should be directed towards a more widespread dispersal of tourist nuclei (PLAN, 1977).

The 1978–1982 plan, after stressing once more the unfavourable spatial impact of tourism in the past which had resulted from the lack of a national zoning model for the long-term development of this sector, proposed that the development and the regionalisation of tourism should be carried out only with the help of specific zoning plans, while for those locations showing signs of saturation a halt in further tourist development was proposed (PLAN, 1979). In the 1981–1985 plan, which constitutes the first plan for the regional development of the country, tourism was seen mainly from the point of view of the regulation of spatial planning problems which tourist development itself had created in the past. Thus, the main goal of this plan was, on the one hand, "the suspension of further tourism development in regions and areas which show marked signs of overcrowding and saturation (e.g. Attiki-Athina, Rhodos, Kerkyra, Kriti and Chalkidiki)" (PLAN, 1980) and, on the other hand, the attempt "to increase the contribution of tourism to the development of problematic regions and to the more general regional development of the country" (PLAN, 1980). In addition, according to the plan, the above goals could be achieved by the parallel aims of "increasing auxiliary accommodation in relation to main accommodation and the correct distribution of such accommodation throughout the country" (PLAN, 1980).

Finally, the 1983–1987 plan, after stressing once again the spatial and environmental problems due to overconcentration of tourist activities in certain parts of the country, set as its main goal the zoning redistribution of tourist demand and supply and the corresponding suspension of tourism development in areas which showed signs of saturation (PLAN, 1983).

3. Conclusion and perspectives

To conclude the presentation of the regionalisation of tourism policies in Greece in the context of regional development, as these have been set out in the various economic development plans, we must mention that a permanent goal of regional policy was the development of tourism because of the belief that "the dispersal of tourism funds throughout the country and especially in regions which have been deprived of other development funds" (PLAN, 1972) is important in the process of increasing regional income in poor and depressed areas.

Examining the parallel growth of private investment in hotel and other facilities in the tourism sector after 1974, where, as the basis of the credit policy, bank criteria were once more applied to loans and where administrative power concerning the size and site of hotel units was removed from NTOG, we may observe the following with respect to the question of their regionalisation:

Continuation of the increase in hotel capacity, with an average annual rate of 6% for the period 1975–1989. Concentration of 80% of the total capacity in 14 Prefectures which, as shown in Table 2, have maintained, for more than thirty years, a relatively similar percentage of the country's hotel capacity (75% in 1959, 71.3% in 1971, 77.9% in 1981 and 78.1% in 1988). This concentration of hotel capacity in 14 Prefectures of the country will become even more pronounced during the next few years if we add to those already existing the hotel construction investments of the 5-year period 1983-1987. Of the investments made in this period, four regions absorbed more than 60% of the total private investment in hotel facilities (Kriti: especially Iraklion and Lassithi, 27%; the south Aegean, especially Rhodos and Cos, 19%; the Ionian Islands, especially Kerkyra, 10%; the north Aegean, especially Samos, 6%). In addition, a further sign of the continuing concentration of hotel accommodation in these regions is the fact that 52% of the architectural plans for hotel installations for the period 1988–1990 that have been approved but have not been built yet concern the above regions.

Evaluating the overall regional policy for tourism in Greece, on the basis of the main goals of all the 5-year plans to date and the rationale of regionalisation reflected in the credit policy for tourism, we may maintain the following:

Today's picture of the regionalisation of tourism activity does not permit us to conclude that the goals set and the measures taken have been successful. It is true that the 14 regions in which 80% of hotel beds are concentrated today do not belong to the highly subsidised zones D and E, but mostly to incentive zone C, where total subsidies (on capital and interest) were from the time of the first development law of 1971 already fairly satisfactory. This fact together with the observation that the three traditional tourist areas (Attiki-Athina, Rhodos and Kerkyra) as well as the two newly rising ones (Kriti, Iraklion and Halkidiki) concentrate and will continue to concentrate more than 50% of total capacity, means that the incentives policy for the regionalisation of tourism, in spite of the extremely high subsidies to the certain areas, was not able to offer more attractive gains to private investors than those offered by the installation of a hotel enterprise in already established tourist areas. This concentration of hotel beds in already established tourist areas can also be explained by the lack of a relatively strict zoning plan. This plan could have constituted the basis for the co-ordination of the incentives policy in promoting tourist projects and of the public investment policy in tourist and more generally in social infrastructure defined by zoning plans and could have offered assurance to private investors that "quality sites" for installation of tourist facilities were available.

The basic problems that face most of the touristically developed or developing areas are more or less problems of quality of the environment than purely economic ones but it is very probable they will turn into main economic problems in the immediate future if they are not addressed today.

To date, tourism development was characterised by a singular dualism, where the increase in tourist superstructure was not coupled by a parallel quantitative and qualitative increase in infrastructure, and also by a lack of a concrete institutional context of tourist policy and particular long-term rules for the production and distribution of tourist consumption. The main environmental consequences of this situation are summarised as follows:

- *Water Supply*. In most cases, the bad exploitation of water resources, the existence of malfunctioning and insufficient water supply networks and the lack of parallel construction of new ones have created great problems for the local authorities of the frontier areas, especially in the islands of the Northern and Southern Aegean and the Ionian Seas. A consequence of this situation is that many tourist lodgings suffer from shortage of water and bad quality of the water supplied during the tourist season.

- *Sewage*. The large increase in tourist consumption at the local level produces in its turn a multiple increase in sewage production, which can be confronted only with large projects of infrastructure at this level. However, the institutional and financial weakness of local authorities to cover the basic needs in infrastructure and, moreover, the fragmentation and dispersion of the legal and illegal tourist accommodation have not allowed the construction of autonomous units of biological cleaning in most of the small and large tourist regions so far. Thus, the anaerobic process of decomposition-absorption meets two large difficulties: (1) Absorption becomes dangerous especially in islands, where the subsoil consists of limestone and, therefore, the risk of groundwater contamination is very high; (2) Due to the large number of sewage pits in many tourist areas, large volumes of sewage are produced daily, for a period of 4 to 6 months every year, which must be disposed of somehow (for treatment or storage in septic tanks).

- *Garbage*. Garbage collection presupposes the existence of cleaning services and also the existence of technically safe receptor areas. The majority of tourist centres in frontier areas do not possess this kind of infrastructure and, thus, there is an increasing danger of contamination of the waterways from uncontrolled garbage disposal and insufficient garbage disposal sites.

- *Master plan*. The lack of master plans and the continuous extension of existing plans in order to legalise illegal structures have also a negative effect on large and small tourist centres. Hundreds of residential enterprises possessing thousands of beds are operating interspersedly and irregularly, situated among discotheques, shops, garages, restaurants and other facilities. This mixing of land uses, which is often accompanied by buildings of ambiguous architectural character, detracts from the image of the whole area. From the viewpoint of tourism development harmonised with the environment now demanded by many foreign tourists, many of these areas may be characterised in the future as unsuitable for tourist areas unless measures to improve this situation are taken.

To close the discussion of the role the regionalisation policy for tourism could play in the context of the regional development of the country as well as of tourism development itself, a few remarks are offered, with the following three basic points in mind: the single European market of 1992, the developments in the countries of Eastern Europe and the particular geographical and geomorphological position of Greece in the Mediterranean.

According to the most recent data, the immense development of intra-European tourism is now an indubitable fact, since it includes 65% of the total of international trans-frontier arrivals. Since 1981 it has brought a highly positive surplus to the "travel balance of payments", with total revenues which amount to 5% of credits from goods and services and which contribute positively to the harmonization of the balance of payments of the Community of the 12. In addition, tourism has a large positive effect on employment, with a total of 7.5 million employed, or 6% of total employment of the European Community. As regards regional development, it may constitute the main element of a counter-policy of attacking unemployment mainly in the problem areas and in those areas of the Community which are suffering from industrial decline. Furthermore, developments in the eastern countries are already creating a further increase of tourist flows to the EC and especially to the Mediterranean, where Greece, with its particular geographical position as well as environmental and cultural character, will constitute one of the main poles of attraction for this new flow of tourism towards the Community of 12. These trends, considered together with a predicted increase in demand for seaside tourism in the Mediterranean (Ginod, 1987), imply a dual task for Greek tourist policy. On the one hand, it must attempt to adjust to the goals of the general European tourist policy towards a further increase in intra-European tourism as well as in international tourism from outside Europe, stressing in particular the geographical position of Greece with the aim of increasing its own share of tourist consumption in the Mediterranean market. In order to succeed, this effort must be accompanied by specific policies designed to solve the particular spatial planning and environmental problems which the regional and tourism policies have created to date. In the context of the new policies, it is inevitable that a certain reorganisation of the "tourist value" will occur in some tourist centres, with all the consequences that the rise or fall in this value may bear. Thus, on the other hand, tourist policy in Greece must attempt, for socio-political reasons, to render the cost of such a reorganisation of the "tourist value" born by tourist centres as small as possible. For this purpose, an extensive programme for the formation of tourist and social infrastructure in the country's tourist centres must be designed and implemented.

Recapitulating, it is stressed again the shift in tourist policy from a focus on foreign currency increase to an emphasis on employment and income increase within the context of the monetary unification of the European Community. This means that tourism policy in Greece must function as part of the country's regional development policy and not only as a policy for the regionalisation of tourism. This new reality requires the scientific community to improve its old instruments and to invent new ones in order to analyse and investigate the regional and local economic and environmental effects of tourism.

References

Franz, A. (1990) *Economic Accounts for Tourism.* Preliminary Draft Report. prepared for OECD, Tourism Committee, Vienna.

Ginod, J. (1987) *Le Tourisme Nautique en Méditerrannee: Les Pays de la C.E.E.* Prepared by the "Bureau d'Études de la Chambre du Commerce et de l'Industrie de Nice et des Alpes-Maritimes" for the EC, Oct. 1987.

Gugg, E. (1972) *Methoden der touristischen Absatzforderung, Méthodes de Recherches. Touristique et leur Application aux Pays et Régions en voie de Développement.* Rapport presenté au 22e Congres de l'AIEST, du 3–9 Septembre 1972, Istanbul. Publications de l'AIEST Bd 12, Bern 1972, pp. 33–37

PLAN, (1968) *Le Plan pour le Développement Économique de la Grèce, 1968–1972.* Ministère de la Coordination, Athens 1968, pp. 94.

PLAN, (1972) *The Plan for the Long-term Development of Greece (1972-1987).* Centre of Planning and Economic research, Part B., Athens 1972 (in Greek), pp. 78–79.

PLAN, (1973) *The Five Year Plan 1973-1977,* unpublished, Athens 1973, pp. 273–276.

PLAN, (1977) *The Plan for Economic and Social Development 1976-1980.* (preliminary draft), Centre of Planning and Economic Research, Athens, 1977 (in Greek), pp. 90–91.

PLAN, (1979) *The Plan for Economic and Social Development, 1978-1982.* (preliminary draft), Centre of Planning and Economic Research, Athens, 1979 (in Greek), pp. 105–106.

PLAN, (1980) *The Plan for the Regional Development 1981-1985.* Centre of Planning and Economic Research Athens, 1980 (in Greek) pp. 47.

PLAN, (1983) *The Plan for Economic and Social Development, 1983-1987.* Ministry of National Economy, Athens, Dec. 1983 (in Greek), pp. 202–207.

Stephen, J. and Smith, I. (1989) Estimating the Local Economic Magnitude of Tourism. In *Tourism Analysis - A Handbook,* Ch. 10, pp. 270-281, Essex.

Zacharatos, G. (1983) *Tourismus und Wirtschaftsstruktur – dargestellt am Beispiel Griechenlands.* Frankfurt a. M., 1983.

TOURIST DEVELOPMENT AND ENVIRONMENTAL PROTECTION IN GREECE

GEORGE CHIOTIS
The Economic University of Athens
Athens
Greece

HARRY COCCOSSIS
University of the Aegean
Karantoni 17, Mytilini, Lesvos 81100
Greece

1. Introduction

This chapter intends to highlight some of the basic policy issues relating to the role of tourism in national and regional development with a particular focus on the strong interrelationships between tourism policies and the environment.

The problems of the environment are examined in the context of national development and the development prospects of regions. The basic question revolves around the role of tourism in Mediterranean countries and particularly in certain regions, which are sensitive to tourism and at the same time sensitive to the preservation of their natural resources and their environmental quality. To illustrate the issues involved in the context of tourist development and environmental protection, the experience of Greece and some of its regions will be used as an example. Special reference is made to the role of the European Community (EC) and international co-operation. The chapter explores the role of tourism in economic development in Greece, the role of tourism in regional development, the relationship between tourism development and environmental problems, future prospects and problems and discusses some of the basic policy issues involved. This presentation is to some extent limited by the lack of a consistent national and tourism policy in Greece, recent political and policy changes and the overall lack of an effective planning framework.

2. The role of tourism in economic development in Greece

Tourism is an important sector of the economy of those countries, which do not have significant margins for development in other alternative sectors or those, which are endowed with rich and easily developable tourist resources. A rich cultural heritage and long tradition in hospitality, a unique natural and built environment and a comfortable climate constitute the major factors, which made Greece a significant tourist attraction pole in the Mediterranean. Following the Second World War, the development of the national economy was based on the construction activity and was not outward looking or export oriented. Tourism and transfer payments became the main foreign exchange sectors. Therefore, tourism's impacts on the Greek economy, society and the environment were significant.

The growing importance of tourism is reflected in the structure of the economy, particularly in the share of tourism in the Gross National Product. In terms of GDP, tourist receipts represent more than 7% of the total, rendering tourism one of the largest sectors. Some 215,000 people are employed in tourist services while an additional 120,000 are considered to depend (indirectly) on tourism as well, representing about 7.2% of total employment. Recent estimates raise the total number of people employed in 1990 in the broader tourist sector to 480,000.

In 1988, total tourist receipts exceeded USD 3.8 billion (ECU 3.2 billion) contributing substantially (almost 50%) to the foreign exchange balance of the country. In 1988, the average per capita expenditure was USD 466.8. The above figures include the "invisibles" such as income from cruising, yachting, purchases through credit cards or repurchases of drachmas by tour operators but do not include income from the "informal" sector of rented rooms, a small percentage (unofficially estimated to 40%) of which is reported as income. The dependence of the Greek economy on tourism demonstrates its vulnerability to external factors and fluctuations in the international tourist demand and underlines the necessity for proper organisation and long-term efficiency of the sector.

Early tourism policy (in the 1950s and 1960s) focused on the attraction of a larger share of the international tourist market through the expansion of services and increase of accommodations. As the sector was just beginning to develop, it became necessary for the state to assume a leading role. New model hotels were built and managed by the National Tourism Organisation of Greece (NTOG). At a later stage (mid-1960s to mid-1970s), the state introduced financial incentives for private investments in new hotel construction and other forms of facilities (camping, bungalows, etc.) to meet the growing demand for tourist accommodation and allowed the use of rental rooms (subject to special licensing) in small coastal settlements to supplement hotel accommodations and offer some additional income to local families. An important characteristic of this period as well as the following and most recent period was the shift of the weight of public investments vis-à-vis private investments from a 1/1.5 to 1/8 (later 1/10) ratio. In the early 1980s, tourism policy favoured, through an incentives scheme, the development of small and medium size usually family-oriented tourist enterprises as a measure for the development of rural areas and as a counterweight to large-scale tourist installations in order to spread the benefits of tourism to a wider base of the population. Complementary forms of tourism such as "social tourism", tourism for the young, etc., were sought to dampen the effects of the fluctuations of the international tourist demand. An important characteristic of the development of tourism in Greece throughout this period is its failure to develop linkages to other sectors of the economy and especially agriculture or industry (i.e. the small handicraft activities), linkages which could be important for the development of certain regions, particularly the islands.

The growth of tourist arrivals has kept a fast pace (over 10% annually since the mid-1960s) particularly in the 1970s. In the last decade, average annual growth has been of the order of 5.3%, exceeding 8 million arrivals in 1987. In 1989, more than 8.5 million tourists visited Greece, most of them coming from European countries (83%) especially Germany, Great Britain and Scandinavia. In the last two years, the average annual growth rate has dropped to approximately 2.5% as a result of the emergence of competition from tourism receiving areas in the Far East, the Pacific and the Caribbean, and economic crises in "home" countries.

According to a special 1984 survey of the Greek National Tourism Organisation, 71% of all visitors to Greece are highly educated, mostly scientists, educators, administrators and students. Over one third of all tourists are between 26 and 40, one third between 16 and 25 and only 2.7% are over 65 years old.

By far, the majority of all visitors arrive by air transport to Greece although a growing number is arriving by private means such as car or yacht. Over 62% of all arrivals from EEC countries and over 45% from all other European countries are realized through charter flights. Limitations of the capacity of the transport system such as air seating capacity in international and domestic flights and lack of direct connections with the major tourist receiving areas of the country are among the factors influencing the reduced growth rate of arrivals in recent years.

In 1989 foreign visitors represented over 75% of a total of over 44 million in terms of overnight stay. According to the special survey of NTOG, the average length of stay in Greece is 14 days. Over 60% of all visitors stay in hotels, 8% in campings, 7% in rented rooms, 4% in relatives or friends and 3% in special luxury resorts.

A total of 423,790 licensed hotel beds and additional 220,000 in rented rooms were available in 1989 while a number of camping sites of a total capacity of over 75000 persons complement the country's tourist accommodations. Luxury and A class accommodations represent over 25% of the total accommodations provided. A particular problem in terms of accommodations is the large number of non-licensed facilities, which evade official controls and taxes.

The need to achieve high levels of tourist attractiveness and the increasing competition in the world tourist market have prompted a reorientation of tourism policy towards an upgrading of the quality of services (NTOG, 1989) in order to maximise tourism's beneficial effects on the national economy. Marketing has been an important tool to this direction as well as the introduction of a number of institutional changes whose aim is to improve policy making for tourism. Examples include the upgrading of the National Tourism Organisation to a Ministry of Tourism and the introduction of time-sharing schemes. The recent change in administration has boosted even further this policy reorientation and shifted the emphasis to a greater reliance on private initiatives in tourism. It is considered that tourism will be a significant sector for economic recovery through the creation of new employment and the attraction of additional foreign exchange but also for promoting environmental protection. The basic goal will be upgrading the quality of tourism in Greece. Tools for the attainment of this goal would include:

- strict rules for the protection of the environment
- legislation to provide safeguards for visitors
- financial incentives for regional development
- training programmes for those employed in the tourist sector

National efforts to improve the quality of tourism services offered involve the following measures:

- adequate infrastructure (roads, ports, airports, etc.)
- high standard accommodation facilities
- limits to the creation of accommodation in the low category facilities
- legalisation and control of non-licensed facilities
- encouragement of luxury hotel facilities with supporting recreation and conference facilities
- creation of special tourist infrastructure (marinas, ski resorts, spas, etc)

Recent changes in the investment schemes provide for substantial tax reductions and increased amortisation rates for the development of marinas, golf courses and convention facilities. Also provisions are made for the highest levels of grants to be available for the development or expansion of luxury hotels or facilities for medical, sport related or winter tourism.

The necessity to develop a coherent tourism policy becomes more pronounced, due to the limited opportunities which are available to develop significant alternative activities in agriculture or industry. This is particularly so in the case of most of the regions of Greece for which tourism retains a significant role, except for the regions of Athens and Thessaloniki and maybe a few others.

3. Tourism and regional development

As tourist assets are not uniformly distributed over the geographical space and as accessibility varies from place to place, the spatial distribution of tourism has been concentrated in a few poles or zones although some dispersion is evident in the past few years. Accessibility and the concentration of tourist services, and, to a lesser extent, attractiveness and employment in tourism, seem to be the most important factors for the spatial differentiation of tourist demand (Komilis, 1986).

Furthermore, the spatial preferences and travel patterns of various nationalities differ significantly. For example, British tourists concentrate in Kerkyra (80%), Dutch in Kriti, Dodecanese and Athina, Scandinavians in the Dodecanese while others are more dispersed, as for example the Italians.

Regionally differentiated incentives were designed to stimulate the development of tourist facilities in less favoured areas through higher grants and interest rate subsidies. Athens and Thessaloniki are normally excluded from these incentives schemes; peripheral border regions acquire the highest incentives, while the rest of the country is divided in two other categories of zones, those near Athina and Thessaloniki and the rest of the country.

In the early period of tourism expansion (late 1960s and early 1970s), a few tourist poles attracted the largest share of tourists and new tourist development, mainly Athina (as a major historic site, as a central node in the national transport network and as the major international gate to Greece) and the islands of Rhodos, Kerkyra and northern Kriti. In the last decade or so, as the number of tourists expanded, more distant locations were integrated in the visit and travel patterns of Greek and foreign visitors and tourism became more dispersed. However, three regions Attiki, South Aegean (mainly Rhodos) and Kriti: still represent 55.4% of the total hotel accommodation of the country and almost 71.24% of total foreign overnight stays. Recent trends suggest, however, a clear preference of visitors for new regions at the expense of the established poles of attraction. The islands of the Northern Aegean (i.e. Samos), the Dodecanese (except Cos and Rhodos), the south of Kriti and the west of Peloponissos are among the regions which seem to benefit the most from tourism in recent years. Kriti, the Aegean and the Ionian islands have experienced in the late seventies and early eighties the fastest pace in terms of new tourist accommodations. Athina is still the major gate of entry for tourists although its share is declining in the past decade or so as regional international airports developed with direct links to several European cities which are centres of international and organised tourism.

Tourism development in some regions meets special problems due to internal and external factors and deficiencies. For example, a particular problem of tourism in some of the regions is the lack of entrepreneurs and appropriately trained personnel, which has affected the level and quality of tourism development. In some areas, tourism development relies mostly on small, usually family-based, firms, which lack the resources and organisational ability to market their services and deal with organised tour operators abroad. Other areas lack the ability to offer high quality services due to the lack of basic infrastructure such as sewage.

Tour operators are also responsible, to some extent, for encouraging the low quality services provided in some areas. Their acquisition of charter airline services in the 1970s and strong competition for booking the flights, often at the expense of other costs, have generated demand for cheaper and lower quality accommodations which, coupled with the lack of regulatory intervention by the state, led to a substantial increase in the provision of lower quality tourist installations.

The new guidelines for regional tourism policy place emphasis on increased efficiency in the touristically developed areas through better organisation of the services supplied and improvements in infrastructure and on the development of a system of effective control and provision of adequate infrastructure for areas which experience rapid tourism development. The new development law (L. 1892/90), which was passed recently provides for grants from 40 to 55% of the productive investment for most areas except Athina and Thessaloniki and special bonuses for modernisation of existing accommodations and for the conversion or repair of listed protected or traditional buildings into hotel accommodations.

4. Tourism development and environmental problems

The relationship between tourism and environmental quality is characterised by dynamic feedback mechanisms (OECD, 1978). Tourism is attracted to high environmental quality and amenity areas. The increased number of visitors and accommodations often degrades the quality of the environment and threatens the natural resources and assets and, thus, tourism is affected in a negative way from a low-quality environment. This is particularly so for areas like Greece which attract tourists because of their natural and man-made assets and resources.

As environmental effects are highly localised, it is difficult to generalise about the impacts of tourism on the environment. Among the most characteristic examples of impacts on the natural environment from tourism observed in Greece are:

- Aesthetic impacts on the natural landscape and cityscape from the large size and scale of some individual tourist facilities particularly those of the earlier periods. Landscape alterations from infrastructure projects have a significant impact.

- Indirect effects from the associated urban development and the lack of local plans for the physical organisation and guidance of such development.

- Risks to natural habitats and resources from uncontrolled tourism development due to the general problems of ineffective land use and urban development controls. In some cases, special measures have been taken designating areas under protection and strict urban development control such as the case of the sea turtles in Zakynthos or as special protection areas in the case of the Monk Seal marine park in the Sporades islands.

- Pollution from wastes such as garbage or untreated sewage discharges in the sea, not necessarily from large scale facilities as all licensed tourist facilities are obliged to provide treatment facilities as well, but mostly from smaller scale developments where there are problems of cumulative effects and ineffective control.

- Significant negative impacts on local water resources due to the high peak demand, uncontrolled land development and threats of contamination from a dense pattern of septic tanks for sewage disposal.

However, it is not the total number of visitors, which is a problem but rather their concentration in space and time. For example, in 1989, more than 19% of all foreign overnight stays occurred in the month of August while almost half of the total overnight stays occurred in the three-month period from July to September and some 80% of all tourist stays were concentrated in the period from May to September.

The temporal concentration of demand creates tremendous stresses on environmental resources, primarily drinking water, and on environmental media which receive eventually the solid and liquid wastes generated directly and indirectly by tourist activity. Seasonal peaks in demand for environmental services create serious financial problems as well, as there are large differences between service requirements in summer and winter, due to the substantial fluctuations in daytime population, rendering problematic the design of satisfactory and cost effective infrastructure.

The impacts of tourism on the environment have changed over the years in scale and intensity and so did the relevant policies. Originally, the negative impacts of tourist facilities on the natural and man-made environment were isolated and of a rather physical nature having to do more with aesthetics and the size and scale of facilities within a given context. As the number of tourist facilities in an area multiplied, their cumulative impacts became more of a concern such as sewage and garbage disposal, congestion, etc. Therefore, policy and the accompanying tools and measures changed in emphasis from physical control through permits for individual projects to area controls with plans for tourism development and/or environmental protection.

In certain respects and for certain policy areas, area control has fared much better than any other measure as, for example, in historic towns or traditional settlements. The NTOG has since the mid-1970s launched a special program for the restoration of traditional settlements on six sites through undertaking the cost of renovation under a long-term lease agreement with property owners. Special incentives are also provided under the national incentives scheme for the repair or conversion of listed, protected or traditional buildings into hostels or hotels provided they are located in specially designated settlements.

Since 1984, tourist installations in rural areas are required to obtain a special Location Permit, which is a type of Planning Permit, before the Suitability and Building Permits are issued, the purpose being to control potential conflicts of land-use in the area. The special permit is given following an examination of existing land-use plans and in accordance to regional policy guidelines in order to protect high productivity agricultural land and the quality of the man-made and natural environment. The criteria employed involve the existence of agricultural land of special value, adequate technical infrastructure, sensitive natural resources, compatible activities, archaeological or historic sites, coastal protection and forest protection.

In terms of planning, there was an attempt in the mid-1970s to develop a National Physical Plan which would designate among other concerns, on the basis of various scenarios, areas for tourism development in an attempt to organise spatially the development of tourist activities. However, the ensuing world-wide economic crisis and uncertainty, together with the associated change in attitudes towards long-term planning in favour of short-term management of resources, shifted the priorities and interests of policy makers away from the Plan which was never approved or implemented. Nowadays, the prospective integration of the European economy and an increasing competition has shifted the attention of policy-makers back to the need for a long-term tourism development strategy.

Although there are no Regional Plans, certain area-wide controls for tourism development are becoming increasingly available. Since 1986, tourism development in certain areas is being put under control on the basis of the share of total accommodations and the reported occupancy rates (Athina, Rhodos, Mykonos, etc.). Also, the Law on the Environment of 1986 allows for the designation of special areas for protection, conservation or the development of "productive activities".

From an institutional point of view, a major obstacle in achieving proper tourism development is the lack of experience and the limited involvement of local authorities in local development issues, including tourism, urban development control and environmental protection. Historical and geographical reasons have favoured the development of a strong centralised national administrative system, which lacks the means to monitor and implement policies at the local level. This weakness, coupled with an absence of local, decentralised administrative structures, has stifled all efforts for efficient policy implementation. In the last decade, substantial improvements have been made towards assisting local governments to overcome their inertia and inefficiency, efforts which have been intensified recently and which will hopefully lead to better results in environmental protection.

Recent initiatives for the control of tourism development involve the designation of additional areas, which can be considered to have exceeded their "capacity" for tourism and areas which need upgrading. Although in theory the policy implications of such an approach are evident, there are tremendous difficulties in technically supporting the specific choices to be made.

Aside from the impacts of tourism on the natural and man-made environment there are also important social and cultural impacts, which ultimately relate to the environmental ones. A very important issue for several tourist areas and regions, in terms not only of local economy but also of social life, has to do with the dominant and overpowering role of tourism and almost total absence of other activities, a phenomenon called often "monoculture". The substantial gains of local people from tourist activities far outweigh the gains from agricultural or fishing activities and, frequently, such traditional activities are abandoned. In the case of developed tourist poles (e.g. Rhodos), social life is often minimal and impoverished in the off-tourist season as the locals travel to Athina or abroad for business or pleasure. For smaller communities, the abandonment of traditional practices has two important effects: a higher dependence on external demand fluctuations and increased vulnerability to the capricious ups and downs of tourism, and abandonment of traditional environmental and land management practices (i.e. alternate cultivation and grazing) with frequently devastating effects on the environment such as overgrazing and erosion and overfishing. The desertification problems of many small islands in the Aegean can be attributed to these factors.

Another potentially negative effect of tourism on social life is the overpresence in one area of tourists from one particular country or culture resulting in dominance of local culture and the often associated loss of local social life. Oddly enough, although some tourists may feel more "at home" and less insecure about their behaviour in a foreign culture, in the long run, the dynamics involved work against the interests of tourists and locals as the former loose the benefits of enriching their experience with another culture, the latter loose their local identity and both may face problems of social tensions, as the experience of some tourist countries suggests.

5. Future prospects and problems

Tourism is expected to become the largest sector in international trade by the year 2000. World-wide tourism activity grows by an annual average rate of 7.1% in terms of arrivals and 12.5% in terms of receipts. The rise of real incomes in the industrialised world, the increase of time available for leisure, the lower travel fares and the influence of mass media are among the factors which are expected to lead to a growth of tourism (Courrier, 1990). The World Tourism Organisation forecasts an increase by 2.5 times of global receipts from tourism (annual average rate of increase of 9%).

The countries of the Mediterranean account for more than one third of the world tourist market for a total of 107.9 million tourists (in 1984). Some of the projections of the UNEP/Blue Plan Unit foresee a doubling in the number of tourists by the year 2000 and a three and a half times increase by the year 2025 under the optimistic scenario, or over 50% increase in the number of tourists under the pessimistic scenario. In almost all of scenarios developed for the future, the tourist potential of southern Europe increases (UNEP, 1988). Although the general trend suggests a rising demand, fluctuations in tourism flows render dubious any concrete projections and, as a result, it is difficult to plan or invest in infrastructure in tourist areas (World Bank and European Investment Bank, 1990). In terms of tourist travel patterns, it is expected that the average length of stay will tend to decrease due to more frequent but shorter holidays, seasonality will be less pronounced (e.g. third-age tourists) and the temporal and geographic distribution of tourists will be more even (UN/ESC/ECE/HBP, 1987).

The European tourist preferences have also changed towards high quality vacations, meaning more space, peace and quiet and more activities (WTO, 1990). Furthermore, some broader patterns of change in the context of the anticipated European market integration are likely to affect southern EC countries and especially Greece. It is rather likely that a great demand will be experienced for vacation houses and other tourist facilities due to the freedom of movement of people and capital, a movement from the industrialised north to the Sunbelt. Also, the expected deregulation of airlines is likely to reduce the costs of travel and increase the number of visitors and tourists. Recent political and socio-economic developments in Eastern European countries are also likely to have some long-term effects in the increase of numbers of tourists to Greece.

The above mentioned developments will certainly intensify the pressures on the environmental resources of the country and its ability to manage them. The favourable combination of climate, clean environment and rich cultural resources is likely to remain a major factor of future growth. The anticipated increase in the number of visitors is likely to strain particularly the coastal areas and islands. Areas already developed touristically are likely to face pressures to expand and upgrade their tourist related infrastructure with special attention to improving the quality of the environment. New areas are also likely to develop into tourist destinations, the primary concerns for these being how to organise and mobilise local resources to attract tourists but also how to maintain their identity and environmental quality, probably through diversification of their product. These pressures are likely to be greater in the very small islands and coastal villages of the Aegean and Ionian seas.

The long-term growing importance of tourism as an economic sector faces some short-term difficulties and problems, mainly due to the basic characteristics of tourism as a world-wide socio-economic phenomenon. These difficulties draw from an increasing concentration of tourist trade in a few world centres and a few large tour operators who can act as an oligopoly and control the market to their own economic benefit at the expense of the receiving countries and the visitors themselves. The vertical integration of tourist services, from transport to hotel management, bears also potential dangers as a few large operators could control the market. Additional problems arise also from the dependence of tourist arrivals on external factors such as foreign exchange rates or random events like terrorism or insecurity.

6. Policy issues

The general patterns of future tourism development discussed above point to some of the key policy issues, which are likely to become of top priority for Greek policy-makers for quite some time in the future. The necessity to introduce long-term tourism planning is one of the first issues to be considered.

In an increasingly competitive and volatile tourist market, as the Mediterranean Region is, it will be necessary to pursue a coherent policy on tourism which would rely on the differentiation of the tourist product offered on the basis of the national cultural and resource-related characteristics. The protection and improvement of the natural and man-made environment is certainly the major axis for such a policy.

The main problem of post-World War II tourism policy in Greece is that, although the comparative advantages of the country were exploited, the model of development which was actually pursued, did not promote a balanced national sectoral policy and regional policy which would maximise the tourism potential and at the same time would avoid the vulnerability of the national economy to external factors, like the strategy followed by some other countries with strong tourism. The development of tourism is, in essence, development of services. The broad patterns of world development and the dynamic evolution of the service sector, particularly the financial and quaternary (R&D) activities, set the framework for the development of tourism in the future. Hence, a promising tourism policy would seek to develop operational linkages of national and regional policies and co-ordination of sectoral policies at each level, specifically in relation to environmental policies.

The type of development model to be followed should relate the socio-economic concerns with the physical and environmental aspects, probably through a set of development scenarios for the nation and strategic plans for each region. The role of regional policy is to assure the co-ordination of the various initiatives taken in the context of sectoral policies and at various levels of decision-making *and* ensure an equitable distribution of financial resources at all levels of planning (Council of Europe, 1988).

In the above-mentioned context, technology offers the opportunity for the harmonisation of policies in space. The prospects for technological development, particularly in the fields of telecommunications and transport (air and sea) services is of utmost importance for Greece as it provides the opportunities for the development of isolated and remote areas, reducing their problems of dependence. Moreover, technology offers the opportunity for integration of tourism and environmental protection, as it is possible to employ technological advances and advanced management techniques to protect the man-made and natural environment upon which tourism depends. These prospects relate mainly to energy conservation and waste management. In this respect, future policy should be such that current deficiencies should be offset and corrected. The EC has approved special programs for the Aegean islands as proposed in the operational programs of the Regional Development Plans, providing for installations of waste treatment facilities in coastal towns with favourable terms of financing.

In the above mentioned context, and in view of the expected diffusion of tourism development throughout the country, it will be necessary to establish efficient local administrative mechanisms to ensure better harmonisation of tourism development with the local environmental, social, economic and cultural conditions. Such efforts should be complemented with extensive cultural heritage and environmental awareness programs, which are needed in order to strengthen the qualitative elements of tourism. These are particularly important for the islands where there is a strong need for cultural upgrading.

In terms of the levels and areas of policy-making relating to tourism, it is necessary to underline the increasing role of the European Community, particularly in regional development through financing of various investment schemes, environmental protection through various programs and transport through co-ordination and regulation, a role complementary to national policy. The development of tourist resources is also facilitated by the promotion on the part of the EC of local and regional development initiatives. These can be complementary to the efforts of the

Greek government for decentralisation which have been particularly intense in the last few years.

The problems and opportunities which Greece faces with regard to tourism are, to a great extent, problems common to many other southern European and Mediterranean countries. International tourism consists of "sectors" or "product packages", in the sense of common features and characteristics of tourist demand. In the case of the Mediterranean, it is not only the similar climatic and physical conditions such as clean beaches and warm weather which are particular in a world-wide sense and common among Mediterranean countries, but also the rich heritage of historic urban centres. Any policy for qualitative improvement by necessity involves upgrading, protecting and enhancing both the natural and urban environment (Argeni, 1990). A problem with respect to tourism development common among several Mediterranean countries is the difficulty – because of international competition – to tax the growing tourist industry to cover infrastructure and environmental improvement costs (World Bank-European Investment Bank, 1990).

The existence of common problems suggests that co-operation between neighbouring countries could offer them significant long-term benefits, combining resources, extending the time periods of visits through better organisation or the development of common networks and broadening their respective markets. This strategy could be also beneficial for the tourists as it will offer them an opportunity to assess and compare the akin cultures of the Aegean and the coasts of Asia Minor and ultimately lead to the establishment of international peace and friendship.

References

Argeni, S. (1990) The urban renovation of mediterranean historic centres. *Observer*, June/July 1990.

Council of Europe (1988) *Conference Européenne des Responsables Régionaux de l' Amenagement du Territoire et du Développement Régional*. Valence (Espagne) 28-30 Avril 1987. Collection Etudes et Travaux, No 5. Council of Europe, Strasbourg.

Courrier, *Special Report on Tourism*. July/August 1990.

Komilis, P. (1986) *Spatial Analysis of Tourism*. Athina, KEPE (in Greek).

NTOG (National Tourism Organization of Greece) (1989) *Annual Report '88–- Forecasts '89*. Athens, Greece.

OECD (1978) *The Economic Aspects of Environmental Protection in Tourism*. Report: Group of experts on environment and tourism. OECD, Paris.

UN/ESC/ECE/HBP (1987) *International Tourism Flows and Patterns*. United Nations Economic and Social Council, Economic Commission for Europe, Committee on Housing, Building and Planning. UN/ECE, Geneva.

UNEP (1988) *Blue Plan Futures of the Mediterranean Basin: Environment and Development 2000-2025*. Sophia-Antipolis, France.

World Bank and European Investment Bank (1990) *The Environmental Program for the Mediterranean*.

WTO (1990) Closing Statement by F. Frangialli, Deputy Secretary-General of WTO (World Tourism Organization) at the Seminar on the Integration of Tourism in Europe. 7-9 May 1990. Istanbul, Turkey.

TOURISM AND THE ENVIRONMENT: IMPACTS AND SOLUTIONS

FRANK J. CONVERY
and
SHEILA FLANAGAN
Tourism Research Unit
Environmental Institute, UCD
Dublin
Ireland

1. Introduction

The environment – natural and man-made – is of vital importance to tourism in Ireland. It represents both the backdrop to many other activities and comprises a major attraction in its own right. The purpose of this chapter is to examine the development of environment-based tourism in Ireland and its possible impact on the landscape. Rural and urban threats to the environment in relation to tourism are discussed and tourism management strategies available for environmental protection are examined and compared with procedures in Northern Ireland.

2. Background – Setting the scene

In 1990, the Irish tourism industry earned £1,552 million ($2,678.5 million) of which £1,139 million ($1,966 million) came from out-of-state visitors and £412 million ($712.3 million) was domestic tourist spending. Total foreign exchange earnings from tourism have shown a considerable increase between 1985 and 1989 (see Figures 1 and 2). The tourism industry also makes an important contribution to employment, much of it being in less economically developed regions of the country, thus helping to equalise the geographical distribution of income.

The 1988–1992 period has been one of unrivalled challenge and potential for Irish tourism:

- The government has established targets for the industry to double tourism numbers and receipts by 1992 and increase employment by 25,000 new jobs.

- Tourism is an excellent economic investment for government and is one of the few strategic options which is labour intensive.

The most recent Irish Tourist Board (Bord Failte) estimate suggests that tourism supported 80,000 job equivalents in 1990. Of these, just over 50,000 were supported by out-of-state expenditure and the remainder by the domestic market (Bord Failte, 1991).

3. Growth of environment-based tourism

Western Europe accounts for 60% of the world's tourism (Irish Tourist Industry Confederation, 1989) and is likely to hold that position for the next 10 years. Ireland is part of Europe and is one of the most underdeveloped and unspoilt parts of the continent, having a distinct comparative advantage in terms of environment-based tourism. The trend in the traditional north-south migration to holiday spots has eased and this gives the Irish tourist product, a green one of beautiful scenery and friendly people, a place in the fast lane for the first time ever.

The development of new products based on the environment (environmental activity tourism) broadens the marketing base and increases Ireland's attractiveness as a destination. This is a key in the drive to increase market share and double tourism receipts and employment.

The physical environment is an important factor in Ireland's tourism product. The tourist industry relies heavily on the country's environment, which is still perceived as being clean, restful and unspoilt. Annual marketing surveys of potential visitors demonstrate that Ireland's scenery is the country's principal attraction as a tourist destination. It comes out top of a list for all holidaymakers, irrespective of their country of origin. Similarly, those surveyed after a visit, list the countryside as the highest-ranking memory of their stay. Other features of an environmental nature score highly, including towns, cities and architectural heritage (Irish Tourism Industry Confederation, 1986). By themselves, people and scenery are not enough to form the basis for good, steady, stable tourism. What Ireland is now engaged in is grafting on a whole range of products, which will be built on environmental excellence to make it a more attractive destination. It is necessary need to get away from the "green desert" image of a nice country of friendly people and poor weather with not much to do.

The importance of the environment in terms of tourism is shown in the most recent Bord Failte (1988) Post-Visit Survey. Every four or five year a post-visit survey of non-ethnic holidaymakers to Ireland is carried out. It provides a great deal of detail on influences on expectations of and attitudes to Irish holidays by non-ethnic holidaymakers. The 1988 report shows that environmental factors are of considerable importance in choosing Ireland as a holiday destination. An examination of certain environment-based activities shows that on a national scale, 46% of those surveyed indicated they had participated in hiking/hillwalking and rambling. By market area, this breaks down as follows: 39% British, 43% North American, 57% European and 42% New Zealand/Australian.

A more significant measure of the importance of these activities is the proportion of holidaymakers who said that their availability was essential to their choosing Ireland for a holiday. Holidaymakers who perceive particular activities to be essential to choosing Ireland for a holiday could be defined as serious participants in these activities, as opposed to the broader participation level. The survey indicates that in the category of serious participants 18% (47,520) indicated that the availability of hill walking/hiking and rambling were paramount to their decision making. Broken down by market area, 16% of the British, 15% of the North American, 24% of the European and 12% of the New Zealand/Australian indicated that they belonged to this category. Within mainland Europe, serious interest in environment-based activities varies on a market by market basis. Hiking, hill walking and rambling enthusiasts are more prevalent in the Netherlands and France but also significant to German holidaymakers. The survey indicates that 264,000 were involved in this category and of these 103,000 were serious about hill walking, etc. The Irish Tourist Board has identified this as a growth product area, suitable for brand marketing and with a good British and European market. According to the Irish Tourist Board, other expanding uses of the environment will be:
- horse riding treks
- climbing bikes/motor bikes
- hang gliding
- deer stalking
- mountain lake fishing
- grouse moor development and shooting
- rock climbing

Detailed information on the monetary value of individual tourist activities is not readily available but the value of marketing a clean environment may be gauged from just two activities, angling and river or lake cruising, which together earned £43 million in 1986.

Angling is a particularly environmentally sensitive tourist resource. Ireland is attractive to salmon anglers – the quality of its angling being superb and the country is one of the world's leading Atlantic salmon producers. A survey, carried out in 1985 by the Salmon Research Trust of Ireland in the west of the country on some rivers, found that, in terms of total expenditure by anglers, each salmon caught by rod and line in this river system had a minimum gross value of £460 ($584). This figure includes hotel and guesthouse accommodation, boat hire and spending in restaurants, craft shops and fishing tackle suppliers. Indeed, the survey found that average expenditure per angler reached £740 ($1278) in the case of visitors from the continent. The 1988 Economic and Social Research Institute Report on an Economic Evaluation of Irish Angling showed that the total value of the angling sector as a whole (inland and marine fisheries) amounted to £72 million ($124 million). It is estimated that angling generates some 1,900 full-time job equivalents through direct, indirect and induced effects in the economy and that it yields direct tax revenue of £15m ($25.9 million) in the form of receipts from VAT, excise duties etc. This illustrates the very significant potential of salmon angling in Ireland, both for the tourist industry and the economy in general.

Thus, were serious damage to occur to inland and sea-based water resources, very large revenues would be in jeopardy. Similar losses would be sustained right across the board if the country's many other environmental resources were damaged beyond the point of restoration.

4. Pressures on natural tourist resources

It is now part of the conventional wisdom that growth in visitor numbers and expenditure depends on activity-based tourism, catering to niche markets, and using the environment as a base. This is a much more complex, albeit interesting, product to produce than the general coach tour and scenery experience. Activities, almost by definition, are clearly part of the local scene, so that the potential for conflict exists. The environments, which are the most appropriate and interesting for activities are also the most fragile, with a very limited capacity to sustain increased use.

One of the real problems is that threats to the physical environment – and hence to the tourism industry – are often so insidious that they are recognised only when it is too late. The process can be slow, seemingly causing little harm, but it is also cumulative and, quite suddenly, a threshold is crossed and it becomes difficult to reverse. Pressures on the environment may be rural- or urban-based or a combination of both.

5. Rural pressures on the tourist environment

The following impacts are probably the most prominent:

Visual pollution. Large dominating hotel buildings are often out of scale and clash with the rural surroundings. The failure to incorporate adequately environmental considerations in the architectural designs of hotels, restaurants and entertainment facilities can lead to consequences which are both environmentally and economically unprofitable. In Ireland, one leisure activity which has been increasing greatly in popularity is camping and caravanning. The number of "pitches" and registered parks has doubled in the last fourteen years, and this sector can now cater for over 37,000 people. There are also many unauthorised sites, where the operators have neglected to apply for planning permission.

For most of the year and throughout most of the country, the volumes involved are so low that little environmental damage is caused. But in the high season, at certain sensitive locations, problems do arise. In terms of visual amenity, the effect of many brightly coloured caravans and tents parked haphazardly on exposed sites can destroy views and prospects over a wide area. This is quite common in the case of amenity areas laid out at public expense and then colonised illegally, often on a semi-permanent basis by those caravanning or camping.

Another form of visual pollution in Ireland affecting the tourist landscape resource is the sporadic development of urban generated housing. It is estimated that in the period 1973–1980 approximately 82,000 one-off houses were built in the countryside, nearly half of them for people who had no functional connection with agriculture. This phenomenon of "bungalow blight" has already altered some of Ireland's most scenic areas. The form it takes may be concentrated in ribbon development on the approach roads to cities and towns or dotted sporadically throughout the countryside. In many cases, such housing is located to take advantage of a particular view, degrading or destroying the very scenic qualities, which attracted the development in the first place.

Sporadic development along coasts in the form of ribbon development has occurred also. Coastal resort development has tended to sprawl along the coastline. This is a response to the need to take advantage of the beach as a primary resource. Ribbon development has also occurred along valleys and scenic routes in inland areas. Minor roads built to accommodate light, rurally generated traffic volumes, have become heavily overloaded and unsafe for activities such as walking, cycling or touring by car.

Overloading of infrastructure. In many resorts, infrastructures are unable to cope with the intensity of tourist variation at peak periods of the year. The result is failures in supply, pollution and health hazards. The disposal of sewage and refuse can also lead to isolated or more widespread pollution and litter problems, while overhead electricity and telephone cables provided to serve one-off housing developments in the countryside can be almost as visually obtrusive as the houses themselves.

In the Lakelands tourist region of Ireland, there were several instances of water pollution caused by inadequate treatment of domestic sewage before it was discharged into receiving waters. This totally inadequate level of infrastructure is not conducive to the proper development of tourism.

Conflicts of access. Tourists may have to pass through farmlands to points of scenic beauty/attraction. The problem of public liability insurance needs to be resolved as a matter of urgency. Irresponsible behaviour by tourists is an associated problem. Damage and destruction of crops, farm buildings and harassment of livestock are often reported. Fires, excessive noise, illegal hunting and fishing and litter also lead to conflicts between farmers and tourists.

Increased erosion from overuse of paths and trails, especially from horse riding can be a problem in sensitive areas. The Irish Tourist Industry Confederation Report shows that more than one million visits were made to the Dublin/Wicklow mountains (an area which covers about 440 km^2 in extent) in the east of the Republic of Ireland in 1985. This puts pressure on land through damage of the natural habitat, erosion, eviction of wildlife, destruction of plant species and general overuse.

Increasing affluence, leisure time and urbanisation have combined to put more pressure on land used for recreational purposes – especially in scenic areas located close to major towns and cities such as Dublin. Such pressures can damage the natural habitat through litter, pollution, erosion, destruction of plant species, eviction of wildlife and general over-uses. When these problems become widespread, the particular area under pressure can be destroyed altogether; participants in the various activities it once supported look further afield then and the cycle begins again at more distant locations.

Farming pressures. Changes in agricultural practices have also had an adverse effect on the rural landscape in recent years. These include the erection, usually without the need to obtain planning permission, of large industrialised farm buildings. Archaeological artefacts have also suffered. In the case of ring forts, for example, the proportion damaged or destroyed since 1841 ranges from 29 per cent in south Donegal to 66 per cent in the Cork Harbour area.

Land improvement through arterial drainage has given farmers better financial returns. In visual terms however, the straightening of river channels and the deposition of dredged spoil on their banks can sometimes be most unfortunate. Ecologically, the adverse effects include the removal of wetland habitats and the disturbance of fish spawning ground.

In recent years, pollution problems, associated with the trend towards intensification in agriculture, have become more acute, especially in pig and poultry production. The disposal of effluent from these large-scale units has resulted in pollution through run-off into streams, rivers and lakes. This problem appears to be most serious in the Irish Lakelands region where the adverse impact of declining water quality costed local tourism interests £1 million in 1982. However, in the late 1980s a reversal in this trend has emerged.

Trees and woodlands. Indiscriminate trimming of hedgerows and felling of "dangerous" roadside trees by the local authorities can have very damaging effects, especially if such works are undertaken during the nesting season. The state has shown a remarkable lack of sensitivity for the widely differing landscapes of Ireland. State-owned forests are characterised by large tracts of single species planting, laid out in regimented patterns. In the last decade, approximately 19,000 acres of state forests were planted each year and around 80 per cent of these comprised only two coniferous species, Sitka Spruce and Lodgepole Pine. Broad-leaved woodlands are mainly in private hands and such woods account for only one-fifth of all afforested areas.

Other threats. The segregation of local residents and tourists can be a problem. The mass tourist may be surrounded by, but not integrated with, local society. Local inhabitants may be prohibited from using tourist facilities. The quarrying of rock, the extraction of sand and gravel and the Irish Peat Board's (Bord na Mona) extensive peat development programme are the most common forms of extractive industry in Ireland. Quarrying can be extremely intrusive, creating unnatural scars on the landscape. The same is true of peat extraction. It has been estimated that only 5 per cent of the country's raised bogs, arguably the most "Irish" of our landscapes, is still in its original condition and it is regarded that even 1 per cent will survive into the future.

6. Urban pressures on the tourist environment

Townscapes. The attraction of the cities, towns and villages in Ireland stem, to a large extent, from the overall impression gleaned from a combination of facades and spaces, changes in levels, contrasts in materials, soft and hard landscaping and street furniture. In many towns and cities, great damage has been done on streetscapes through thoughtless replacement of traditional shopfronts with mass produced aluminium and plastic, usually in brash colours. In Dublin, it has been estimated that at least 50 Georgian houses listed for preservation in the City Development Plan have been subjected to material alteration over the early 1980s. Of the Georgian houses included in the less stringent "protected" category, scores have been redeveloped or replaced by "replicas" – usually pastiche attempts to echo the past.

Dereliction. Closely linked with townscape is the issue of dereliction, which regularly features among the unfavourable comments made by tourists returning home after visits to Ireland. Tourists tend to visit traditional city, town and village central areas for their shopping needs since these environments form part of the uniquely Irish atmosphere which attracted them to Ireland in the first place. In recent years, however, there has been a proliferation of purpose-built shopping centres, usually at the outskirts of cities and towns which compete with the traditional town centres rendering downtown shops non-viable and forcing them to close. The abandoned commercial buildings eventually deteriorate and become derelict.

Other problems. The greatest change that has affected the central areas of the cities in recent times has been the intrusion of vehicular traffic. Apart from the air and noise pollution generated, existing streets, some of them medieval in origin, have been widened, narrow junctions reamed out, new routes and car parks created. In tourist resort areas, traffic congestion has emerged as one of the most serious consequences of resort development. In a poll conducted by the Swiss Tourism

Federation in various Swiss holiday resorts, visitors actually rated 'traffic chaos' as the greatest problem facing resorts and the one requiring the most urgent solution.

7. How can use conflicts and environmental problems be avoided?

In order to bring about a rational utilisation of tourist resources, it must be incumbent upon all from the beginning to take into account the national priorities and the needs of others, especially the resident local populations and their traditional activities upon which they depend for a livelihood. Otherwise tourism will continue to be seen as an activity which is imposed from outside upon the local community, put in place for the benefit only of tourists. On the other hand, moves by the Department of the Environment, both in Northern Ireland and the Republic, to curtail developments, such as location of marinas in certain areas sometimes meets with objections as such projects are economically beneficial, creating employment and prosperity in the region. It is often difficult to get the balance right between economic prosperity and environmental protection.

Local people living in areas of tourist attraction – be it the countryside or a coastal environment – need to be, in effect, compensated for the intrusions which tourism inevitably brings. It is better if compensation comes in the form of a return on investment in the tourism sector than in the form of a "handout", although there is certainly a place for the latter if revenue gaining opportunities must be significantly compromised in order to meet tourism objectives. Another complementary strategy worth of consideration in this context is the concept of sustainable tourism.

8. What is sustainable tourism?

The three basic attributes of sustainable tourism are:

For the host area, it should provide carefully planned economic growth with satisfying jobs, without dominating the economy. It must not abuse the natural environment and should be architecturally respectable. Decision-making should be local and traditional values and societies should be maintained. The benefits of tourism should be diffused through many communities and not concentrated on a narrow coastal strip or scenic valley (Developing Sustainable Rural Tourism, April 1990, Bernard Lane, University of Bristol).

For the holiday makers it should provide a good value, harmonious and satisfying holiday experience. The experience must respect the intelligence of both the visitor and the host population.

For the operator, responsibility will be a key to success. Responsibility entails accepting and building up the role of the go-between – explaining the need for the long term ecological care of the tourism resource- scenery, place, people, fauna and flora – to both the host people and the visitors.

9. How can sustainable tourism be implemented?

Sustainable tourism is an approach rather than a series of laws. There are, however, four interlocking strands, which are central in helping the evolution of sustainable tourism.

1. Education is important. The philosophy of sustainable tourism must reach a wide audience including politicians, planners, tourism operators, developers and tourists themselves.
2. Local tourism management strategies.These strategies must be designed in consultation with local people, businesses, planning authorities, ecologists and tourist boards.
3. Rural tourism forums. Covering groups of villages or small cultural regions are effective methods of bringing together accommodation providers, operators of local attractions, voluntary groups and the authorities together. The forums can discuss the implications of the sustainable tourism approach, produce marketing material, administer training schemes and act as a sounding board for updating the Tourism Management Strategies.
4. Information exchange. In rural tourism development and management projects is important if locally based groups are to take full opportunity of the possibilities open to them. In Britain, the work of the "Country Village Weekend Breaks" group has important lessons for many rural communities. The Ballyhoura Failte project is an impressive example of the concept at work in Ireland. Ballyhoura (in Co. Limerick) seems to be exactly what the theorists are advocating. It involves the local community taking a bottom-up approach and promotes the natural advantages of the area.

The work of the Irish Resource Development Trust in Waterville, Co. Kerry (South West Ireland), is another interesting case in point. One of the main aims of the trust is to develop a model and promote it nationally as a structure for achieving sustainable employment in rural Ireland. Local area companies are being promoted by the Trust in association with County Development Teams. It is about supporting people who want to help themselves. The model is a 'third sector' company model and is a financially expressed partnership between the community, public officials and the commercial sector. The aim is to bring together competent people from the community, the commercial sector and the public service to provide a comprehensive local initiative for enterprise and investment, to bring new resources and to release and support creative energies and skills. As in the case of Bernard

Lane's Sustainable Rural Tourism model in Britain, this Irish model for Rural Development advocates a "bottom-up" approach.

Waterville in South West Kerry is a village of some 1200. IRD Waterville Ltd was established in 1987 funded by the local community [over £20,000 per annum ($34,538)], the Irish Resource Development Trust and the Central Development Committee of the Department of Finance. Finance for the Trust's backing of the Waterville model is provided by the ESB, Allied Irish Banks and the Gulbenkian Foundation.

In 24 months IRD Waterville had:
- appointed a manager, established a tourist office, developed tourism products, specifically village refurbishment, trout and salmon fisheries enhancement
- implemented a successful marketing and direct selling programme to increase visitor bednights and stimulate private investment in hotels and leisure facilities and other business
- established a craft market retailing the products of 21 craft workers
- helped established Sub Aqua Diving businesses and Coach Tour Courier operations
- piloted a vegetable growing scheme
- helped establish a Salmon Farming venture

Additional spin-offs include
- employment from tourism
- increased income for 21 craft workers
- 3,500 bed nights, off peak times, in the first season

A programme to develop this model nationally offers a coherent, new and realistic approach to rural development which can become self-sustaining.

The concept of integrated rural development or sustainable rural approach to tourism promotion must seem somewhat utopian, but the approach makes a lot of sense especially in areas where the environment is particularly sensitive and likely to be disturbed easily by tourism development.

10. Epilogue

The environment in Ireland is an essential ingredient of the tourism product. Already one of Europe's largest package tour operators, Thomson Holidays has set up a high level working party to consider how it can help the trend towards responsible tourism. The Spanish National Tourist Office is reconsidering its approach to growth and is beginning a new "Green Spain" campaign. Ireland traditionally the greenest of all isles should be able to capitalise, in a responsible way, on these new and welcome trends.

References

Antrim Coast and Glens (1988) *Area of Outstanding Natural Beauty, Guide to Designation.* Department of the Environment, Northern Ireland, July 1988.

Bord Failte (1988) *Borde Failte Post-Visit Survey.* Dublin

Bord Failte (1989) *Developing for Growth – A Framework Development Plan for Irish Tourism.* Dublin.

Bord Failte (1991) *Bord Failte Tourism Facts 1990.* Dublin.

Deane, B. M. (1986) *An Employment Growth Area: The Tourism Industry.* Paper presented to Conference on Unemployment, July 1986.

Department of the Environment (1990) *An Environment Action Programme.* Dublin.

Department of the Environment for Northern Ireland. *Protecting Your Environment - A Guide.*

Department of Agriculture for Northern Ireland (1990) *Countryside Management - The Dani Strategy.*

Greer, J. V. and M.R. Murray (1988) *A Recreation Strategy for the Mourne Area of Outstanding Natural Beauty.* The Sports Council for Northern Ireland, June, 1988.

Irish Tourist Industry Confederation (1986) *Report on Tourism and the Environment.* October 1986.

Irish Tourist Industry Confederation (1989) *Doubling Irish Tourism – A Market-Led Strategy.* Dublin.

The Stationary Office (1987) *Improving the Performance of Irish Tourism.* Dublin.

The Stationary Office (1989) *Ireland - National Development Plan 1989-1993.* Dublin.

Whelan, B.J. and G. Marsh (1988) *An Economic Evaluation of Irish Angling.* Economic and Social Research Institute and the Central Fisheries Board, Dublin.

AGROTOURISM AND THE RURAL ENVIRONMENT: CONSTRAINTS AND OPPORTUNITIES IN THE MEDITERRANEAN LESS-FAVOURED AREAS

THEODOSIA ANTHOPOULOU
Department of Social Policy
Pantion University
Athens
Greece

1. Introduction

The environmental problems of the Mediterranean areas are very important, due to the continuous growth in tourism and the concentration of economic activities along the coastline. Meanwhile, rural regions in the hinterland, mainly on the mountains and hills, suffer from population decline and the abandonment of productive, agricultural activities. Agrotourism projects have been introduced within the framework of European Union policies for sustainable development in the less favoured, not to say, fragile, rural zones. They aim to exploit the natural and socio-cultural heritage (rural architecture, gastronomy, folklore), motivating country people to keep their farming activities and allowing them to earn complementary income. This extra income is assumed to offer partial remedy to the seasonal or permanent under-employment of family members, mainly women or future successors of the farm; it can even be essential for the socio-economic reproduction of farm household and, therefore, for the maintenance of agriculture in small areas. The main question this paper asks is whether agrotourism can contribute to the preservation of the rural environment.

2. New agricultural functions and prospects for the mediterranean less-favoured regions in the framework of the New Common Agricultural Policy

After the 1992 reform, the essential aim of the European Common Agricultural Policy (CAP) has been to manage volumes of food products in order to stabilise the agricultural market, and at the same time try to reduce the negative effects of intensive farming on the environment (like excessive fertilising, use of pesticides, dangerous fungicides). The extensification of productive agricultural systems aims at a less input-consuming agriculture or an agricultural more traditional in its practices and, consequently, more adapted to the local environment (Mormont, 1996; Whitby, 1996). But in the Mediterranean mountainous and less-favoured zones, agricultural production is already extensive and rural societies need a revival, which must be supported by a wide range of economic activities linked to their region and their know-how.

The reform of the CAP has been based on an economistic approach (such as stabilisation of agricultural prices to the rate of the world market, limitation of public expenditures, and so on) and, hence, has not taken into account the variety of local situations within Europe's territory. The new CAP continues to encourage the specialisation and concentration of farming activities in the most productive and competitive rural regions without taking care of disadvantaged areas due to physical or other reasons (De Casabianca, 1997).

Rural Mediterranean regions with weak environmental potential and low yields, like the mountainous and less-favoured zones (Dir. 75/268/CEE), 'remain penalised' by the market law in the sense that they are not able to compete with the high productivity regions. They are always likely to be highly socio-economically marginalised due to inefficient production conditions (like small and compartmentalised land properties, low agronomic potential, lack of finance capital, and so on). Consequently, they become subject to the risk of environmental desertification (abandonment of farming activities, soil erosion, land degradation, fires, etc.). Many cases in the Pindos Mountains in Greece, Extramadura's high regions in Spain, or in the mountainous hinterlands of the Mediterranean coast in France, can be mentioned where population densities are low. Areas where farming activities have been abandoned become very sensitive to the risk of fire. At the same time, the absence of other activities apart from agriculture have created severe social problems and especially unemployment (MEDEF, 1992).

Diversification of farming into activities other than those producing raw materials, like agrotourism, the processing and on-farm sale of agricultural produces, is presented as an alternative for rural planning and sustainable development, especially in the Mediterranean mountainous and disadvantaged areas (CE, 1994). The diversification of the activities in these areas together with quality strategies aims at revealing regional competences and at mobilising local human and natural resources. Today these areas are facing new prospects, which are based on a 're-territorialised' model emphasising 'economies of quality'.

Overall, agriculture in the Mediterranean mountainous and disadvantaged regions cannot be considered as an economic activity capable of securing the socio-economic reproduction and well being of farm households. Nevertheless, its importance as an economic activity is unquestionable. Firstly, although agriculture contributes to the environmental management of fragile ecosystems, this function is not enough to justify and finance agricultural activity in an area. Secondly, because even though agriculture is not very competitive in such disadvantaged rural or ecological contexts, as the Mediterranean ones, it possesses a relatively important potential in less-favoured areas likely to support other activities. Tourism is a typical example as it finds essential assets in the mobilisation of regional gastronomic resources as well as in the more humanised landscapes of rural areas (de Casabianca, 1997).

In this theoretical perspective, rural tourism, and more specifically agrotourism, must function in accordance with agriculture in the mountainous and less-favoured zones. This means that one activity must feed and maintain the other by insuring extra income to the farmers and other local actors involved. Agriculture is called more and more to develop and offer various services and leisure activities, apart from its traditional role in producing foodstuff, mainly in the less competitive areas in terms of economies of scale.

3. Agrotourism: a means of developing and preserving the environment in the Mediterranean mountainous and less-favoured regions

Although agrotourism is highly recommended by specialists and practitioners of local development in the framework of agro-environmental policies, its definition remains often mistaken with that of rural tourism. By agrotourism, we should understand "all tourist services, like accommodation, catering and leisure activities, given by farms, generally family farms, with agriculture as main activity or even secondary. Agrotourism is, therefore, one of the components of rural tourism; i.e. tourism in rural areas" (Bazin and Roux, 1997).

It was not before the 1980s that interest in agrotourism began to develop. This was due, partly, to a dynamic urban demand for holidays and leisure in rural areas and, partly, to the offer of tourist services by farmers who were trying hard to diversify and improve their income in the face of agricultural market crisis. At the same time, agrotourism was highly stimulated by the European socio structural funds in favour of the disadvantaged and/or fragile areas (objectives 1 and 5a) in order to develop and diversify activities in rural areas. This is considered an important tool for rural planning and a stimulus of local economies, especially because of its positive effects on employment (LEADER Dossiers, 1993).

Agrotourism is an alternative to mass tourism, which is seen as harmful to the natural environment (destruction of local crops, introduction of foreign plants, discarded refuse, coastal water pollution, and so on) and the landscape (expansion and canalisation of urban areas, destruction of architectural heritage, etc.). Moreover, it has been observed that the local population does not have direct or indirect economic benefits from mass tourism because of foodstuff and manpower imports used in the various tourist services. On a regional scale, a spatial polarisation phenomenon can be observed with various economic and tourist activities concentrating in very attractive tourist areas to the detriment of the farming hinterlands. This is, par excellence, the case in all the Mediterranean coastal or mountainous areas which are well equipped in recreation and accommodation facilities, often oriented to mass tourism. This movement leads to a massive immigration from neighbouring areas, harming their socio-economic structure. Tourist expansion also leads to saturation of coastal areas and, consequently, to the spread of tourist nuisance to interior rural areas. This can cause landscape degradation, as in the Provence area in France or in the Greek islands (CE, 1994). As mass tourism is not associated with regional particularities and quality of local products, the major tour-operators can easily change tourist streams and tours, in which case the local actors involved (farmers, hoteliers, restaurant owners, etc.) can hardly react.

Forms of rural tourism which are limited to simple accommodation in small-sized hotels in rural areas or to 'rooms-to-let' by farmers or other parties do not seem to support rural incomes and promote the maintenance of the natural and built environment. Similarly, polarisation phenomena are observed in these cases as well. As far as supply is concerned, rural areas have not generally developed a structured marketing approach for rural tourism products; at the same time, country people are not the only suppliers of services such as accommodation and catering. In France, only one third of all the 'gîtes ruraux' belongs to farmers (Perrier-Cornet and Capt, 1995). In Paros (one of the Cyclades islands in Greece) surveys in tourist villages have revealed that among the owners of 'rooms-to-let' in 'rural areas' (terms used by the National Tourism Organisation), the percentage of those who have a farming activity does not exceed 10 per cent, whereas less than half of these owners are permanent inhabitants of the island (Anthopoulou, 1997). These forms of 'rural tourism', developed by entrepreneurs external to rural areas and dissociated in reality from the local culture, work along mass tourism stereotypes received in the hotel industry. As a result, the risk of degradation of the natural environmental and the rural heritage exists as a consequence of bad management practices.

On the contrary, agrotourism, which combines agricultural production with the provision of tourist services and leisure activities in a family farm surrounding can be spread to rural areas as a whole and create positive externalities on the local economy. A relevant survey conducted on the Highlands of Scotland, concerning the impacts of different types of tourism on the local economy (analysis of direct, indirect and induced impacts), has shown that soft/land-based tourism is better integrated in the local and regional economy than mass tourism, generating higher income and more employment per unit of visitor expenditure, even though expenses per tourist are higher in mass tourism than in soft tourism (Slee *et al.*, 1997).

Moreover, in the Mediterranean mountainous and disadvantaged zones, which are sensitive to environmental hazards, agrotourism is considered as a tool for land management and for opening up these areas to external (positive) socio-economic influences. On the one hand, manual (as opposed to mechanised) work on land and the associated techniques (terracing, small-scale hydraulic works, and so on) contribute to the maintenance of the physiognomy of the rural environment. On the other hand, small-scale, high quality farm production, employing traditional practices, finds an important market in tourist activities.

In addition, the Mediterranean areas provide an excellent natural habitat for agrotourism to thrive: the biodiversity in their ecosystems, the originality of their landscapes touched by man's age-old presence and the rich architectural heritage of the countryside impart a sense of authenticity, which is greatly demanded and appreciated by tourists. On the contrary, in rural areas of intensive farming, the trivialisation of the rural landscape and environmental degradation can lead to a decline in tourist demand. Survey results conducted on 'gîtes ruraux' in Brittany, in north-west France, in order to identify the impacts of intensive farming on the profitability of rural tourism accommodation enterprises (using the technique of hedonic prices), have shown that the demand and price for gîtes are negatively affected by intensive fodder crops and livestock systems (pigs, poultry and cattle) (Le Goffre and Delache, 1997). As an OECD study has noted, every rural area cannot succeed in tourism: an interesting landscape or animal life are necessary but not sufficient conditions for success (Muheim, 1995).

Finally, the Mediterranean hinterland located far from areas urbanised by seaside tourism or damaged by intensive irrigated farming, can provide important assets for the development of agrotourism, if their natural and human resources are utilised in accordance with their rural character.

4. The particularities and assets of agrotourism in the Mediterranean rural areas

Agrotourism has a long tradition in some European countries, for example in Austria and in Switzerland, due to particular public policies in favour of its development. For a long time in these countries, there was the political will to associate agricultural activities with welcome and accommodation facilities in farms in order to support mountain farming. Sometimes, measures for the protection of properties were taken to guarantee that country people would control their land against anarchic development of second residence, as it was the case in Tyrol (Austria). In other northern European countries, as for example in the Netherlands, the United Kingdom, and Germany, agrotourism activities have constantly grown over the past few years, whereas farmers try to offer more than the conventional services to visitors (sport and leisure activities, local gastronomy, on-farm sale of agricultural produce, trekking, educational farms for children, and so on). In Switzerland and Norway, 20 per cent of the farms are dedicated to agrotourism, in Austria 10 per cent, in the United Kingdom 8 per cent, less in France – about 2 per cent, and even lesser in Southern European countries as in Spain (only 0.5 per cent), Greece and Portugal (Bazin and Roux, 1997).

Indeed, agrotourism has developed only modestly in Southern Europe. Despite the fact that the Mediterranean areas possess undeniable tourist assets – diverse landscapes, gastronomy, architecture, historical sites and monuments, mild climate which makes possible a longer tourist season into the year – agrotourism has only recently started to develop through the economic incentives of the CAP socio-structural measures (as the Dir. CEE/2328/91 and the Programme LEADER). Which reasons can explain this delay in the Mediterranean countries compared to the countries in northern Europe? And, what are the expected results in the Mediterranean mountainous and disadvantaged zones since agrotourism is considered as an important option for local economic development?

At first, it is noted that, despite the positive feedback effects of agrotourism on local economies, the experience in the European Union shows that it is difficult for farmers to develop extra-agricultural activities and, therefore, to become managers of small firms offering various services, like agrotourism. This means that farm holders will have to make a radical social and occupational change in moving from an occupation centred on the control of technical processes, as it is the case of agricultural activities, to an occupation demanding more open social relationships and structures capable of offering services to specific tourist clienteles (Perrier-Carnet and Capt, 1995). Farmers and other rural actors involved need time to accept the idea, to get training and to adopt new technical and organisational innovations in agrotourism projects. This occupational shift and the ability for technical and social adaptation seem to be much more difficult to achieve in mountainous and disadvantaged zones, where rural societies were, and still are, isolated and follow more traditional management practices. Beyond this general observation which concerns all rural areas, the particularities of the agrarian structures and socio-economic evolution of Mediterranean rural areas, and especially of those lagging behind in development, can explain the opportunities and limits for agrotourism development in southern Europe. In general, the following features are common to most of these areas:

As far as their agrarian structure is concerned, contrary to the model of 'isolated farms' encountered in the rural areas of northern Europe (an ancient rural structure which resulted from the era of the enclosed countryside), the Mediterranean farm, viewed as a unit of production, is generally separated from the rural settlements (Lebau, 1992). In the Mediterranean region, rural settlements are usually concentrated, whereas the fields are scattered around the villages. In this case, the direct association between tourist accommodation (which is located in a compact village) and agricultural activities of the farmer - like cultivation of the land and animal husbandry - which take place at distant fields is not evident in the context of agrotourism. The same applies to the integration of various tourist services into agricultural activities and rural life in general. Even for the farmer, there is a certain separation between his occupation within the farm and his social and family life, and, as a result, a kind of social individualism. That is why it may be difficult to link these two functions, productive and social, within an 'agrotourism product', and so attain the goal of preservation of the rural environment through an evaluation of the rural and built heritage.

As far as society and human resources are concerned, the Mediterranean hinterlands, particularly the less-favoured and disadvantaged areas, have suffered from out-migration the result being the ageing of the population. This partly explains why local actors lack innovative initiatives for reviving their small region;

why farmers are reluctant towards new forms of economic activities, like agrotourism and, finally, why they meet difficulties in seeking to broaden and reorganise their agricultural occupation around the provision of services to specific customers.

But, from the supply point of view, farmers are not the only actors to claim the provision of tourist services in rural areas. If farmers cannot augment their activities, small entrepreneurs of a farming or other origin may invest as well in agrotourism. They build small-sized hotels or 'rooms to let' which sometimes offer catering and leisure activities of the 'agrotourism' type, but without a real link with local farming production and the rural culture. This can lead to an occupational abuse, as this type of rural tourism does not necessarily contribute to environmental protection in the sense that it may not protect the rural landscape, ecosystems and the architectural heritage.

As far as technological and economic structures are concerned, less favoured rural areas in Southern Europe suffer also from a lack of public infrastructure, which would support local development endeavours, such as the development of agrotourism. This infrastructural deficiency partly explains why farmers are discouraged to invest in such projects.

On the other hand, at the level of the family unit, the Mediterranean agricultural holdings in less-favoured areas suffer from significant depreciation and economic weaknesses. These areas cannot afford to invest in tourism – e.g. to restore old buildings, build new premises, buy home equipment – without technical and economic assistance from the public and the private sector.

Moreover, in these countries, where the institutional and organisational framework governing farming activities is relatively deficient, farmers are unable to modernise and diversify their activities due to lack of information networks and diffusion of technical 'know-how' on both regional and national scales. Therefore, the absence of specific public policies in favour of agrotourism development which will relate tourist services to agricultural activities becomes more serious in the Mediterranean rural areas, as farmers are often not sufficiently informed and organised to undertake the required collective actions in their territories.

As far as demand for agrotourism services is concerned, this is not high in the Mediterranean rural areas. There are two essential reasons for this situation. Firstly, rural-urban migration is still a recent experience (after the Second World War), and, consequently, rural immigrants in urban areas are still in touch with their parents and places of origin. Usually, they keep their patrimonial home as second residence for holidays or as main home for their retirement. This firm relationship with their village of origin is the reason why urban people do not feel an urgent need to trace their roots or seek the experience of country authenticity through agrotourism. In reality, they have never lost contact with their rural culture. Secondly, in the Mediterranean countries, the dominant habit of seaside holidays concentrates tourist interest on the littoral zones to the detriment of the farming hinterlands.

Finally, it appears that technical, economic and organisational weaknesses as well as the traditional preference of holiday-makers for coastal resorts make that agrotourism is developing in areas where there are already many tourist advantages. This is the case with coasts (summer vacations) and the mountains (winter sports). Generally, these areas are considered rural and they are often characterised as disadvantaged according to the criteria of the European Union as they possess natural handicaps. Therefore, they can benefit from European Union funding to undertake local development actions, including agrotourism. Alternatively, in the coastal or mountainous areas, agrotourism enterprises can benefit from public infrastructure works as well as from tourist flows generated by tour-operators; i.e. they can benefit from the positive externalities of existing or new infrastructure and tourist services. For example, in France, the highest proportion of agricultural holdings practicing agrotourism (between 5 to 10 per cent) is recorded in the Mediterranean arc which includes the piedmonts and foothills of the Pyrénées, the Massif Central and the Alps (Bazin and Roux, 1997). In Greece, analysis of the Ministry of Agriculture statistics revealed that two thirds of the total of 835 agrotourism units which got subsidies from European Union funding programs (PIM, Dir. 2328/91/CEE) are located in the islands, 17.5 per cent in the mountains, 9 per cent on the mainland coastline.

This uneven spatial distribution of agrotourism activity indicates disorientation in the initial goal of regional and local actions for the revitalisation of rural, impoverished areas, which to diffuse territorial development through the mobilisation of local, natural and human resources. The concentration of agrotourism units in the advantaged tourist areas reproduces the same productivistic model, which favours the establishment of economic activities in the most competitive regions. Moreover, this model leads to the same secondary, negative effects on the environment (e.g. overpopulation, waste disposal problems and coastal water pollution) as conventional tourism, which is based on the 'sea and the sun' dominant image and ignores the rural hinterlands.

The concentration of agrotourism in developed regions implies that, in the long term, the equipped buildings will be used as secondary residence for the farm holder during his retirement or they will be passed on as main homes to his successors. It also implies the risk of the abandonment of the agricultural activity after amortisation of the agrotourism investment supported by European Union funding. This can lead to the separation of the two functions, which are the pillars of agrotourism and the progressive transformation of agrotourism holdings into exclusively tourist units without taking care of the environmental function of agriculture. It should be mentioned also that, in the Mediterranean zones, there is traditionally high competition for land between tourism and farming which pushes land prices to forbidding levels and harms agriculture, in the final analysis, as tourism is more profitable in the short term than farming.

In the context described above, three main types of agrotourism units can be distinguished in the Mediterranean areas, and, particularly, in Greece:

Holdings which offer veritable agrotourism services, the heads of which have agriculture as their principle occupation, offer primary produces or home-made products, and, therefore, allow the visitor to discover and experience the rural heritage and way of life. These farms can potentially participate in the efforts of the European Union policies for sustainable development in declining or fragile rural areas as agriculture plays an important part in land management and the preservation of the environment.

Intermediate holdings, the heads of which are country people with farming as a secondary occupation; e.g. retired people, petty traders and workers who use agrotourism to increase their income. Usually, the agricultural activity is of the traditional, extensive, Mediterranean type (olive trees, wine-growing areas, etc.) which, whether it participates or not to the agrotourism activity, can be an alibi for the inclusion of 'farm holdings' into the relevant European Union programmes. The support offered for such marginal complementary farm activities (for hobby or, more often, as an alibi) can introduce tourists into rural surroundings and contribute to the preservation of the environment. But, this type of agrotourist holdings conceals the risk of the abandonment of agriculture when it becomes unprofitable, or when the agrotourist unit is passed to successors.

Holdings with a diverted agrotourism vocation, the heads of which are not farmers, but country inhabitants, migrants returning to their villages, or retired people who do not own farmland. These simply benefit from public incentives to invest in farm tourism, aiming, in the long term, at an exclusive tourist use by themselves or their successors. In this case, there is no link between tourism and farm activity, and so no contribution of agrotourism to environmental preservation.

This typology will be used in the case study of agrotourism in Greece, which will be presented in Section 6.

5. The introduction and organisation of agrotourism in Greece

In Greece, the notion of rural tourism started to develop in the 1980s in the context of the discussion about 'endogenous development', launched by the five-year plans of the Greek socialist government for the revitalisation of the Greek countryside. This discussion was a response to the provision of European Union funding in favour of economically less-developed areas. It is noted that Greece became a member of the European Union in 1981 and that more than 80 per cent of its territory contains mountainous and disadvantaged areas (Dir. EEC/75/268). At the same time, major problems caused by the unruly development of mass tourism since the 1970s have pointed to the urgent need to protect the sites of high ecological and historical value by means of alternative types of tourism.

The notions of rural and agrotourism are still confused, however. According to the Ministry of Agriculture, which is the national organisation in charge of agrotourism development in Greece and the management of European funds: "Farm tourism consists of various small-scale rural activities, of a family or co-operative structure, which are run by people working in agriculture to offer them alternatives in terms of employment and to assist them increase their income. Furnished rooms or apartments, family restaurants, sport and leisure activities, arts and crafts workshops, and processing of farm products are state-aided investments". It is noted that, even by definition, welcome and stay facilities can exist independently of rural family housing; in that case, fully equipped or new buildings do not include traditional architectural or rural home functional elements. On the contrary, the model of a simple 'room-to-let' or furnished apartment is reproduced which can be encountered in any tourist area of the Cyclades Islands or in Crete. At the same time, there is no requirement for the agrotourist holder to include breakfast and/or a meal to his services, using the products of his farm or at least of his area. However, it is the provision of agrotourist products using farm produce, involving the rural environment and based on local traditions that really makes the difference between agrotourism and rural tourism. Apart from a false 'agrotourism product', this situation implies also losses of added value on the agricultural production of the farms and of the positive multiplier effects on the local economy and the environment.

In this context, the critical question is to what extent is rural tourism, and even more agrotourism, in Greece structured around local produces and 'know-how' and whether the associated leisure services are tied to rural life. And, consequently, to what extent can agrotourism development contribute to the preservation of the natural, built and cultural environment of the country? In this section, the two main types of agrotourism facilities, encouraged by public organisations in the framework of the European Union programmes and initiatives (PIM, Dir. EEC/797/85, Dir. EEC/2328/91, LEADER) are examined: (1) 'rooms-to-let' by country people or farmers and small hotel accommodation in rural areas and (2) the women's co-operatives in agrotourism.

According to the records of the Greek Ministry of Agriculture on agrotourism accommodation, there were 835 owners of 'rooms-to-let' in 1997 (Ministry of Agriculture, 1997). Of these owners, 298 (35.7 per cent) offer only a simple room separated from their house, 386 (46.2 per cent) provide also breakfast, 56 (6.7 per cent) provide breakfast and meals and, finally, 95 (11.4 per cent) rent rooms with kitchens to share or equipped studios. Based on these data, it is easily noticed that this type of welcome and accommodation in a rural/farm family is far being from the so-called 'veritable agrotourism services', which combine stay in a family environment with enjoyment of local produces and gastronomic specialities. More than one third of the 'agrotourism owners' offers only accommodation, dissociated from the rural family life, less than half of them offer breakfast, whereas a not insignificant percentage of them offers impersonal services unrelated to local traditions, as it concerns fully-equipped, independent apartments.

The agrotourism women's co-operatives are the first attempt to set up integrated agrotourism services in the sense that they combine accommodation in a rural family with the provision of farm products and craft items made by local women. This original type of agrotourism also expresses the political will to support women farmers while promoting alternative quality tourism in less-developed rural areas. The conception of this action belongs to the Greek State Secretariat for the Equality of Sexes which was created in 1983 within the Ministry of the Presidency. In the dominant ideological context of the socialist government, at that time, the Secretary for the Equality of Sexes was looking to develop new forms of economic activities which would lead women living in rural areas to a social and economic emancipation by exploiting their know-how (Anthopoulou, 1995). The decision to create agrotourism women's co-operatives has been reinforced also by the following facts. Most of the farmwomen are seen as simple family assistants in the agricultural holding. Without having an occupational status nor personal income, in spite of their contributing to the various tasks on the farm, their involvement in professional bodies or social institutions (co-operatives, associations, local authorities, and so on) remains very limited. Thus, their participation in an exclusively female co-operative structure would be easier compared with a mixed male-female structure (Iakovidou, 1995). In this way, agrotourism appeared to be the best mechanism to assist women farmers to overcome their social and economic isolation. Moreover, the establishment of an agrotourism programme does not require very high capital investment.

In the framework of women's co-operatives, created by the State Secretary for the Equality of Sexes and supervised by public authorities (the Ministry of Agriculture, the National Tourism Organisation and the Agricultural Bank), agrotourism was developed in several localities thanks to the dynamism of rural women. Several studies on women's co-operatives (Iakovidou, 1995) show how successful this decision has been compared with the Secretary's initial goals: women's income has increased, their creative spirit has improved thanks to working together, the possibilities of cultural exchanges grew and a relative socio-economic independence was created.

However, the analysis of agrotourism units shows that, in a certain way, the services offered have moved away from agricultural activities. To investigate this situation in depth, the case of the women's co-operative of Petra village in the island of Lesvos is examined, which has been one of the first four Greek agrotourism co-operatives established in Greece in 1983.

6. The case of the agrotourism women's co-operative in Petra (island of Lesvos)

Petra is a coastal rural district located at the north of Lesvos, Greece, 60 km from the capital of Mytilini. Of its total area of 1300 ha, agricultural land occupies 670 ha with 271 covered by olive groves and 240 ha by annual crops (fodder, cereals and fallow land).

At present the women's co-operative counts 36 members disposing 210 beds. It also owns a restaurant, which offers local food. Among the 36 members, 21 are actively employed (19 inhabit Petra and 2 Athens), and 15 are not (11 inhabit Petra,

2 are in Mytilini, 2 in Athens). A questionnaire survey was conducted in May 1997[1] to study the characteristics of this co-operative. Based on the analysis of the responses of 19 active members of the co-operative, three main types of agrotourism units could be distinguished (see also Section 4).

6.1. 'VERITABLE' AGROTOURISM UNITS

Six 'veritable agrotourism' units, the heads of which are women farmers or farmers' housewives, seasonally occupied in agriculture. Their mean age is 54.3 years (age between 34 and 68 years). All of them are retired or widows. They own an average of 3.5 rooms (from 1 to 5 rooms) and 7.7 beds (from 2 to 12 beds) in buildings which are separated from their homes but are either in the same premises or elsewhere in the village of Petra. Only two women (the oldest ones, 68 years old) own old farm buildings situated in the middle of their fields, which have been, restored (stall with barn): these are indeed veritable agrotourism ensembles. But as their buildings are outside the village centre and relatively far from the beach (500 to 700 metres), these women have difficulties to rent their rooms due to their competitors located close to the beach. The rural family possesses small areas planted with olive trees (from 0.5 to 2.5 ha; 25 to 300 trees) and vegetables (tomatoes, eggplants, pumpkins, string beans, and so on) grown for family consumption, tourists and sometimes sold to local taverns. On some areas, clover is grown for herds (max.1.5 ha). A few domestic animals are maintained (1 to 10 sheep and goats plus some chickens). Home-made cheese, yoghurt and, sometimes, raw milk are offered to tourists. In 4 cases, the husbands or sons of the women of the co-operative raise mainly sheep (50 to 150 animals) and may own a tractor as well. Usually, the women offer breakfast made with their own home-made products (jams, cheese, yoghurts, creams, eggs and bread). Occasionally, they offer meals and sell crafts and products from their region (olive oil, cheese, pasta, embroidery, tapestry, and so on). One of them teaches how to cook, gives lessons of folk dancing and is involved in countryside discovery excursions organised by the co-operative.

Women confess that they are happy with their involvement in agrotourism because they can contribute to the family income, and also they have the opportunity to meet other cultures. However, they would like to have permanent assistance from public organisations as regards on-the-job training and promotion of their products in the tourist market. They would like also to be protected against mass tourism, which develops quickly in their area detracting from its rural image and peace. Moreover, tour operators offer package holidays with tempting prices

---

[1] This survey was conducted in the framework of the Greek-French 'Programme PLATON 96' entitled 'Étude des interactions entre la valorisation locale des produits agricoles et le développement du tourisme rural dans les zones de montagne et défavorisées méditerranéennes' under the scientific responsibility of the University of the Aegean - Mytilini (dir. Th. Anthopoulou) and INRA-Paris (dir. B. Roux).

(plane tickets plus accommodation) which compete with the women's co-operative whose prices are very attractive but they do not have the possibility to be organised through an international tourist agency.

Five out of the six women find that agriculture and tourist activities are complementary, as agriculture provides foodstuff as well as a rural image for tourists searching for authentic rural environments. Moreover, none of these activities by themselves can provide a sufficient income for the rural household and, finally, there is no constraint in their timetable, as both activities are seasonal. Collection of olive and ewe milking are practised in the winter, whereas tourists visit the area from the end of spring to the beginning of autumn (May to September). Finally, none of these women considers abandoning agriculture because it provides extra income and products for their own consumption. Moreover, they are sentimentally attached to the land of their ancestors.

6.2. INTERMEDIATE AGROTOURISM UNITS

Six 'intermediate units', in the sense that the women who are the heads of the agrotourism holdings have a principal extra-farm occupation whereas their farming activities are secondary. Their mean age is 48.5 years (between 32 and 69 years). One of them is retired and one is a widow. The others work in firms related to Petra's recent tourist development: three are cooks and waitresses in restaurants, one is a part-time worker in the local cultural museum, and the last runs the co-operative's restaurant after having rented the business and the equipped premises.

They own an average of 3.2 rooms (from 1 to 8 rooms) and 6.5 beds (from 3 to 16 beds) in buildings, which are separated from their home, or they are in the same premises. In two cases, the women rent (6) furnished apartments/studios.

The agricultural holding is limited to a few olive trees (from 10 to 150 trees) and sometimes a goat or some chickens. Therefore, it is a small-sized farm producing mainly for family consumption and very little for the tourist market. However, the women try to offer local produces, which are bought locally or are provided by their parents who are farmers, like cheese, eggs and vegetables. Also they sell home-made products and craft like pasta, jams and embroidery. Three of the women teach how to cook. All of them consider their jobs interesting because they offer them economic and social emancipation. With the exception of one woman, the others claim that they do not wish to abandon the cultivation of olive trees, mainly for sentimental reasons, and they consider it as part of the agrotourism unit even though it is only a marginal crop.

6.3. DIVERTED AGROTOURISM UNITS

Seven units with a diverted agrotourist vocation, in the sense that the heads of the agrotourism holdings do not work in agriculture since they do not own any farmland. If they own a few olive trees, usually they rent them. Their mean age is 48.9 years (age between 23 and 75 years). This broad age range is due to the fact that younger women have inherited the agrotourism units (rooms to let) from their mothers or mothers-in-law whereas older women, if they are still members of the co-operative, have transferred their lands to their children or they have rented them. Five of them are housewives; two are widows and one is retired. One is employed in a hotel and the other works in her husband's shop. We should mention that the husbands of all these women are occupied in extra-farm activities as employees or in free-lance professions.

The women own an average of 2.1 rooms (from 1 to 4 rooms) and 4.4 beds (from 2 to 8 beds) in buildings, which are separated from their home or are in the same premises. As these women do not own farms, they buy all the products for tourists from local producers and, sometimes, they make traditional craft to add to their agrotourist income. However, all women find that tourism is an activity complementary to agriculture. But all of them have left agriculture or they do not wish to buy agricultural land (if they do not own it already). In this third type of agrotourism unit, the dissociation of the two activities – agriculture and tourism – is clear and its contribution to environmental preservation through the maintenance of farming is questionable.

In all units, which have been examined, there are three different types of agrotourism units and, therefore, three different cases of interaction between family agriculture and tourist services.

In the first case of 'veritable agrotourism holdings', both activities are in agreement with each other, in the sense that one is supplying the other with material and immaterial products, like agricultural produces, landscape, traditions, cultural elements, etc. Both activities have a family microstructure and they are complementary to each other from the economic point of view, as none of them can provide sufficient income by itself to the rural household, whereas there is no competition in their time-share. The preservation of traditional Mediterranean farming is contributing to the preservation of the environment and the natural resources.

In the second case of 'intermediate agrotourism holdings', agriculture is a residual activity, economically marginal to the agrotourism unit. Consequently, it might be abandoned when the agrotourism unit is passed on to the younger family successors, especially when the actual main occupation and incomes of family members have no relation to agriculture. This marginal agriculture which is destined more for family consumption than for agrotourism activities seems to have a limited, short-term contribution to environmental maintenance.

In the third case of 'diverted agrotourism vocation units', farming is absent at all; it simply offers accommodation for tourists in rooms or apartments for rent and their owners, members of the co-operative, do no engage in agricultural activities. As a result, there is no link between agriculture and tourism development, except that the women of the co-operative generally tend to sell and promote local, farm and craft products.

Thus, despite the success of the initial goals of setting-up the women's co-operative in Petra, socio-economical emancipation of the rural women and revival of local economy, agriculture itself as an essential component of the agrotourist unit has a limited place and runs the risk even to be abandoned. In this context, the important role of agrotourism for the preservation of the environment and the rural landscape through agriculture activities is diminished, from the moment that tourism develops and becomes socio-economically more interesting compared to the farming activity. The reduction of the environmental importance of agriculture seems to be more serious in the Mediterranean countries as there is no specific agrotourism policy at the national and regional level and the technical and organisational framework is not enough to direct and support agrotourist holdings.

7. Conclusions

In the mountainous and less-favoured areas, agrotourism is viewed as an important tool to promote sustainable development in two respects: it provides extra income to agricultural holdings and contributes to the preservation of the environment in fragile rural areas through land cultivation. At the same time, there is an increased urban demand for quality food and leisure services connected to the rural heritage. The setting-up of agrotourism initiatives, individually or collectively, was made possible due to the economic assistance offered by the European Socio-structural Funds, in the form of subsidies of the Dir. EEC/797/85 and EEC/2328/91, Programme LEADER, and so on. However, agrotourism is hardly undertaken by farmers and other rural people in order to enlarge and differentiate agricultural activities, as it demands professional abilities concerning the provision of specific services which are radically different from the local know-how and agricultural production techniques. In the Mediterranean areas, the undeniable assets of the rural heritage offer a favourable context for the development of agrotourism. But, several constraints due to the historical, physical and socio-economic particularities of the region, make progress slow. As far as demand is concerned, the fact that the inhabitants of Mediterranean cities have often kept the contact with their location of origin is a disincentive to look for contact with nature and its products through agrotourism.

The stakes between the natural assets of the Mediterranean environment (mild climate, proximity to the sea, rich biodiversity, diverse landscapes, etc.) and the socio-structural handicaps (i.e. small farms, low farm yields, traditional techniques, etc.) create a high competition between tourism and agriculture in terms of economy and land use. This competition implies the risk of the gradual abandonment of agriculture and the conversion of the farm premises into simple 'rooms-to-let' in a rural area by the younger farm successors after the actual rural holders have retired. This separation between farming and provision of agrotourism services is mainly observed in the Mediterranean regions with weak economies and not sufficient legislative and structural support for economic and technical assistance to agrotourism holders and other involved services.

In Greece, the analysis of a case study, the women's co-operative in Petra, permitted the distinction of three types of agrotourism units: (1) 'veritable agrotourism units', based on family agriculture and the offer of farm produces and crafts, (2) 'intermediate agrotourism units' which maintain a small scale agriculture which will probably be abandoned by the successors who have no economic interest and farm experience, and (3) 'agrotourism units with a diverted farm vocation', whose holders do not possess farmland or they are not occupied in farming; their unit lives on the fringe of the rural area and culture.

In that Mediterranean context, and more precisely in Greece, where agriculture as one of the main agrotourism pillars is not very competitive, while mass tourism has damaged the environment very badly, these new rural professionals need legal, technical and financial assistance in order to define and protect their 'agrotourism product' and strengthen their professional status. Also, agrotourism would not make significant progress without co-ordination among the various local actions undertaken for the diversification of rural activities and the promotion of sustainable development. In addition, agreement is needed between the various organisations and actors intervening at the national and regional levels through the management of the European Funds and offering technical assistance. Finally, an agrotourism development policy for the Mediterranean mountainous and less-favoured areas is deemed necessary which would take into account the need to keep both the population and the agricultural activities which are gradually abandoned in order to preserve the fragile Mediterranean environment.

References

Anthopoulou, Th. (1995) La coopérative féminine d'agrotourisme à Petra dans l'île de Lesvos. In J. Battesti (ed.), *La RDT dans les Zones Rurales et les Régions Insulaires de Union Européenne*, Commission Européenne – DGXII. Bruxelles, pp. 86–89.

Anthopoulou, Th. (1997) L'enjeu de petites économies en milieu insulaire grec: l' île de Paros (Cyclades). In A.L. Sanguin (ed.), *Vivre dans une Île. Une Géopolitique des Insularités*, L'Harmattan. Paris, pp. 241–252.

Bazin, G. and B. Roux, (1997) L' agritourisme: un atout pour les zones rurales difficiles méditerranéennes? In B. Roux and D.Guerraoui (eds), *Les Zones Difficiles Méditerranéennes*, L'Harmattan. Paris, pp. 339–361.

Coulombe, P. (ed.) (1986) *Le Tourisme contre l'Agriculture?* ADEF, Paris.

Commission Européenne (1994) *Coopération pour l'Aménagement du Territoire Européen*, Bruxelles.

De Casabianca, F. (1997) *Pour l'Élaboration d'une Politique Agricole et Environnementale Adaptée aux Îles Méditerranéennes*. INRA-Corse.

Iakovidou, O. (1995) L'agrotourisme en Grècque: Le modèle des coopératives agritouristiques féminines. In ENITA (ed), *Rencontres Internationales sur le Tourisme en Espace Rural. L'Organisation des Partenariats*, Collections - Actes, Clermont-Ferrant.

Leader Dossiers (1993) *Tourism at the Service of Rural Development*, LEADER Coordinating Unit/AEIDL. Bruxelles.

Lebeau, R. (1992) *Les Grands Types des Structures Agraires dans le Monde*. Ed. Masson, Paris.

Le Goffe, P. and X. Delache (1997) Impacts de l'agriculture sur le tourisme. Une application des prix hédonistes. *Economie Rurale* **239**, 3–10.

MEDEF (Dir:G. Bazin and B. Roux) (1992) *Les Facteurs de Résistance à la Marginalisation des Zones de Montagne et Défavorisées Méditerranéennes Communautaires* Office des Publications Officielles des Communautés Européennes, Luxemburg.

Ministry of Agriculture (1997) *Agrotourism : Vacation in the Countryside*. Athens.

Mormont, M. (1996) Agriculture et environnement: pour une sociologie des dispositifs. *Economie Rurale* **236**, 28–36.

Muheim, P. (1995) Tourisme et développement rural, une opportunité à mieux exploiter selon l'OCDE. *L'Information Agricole* **682**, 25–26.

Perrier-Cornet, P. and D. Capt (1995) Les agriculteurs face à la nouvelle PAC. Quelles perspectives pour quels territoires? *Economie Rurale* **225**, 22–27.

Slee, B., Farr, H. and P. Snowdon (1997) The economic impact of alternative types of rural tourism. *Journal of Agricultural Economics* **48**:2, 179–191.

Whitby, M. (ed.) (1996) *The European Environment and CAP Reform. Policies and Prospects for Conservation*. CAB International, Oxon.

INDEX

ENVIRONMENT & ASSESSMENT

KLUWER ACADEMIC PUBLISHERS – DORDRECHT / BOSTON / LONDON